FUTURE POLITICS

Politics in the twentieth century was dominated by a single question: how much of our collective life should be determined by the state, and what should be left to the market and civil society?

Now the debate is different: to what extent should our lives be directed and controlled by powerful digital systems—and on what terms?

Digital technologies—from artificial intelligence to blockchain, from robotics to virtual reality—are transforming the way we live together. Those who control the most powerful technologies are increasingly able to control the rest of us. As time goes on, these powerful entities—usually big tech firms and the state—will set the limits of our liberty, decreeing what may be done and what is forbidden. Their algorithms will determine vital questions of social justice. In their hands, democracy will flourish or decay.

A landmark work of political theory, *Future Politics* challenges readers to rethink what it means to be free or equal, what it means to have power or property, and what it means for a political system to be just or democratic. In a time of rapid and relentless changes, it is a book about how we can—and must—regain control.

Jamie Susskind is an author and barrister. A graduate of Oxford University, he has held fellowships at Harvard University's Edmond J. Safra Center for Ethics and Berkman Klein Center for Internet and Society, and at Cambridge University's Leverhulme Centre for the Future of Intelligence.

JAMIE SUSSKIND

LIVING TOGETHER IN A WORLD TRANSFORMED BY TECH

FUTURE POLITICS

OXFORD
UNIVERSITY PRESS

OXFORD
UNIVERSITY PRESS

Great Clarendon Street, Oxford, OX2 6DP,
United Kingdom

Oxford University Press is a department of the University of Oxford.
It furthers the University's objective of excellence in research, scholarship,
and education by publishing worldwide. Oxford is a registered trade mark of
Oxford University Press in the UK and in certain other countries

First Edition published in 2018
First published in paperback 2020

Impression: 1

Published in the United States of America by Oxford University Press
198 Madison Avenue, New York, NY 10016, United States of America

British Library Cataloguing in Publication Data

Data available

Library of Congress Cataloging in Publication Data

Data available

ISBN 978–0–19–882561–6 (Hbk.)
ISBN 978–0–19–884892–9 (Pbk.)

Printed and bound in Great Britain by
Clays Ltd, Elcograf S.p.A.

This book is dedicated to my parents,
Michelle and Richard

PREFACE TO THE PAPERBACK EDITION

A year is a long time in politics and an even longer time in technology. Since *Future Politics* was published, a political awakening has taken place in the public consciousness, in Silicon Valley, and in what used to be called the corridors of power. The last year has seen the emergence of a global framework for the governance of artificial intelligence, calls by public authorities for the tech giants to be broken up, and a plea by the founder of Facebook for greater regulation of the internet. These stories, and countless others like them, mark the beginning of an era in which the defining political issue is how far, and on what terms, our lives should be governed by powerful digital systems.

I have been grateful for the opportunity to make one or two small corrections to this edition of *Future Politics*, but the fundamental premise remains unchanged and more urgent than ever: we are not yet ready—intellectually, philosophically, or morally—for the world we are creating. There is still so much to do.

Jamie Susskind
London
March 2020

ACKNOWLEDGEMENTS

I could not have written this book without the help of my friends, colleagues, and family.

Most of *Future Politics* was completed during a Fellowship at Harvard University's Berkman Klein Center for Internet and Society. The Center is a special place, and its staff and Fellows are a constant source of inspiration. I am also indebted to the staff and members of Littleton Chambers, who have tolerated my absences and supported my work with great patience and loyalty.

I have been lucky to find, in Dominic Byatt, the ideal editor: always insightful, frequently critical, and generally indulgent of my foibles. My agent, Caroline Michel, has looked after me from the beginning and I am always thankful to have her on my side. It has been a pleasure to work with the fine teams at Oxford University Press and Peters Fraser + Dunlop: Tim Binding, Alexandra Cliff, Tessa David, Kate Farquhar-Thomson, Phil Henderson, Dan Herron, Erin Meehan, Laurie Robertson, Sarah Russo, and Olivia Wells. Chris Summerville expertly copy-edited the final text.

I am grateful to Luciano Floridi, Vicki Nash, and Susannah Otter, whose early guidance helped to get the project off the ground. Fred Popplewell's research was enormously useful. I have benefitted deeply from conversations with Yochai Benkler, Alex Canfor-Dumas, Amber Case, Matt Clifford, David Cox, Primavera De Filippi, Gabriella Fee, Howard Gardner, Josh Glancy, Philip Howard, Laurence Lessig, Andrew Perlman, Michael Sandel, Bruce Schneier, Carina Namih, Beth Simone Noveck, David Weinberger, Owain Williams, Ellen Winner, Tom Woodward, and Jonathan Zittrain. David Wilkins has been an invaluable source of wisdom and counsel.

Suzanne Ashman, James Boyle, Lizzie Gaisman, Nicholas Gaisman, Dorita Gilinski, Philip Howard, and Martha Minow offered invaluable commentary on various drafts, and Ifeoma Ajunwa and Maxine Mackintosh provided feedback that dramatically improved Part V. I will always be grateful to Olivia Wollenberg for her encouragement and advice as the writing process got underway.

Several readers were kind enough to read the manuscript in full: Tony Blair, Gordon Brown, Matt Clifford, Alex Canfor-Dumas, Kim FitzGerald, Matthew Flinders, Howard Gardner, Beeban Kidron, Laurence Mills, Marius Ostrowski, Fred Popplewell, Susannah Prichard, Dov Seidman, Daniel Sleat, Owain Williams, and Tom Woodward. Their many (many) points of criticism have made it a much better book. Throughout, Pavithra Mahesh encouraged me, for better or worse, to write with my own voice. Philippa Greer made a big difference in a short space of time.

The great Matt Orton has been an unfailing source of writerly wisdom, feedback, and motivation. Chris and Diana Orton kindly let me squat in their Edinburgh cottage while I finished writing.

If there is any good political theory in these pages, it is down to Simon Caney, who fired my passion for the discipline as an undergraduate and, ten years later, reviewed the manuscript of this book with his customary rigour.

Finally, I wish to thank my family. I am always amazed by how fiercely my mum, Michelle, believes in my work. I would be nowhere without her love and support. She is my surest source of strength and encouragement. My sister Ali has been a rock: I trust her judgement entirely and have leaned on it countless times. Her painstaking labours in the closing stages helped me to get the manuscript over the line (just about) in time.

Writing about the future has become a kind of weird family business for the Susskinds. Some readers will know that my dad, Richard Susskind, has been writing about the impact of technology since the early 1980s; and that he and my brother Daniel Susskind (an economist) co-authored a book called *The Future of the Professions* in 2015. If you read that book, you will see the extent of my intellectual debt to them both. Daniel has been my most thoughtful and generous reader, as well as my fiercest supporter.

I do not have the words to describe what I owe to my dad or how grateful I am to him. It's been the greatest joy and good fortune of my life to have him as my best friend, mentor, and guide. We've been in constant conversation for nearly thirty years, and for the last few we've mostly been talking about this book. His influence and inspiration are present on every page.

Jamie Susskind
London
March 2020

COPYRIGHT NOTIFICATIONS

CONTENTS

PART IV FUTURE DEMOCRACY

PART V FUTURE JUSTICE

PART VI FUTURE POLITICS

'A new political science is needed for a world itself quite new'

Alexis de Tocqueville, *Democracy in America* **(1835)**

Introduction

The future stalks us. It is always waiting, barely out of sight, lurking around the corner or over the next rise. We can never be sure what form it will take. Often it catches us entirely unprepared.

Nowadays, many of us share the sense that we are approaching a time of great upheaval. The world seems to be changing faster than we can grasp. Often we struggle to explain political events that would have been unimaginable just a few years ago. Sometimes we don't even have the words to describe them. Inwardly, we know that this is just the beginning.

The premise of this book is that relentless advances in science and technology are set to transform the way we live together, with consequences for politics that are profound and frightening in equal measure. We are not yet ready—intellectually, philosophically, or morally—for the world we are creating. In the next few decades, old ways of thinking that have served us well for hundreds, even thousands, of years, will be called into question. New debates, controversies, movements, and ideologies will come to the fore. Some of our most deeply held assumptions will be revised or abandoned altogether. Together we will need to re-imagine what it means to be free or equal, what it means to have power or property, and even what it means for a political system to be democratic. Politics in the future will be quite unlike politics in the past.

Politics in the twentieth century was dominated by a central question: how much of our collective life should be determined by the state, and what should be left to the market and civil society? For the generation now approaching political maturity, the debate will be different: to what extent should our lives be directed and controlled by powerful digital systems—and on what terms? This question is at the heart of *Future Politics*.

In the next few decades, it is predicted, we'll develop computing systems of astonishing capability, some of which will rival and surpass humans across a wide range of functions, even without achieving an 'intelligence' like ours. Before long, these systems will cease to resemble computers. They'll be embedded in the physical world, hidden in structures and objects that we never used to regard as technology. More and more information about human beings— what we do, where we go, what we think, what we say, how we feel—will be captured and recorded as data, then sorted, stored, and processed digitally. In the long run, the distinctions between human and machine, online and offline, virtual and real, will fade into the background.

This transformation will bring some great benefits for civilization. Our lives will be enriched by new ways of playing, working, travelling, shopping, learning, creating, expressing ourselves, staying in touch, meeting strangers, coordinating action, keeping fit, and finding meaning. In the long run, we may be able to augment our minds and bodies beyond recognition, freeing ourselves from the limitations of our human biology.

At the same time, however, some technologies will come to hold great power over us. Some will be able to force us to behave a certain way, like (to take a basic example) self-driving vehicles that simply refuse to drive over the speed limit. Others will be powerful because of the information they gather about us. Merely knowing we are being watched makes us less likely to do things perceived as shameful, sinful, or wrong. Still other technologies will filter what we see of the world, prescribing what we know, shaping the way we think, influencing how we feel, and thereby determining how we act.

Those who control these technologies will increasingly control the rest of us. They'll have *power*, meaning they'll have a stable and wide-ranging capacity to get us to do things of significance that we wouldn't otherwise do. Increasingly, they'll set the limits of our *liberty*, decreeing what may be done and what is forbidden. They'll determine the future of *democracy*, causing it to flourish or decay. And their algorithms will decide vital questions of *social justice*, allocating social goods and sorting us into hierarchies of status and esteem.

The upshot is that political authorities—generally states—will have more instruments of control at their disposal than ever before, and big tech firms will also come to enjoy power on a scale that dwarfs any other economic entity in modern times. To cope with these new challenges, we'll need a radical upgrade of our political ideas. The great English philosopher John Stuart Mill wrote in his *Autobiography* of 1873 that, 'no great improvements in the lot of mankind are possible, until a great change takes place in the fundamental constitution of their modes of thought.'[1]

It is time for the next great change.

The Next Great Change

We already live in a time of deep political unease. Every day the news is of bloody civil war, mass displacement of peoples, ethnic nationalism, sectarian violence, religious extremism, climate change, economic turbulence, disorienting globalization, rising inequality, and an array of other challenges too dismal to mention. It seems like the world isn't in great shape—and that our public discourse has sunk to the occasion. Political élites are widely distrusted and despised. Two of the most consequential exercises in mass democracy in the English-speaking world, the 2016 US presidential election and UK Brexit referendum, were rancorous even by the usual unhappy standards, with opposing factions vying not just to defeat their rivals but to destroy them. Each was won by the side that promised to tear down the old order. Neither brought closure or satisfaction. Increasingly, as Barack Obama noted at the end of his presidency, 'everything is true, and nothing is true'.[2] It's getting harder for ordinary citizens (of any political allegiance) to separate fact from fraud, reality from rumour, signal from noise. Many have given up trying. The temptation is to hunker down and weather the present storm without thinking too hard about the future.

That would be a mistake.

If mainstream predictions about the future of technology are close to the mark, then the transformation on the horizon could be at least as important for humankind as the industrial revolution, the agricultural revolution, or even the invention of language. Many of today's problems will be dwarfed by comparison. Think about the effect that technology has already had on our lives—how we work, communicate, treat our illnesses, exercise, eat, study, and socialize—and then remember that in historical perspective, the digital age is only a few seconds old. Fully 99.5 per cent of human existence was spent in the Palaeolithic era, which began about 3 million years ago when humans began using primitive tools. That era ended about 12,000 years ago with the last ice age.[3] During this long twilight period, people noticed almost no cultural change at all. 'The human world that individuals entered at birth was the same as the one they left at death'.[4] If you consider that the earliest human civilizations emerged some 5,000 years ago, then the seventy or so years that we have lived with modern computing machines, the thirty or so we have had the world wide web, and the decade we've spent with smartphones don't seem very long at all. And while time passes linearly, many developments in digital technology are occurring exponentially, the rate of change accelerating with each passing year.

We have no evidence from the future, so trying to predict it is inherently risky and difficult. I admire those who try to do so in a rigorous way, and I have borrowed extensively from their work in this book. But to be realistic, we should start by acknowledging that such predictions often badly miss the mark. Much of the future anticipated in these pages will probably never come to pass, and other developments, utterly unforeseen, will emerge to surprise us instead. That said, I believe it is possible to make sensible, informed guesses about what the future might look like, based on what we know of the current trends in science, technology, and politics. The biggest risk would be not to try to anticipate the future at all.

The story is told of an encounter between the Victorian statesman William Gladstone and the pioneering scientist Michael Faraday. Faraday was trying to explain his groundbreaking work on electricity to Gladstone, but Gladstone seemed unimpressed. 'But what use is it?' he asked, with growing frustration; 'What *use* is it?'

'Why sir,' replied Faraday, reaching the end of his patience, 'there is every possibility that you will soon be able to tax it.'

Many innovators, like Faraday, find it hard to explain the social and practical implications of their work. And the rest of us, like Gladstone, are too often dismissive of technologies we don't yet understand. It can be hard to see the political significance of inventions that, at first glance, seem to have nothing to do with politics. When confronted with a new gadget or app, we tend not to think first of all about its implications for the political system. Instead we want to know: what does it do? How much does it cost? Where can I get one? This isn't surprising. In general, technology is something we encounter most often as consumers. But this rather narrow attitude now needs to change. We must apply the same scrutiny and scepticism to the new technologies of power that we have always brought to powerful politicians. Technology affects us not just as consumers but as citizens. In the twenty-first century, *the digital is political.*

This book is partly for Gladstones who want to understand more about technology and partly for Faradays who want to see more clearly the political significance of their work. But mainly it's for ordinary citizens who want to understand the future a bit better— so if nothing else they can hold the Gladstones and the Faradays to account.

Philosophical Engineers

Consider the following passage:

> Here's to the crazy ones. The misfits. The rebels. The troublemakers. The round pegs in the square holes. The ones who see things differently. They're not fond of rules. And they have no respect for the status quo. You can quote them, disagree with them, glorify or vilify them. About the only thing you can't do is ignore them. Because they change things. They push the human race forward. And while some may see them as the crazy ones, we see genius. Because the people who are crazy enough to think they can change the world are the ones who do.

These are not the words of a politician. They're from the voiceover to 'Think Different', a 1997 Apple advertisement featuring footage of iconic rebels including Mahatma Gandhi and Martin Luther King. The ad embodies a worldview, widely held among tech entrepreneurs, that their work is of philosophical as well as commercial importance. 'It is commonplace in Silicon Valley,' explains Jaron Lanier, 'for very young people with a startup in a garage to announce that their goal is to change human culture globally and profoundly, within a few years, and that they aren't ready yet to worry about money, because acquiring a great fortune is a petty matter that will take care of itself.'[5] There is something attractive about this way of thinking, partly because it suggests that tech companies might not be as rapacious as they are sometimes made out to be. And the basic premise is right: digital technologies *do* indeed have an astounding capacity to change the world. Compare the following statements:

> 'The philosophers have only *interpreted* the world in various ways; the point is to *change* it.'

> 'We are not *analysing* a world, we are *building* it.'

The first is from Karl Marx's 1845 *Theses on Feuerbach*.[6] It served as a rallying cry for political revolutionaries for more than a century after its publication. The second is from Tim Berners-Lee, the mild-mannered inventor of the world wide web.[7] Marx and Berners-Lee could scarcely be more different in their politics, temperament, or choice of facial hair. But what they have in common—in addition to having changed the course of human history—is a belief in the distinction between *making* change and merely *thinking* about it or *studying* it. On this view, far from being a spectral presence out of our control, the future is something we design and build.

'We are not experimental philosophers,' says Berners-Lee, 'we are philosophical engineers.'[8] It's a practical and hands-on way of looking at life, one more familiar to builders and inventors than to tweedy academics or beturtlenecked philosophers. It also happens to be the defining mindset of our age. Today, the most important revolutions are taking place not in philosophy departments, nor even in parliaments and city squares, but in laboratories, research facilities, tech firms, and data centres. Most involve developments in digital technology. Yet these extraordinary advances are taking place in a climate of alarming cultural and intellectual isolation. With a few exceptions, there is a gulf between the arts and the sciences.

Political philosophy and social policy rarely appear in degree pro-
grammes for science, technology, engineering, and mathematics.
And if you ask the average liberal arts student how a computer
works, you are unlikely to get a sophisticated response.

In tech firms themselves, few engineers are tasked with thinking
hard about the systemic consequences of their work. Most are
given discrete technical problems to solve. Innovation in the tech
sector is ultimately driven by profit, even if investors are prepared
to take a 'good idea first, profits later' approach. This is not a criticism:
it's just that there's no reason why making money and improving
the world will always be the same thing. In fact, as many of the
examples in this book show, there's plenty of evidence to suggest
that digital technology is too often designed from the perspective
of the powerful and privileged.

As time goes on, we will need more philosophical engineers
worthy of the name. And it will become even more important for
the rest of us to engage critically with the work of tech firms,
not least because tech working culture is notorious for its lack of
diversity. Roughly nine out of every ten Silicon Valley executives
are men.[9] Despite the fact that African-Americans make up about
10 per cent of computer science graduates and 14 per cent of the
overall workforce, they make up less than 3 per cent of computing
roles in Silicon Valley.[10] And many in the tech community hold
strong political views that are way outside the mainstream. More
than 44 per cent of Bitcoin adopters in 2013, for instance, professed
to be 'libertarian or anarcho-capitalists who favour elimination
of the state'.[11]

As I will argue, we put so much at risk when we delegate matters
of political importance to the tiny group that happens to be tasked
with developing digital technologies at a given time. That's true
whether you admire the philosophical engineers of Silicon Valley
or you think that most 'tech bros' have the political sophistication

of a transistor. We need an intellectual framework that can help us to think clearly and critically about the political consequences of digital innovation. This book hopes to contribute to such a framework, using the ideas and methods of political theory.

The Promise of Political Theory

The purpose of philosophy, says Isaiah Berlin, is always the same: to assist humans 'to understand themselves and thus operate in the open, and not wildly, in the dark.'[12] That's our goal too. Political theory aims to understand politics through the concepts we use to speak about it.[13] What is *power*? When should *freedom* be curtailed and on what basis? Does *democracy* require that everyone has an *equal* ability to shape the political process? What is a *just* distribution of society's resources? These are the sorts of questions that political theorists try to answer. The discipline has a long and rich history. From Plato and Aristotle in the academies of ancient Greece to Thomas Hobbes and Jean-Jacques Rousseau in the tumult of early modern Europe, to the giants of twentieth-century political thought like Hannah Arendt and John Rawls, western political thinkers have long tried to clarify and critique the world around them, asking why it is the way it is—and whether it could or should be different.

For several reasons, political theory is well-suited to examining the interplay of technology and politics. First, the canon of political thought contains wisdom that has outlived civilizations. It can shed light on our future predicaments and help us to identify what's at stake. We'd be foolish not to plunder the trove of ideas already available to us, even if we ultimately decide that some of those ideas need an upgrade or a reboot. Political theory also offers *methods* of thinking about the world that help us to raise the level of debate above assertion and prejudice.

To my mind, the best thing about political theory is that it deals with the *big* themes and questions of politics. It offers a panoramic view of the political forest where other approaches might get lost in the trees (or stuck in the branches). That's necessary, in our case, to do justice to the subject-matter. If we think that technology could have a fundamental impact on the human condition, then our analysis of that impact should be fundamental too. That's why this book is about four of the most basic political concepts of all:

Power: How the strong dominate the weak

Liberty: What is allowed and what is prohibited

Democracy: How the people can rule

Social Justice: What duties we owe to each other

In a time of great change, I suggest, it pays to go back to first principles and think about these concepts quite apart from any particular legal regime. That way we might be able to imagine a superior system to the one we have inherited.

Political theory is also useful because it allows us to think critically not just about politics but also about how we *think* and *speak* about politics. Concepts are the 'keyholes through which we inevitably see and perceive reality'.[14] When I want to say something to my neighbour about politics, I don't need to start from scratch. I know that if I say a process is 'undemocratic' then she will have a pretty good idea of what I mean and the connotations I wish to convey, without any need for me to explain what democracy is and why it should be considered a good thing. That's because we are members of the same linguistic community, sharing a 'common stock of concepts' drawn from our shared history and mythology.[15] It's convenient.

On the other hand, what we want to say about politics can sometimes be limited by the poverty of the words at our disposal.

Some things seem unsayable, or unthinkable, because the common stock of concepts hasn't yet developed to articulate them. 'The limits of my language,' says Ludwig Wittgenstein, 'mean the limits of my world.'[16]

What this means in political terms is that even if we could see the future clearly, we might not have the words to describe it. It's why, so often, we limit our vision of the future to a turbo-charged version of the world we already live in. 'If I had asked people what they wanted,' said Henry Ford, the first mass-producer of automobiles, 'they would have said faster horses.' Ford recognized that it can be hard to conceive of a system radically different from our own. Failure to keep our language up-to-date only makes it harder.

Why I Wrote this Book

My first real taste of political theory was at university, where I fell in love with the discipline under the watchful eye of some wonderful professors. It sparked an obsession that has stayed with me since. (I acknowledge in hindsight that my most successful undergraduate romance may well have been a passionate but doomed affair with the German philosopher G. W. F. Hegel.)

Passion aside, something troubled me about me the discipline of political theory. Political theorists seemed to pride themselves on thinking deeply about the *history* of political ideas but, with some exceptions, were almost entirely uninterested in their *future*. I found this strange: why would the same scholars—so sensitive to context when writing about the past—discuss politics as if the world will be the same in 2050 as it was in 1950? It seemed that a good deal of very clever political theory was of little practical application because it did not engage with the emerging realities of our time. When I thought about politics in the future, I thought about

Orwell, Huxley, Wells—all novelists from the early twentieth century rather than theorists from the twenty-first. It turns out that I wasn't alone: since the election of Donald Trump to the US presidency in late 2016, Orwell's *Nineteen Eighty-Four* has surged up the best seller lists. But this prompts the question: if we want to understand the world as it will be in 2050, should we really have to rely on a work of fiction from 1949?

After I left university and became involved in my own modest political causes, my niggling sense of unease—that political theory might be unable, or unwilling, to address the looming challenges of my generation—became a more urgent concern. What if developments in technology were to happen so fast that we lacked the intellectual apparatus to make sense of them? What if, unthinkingly, we were about to unleash a future that we couldn't understand, let alone control?

I wanted answers, and that's why I began working on this book.

The Politics of Technology

Before ploughing on, let's begin with a simple question: what is the connection between digital technology and politics?

Technology in General

New technologies make it *possible* to do things that previously couldn't be done; and they make it *easier* to do some things we could already do.[17] This is their basic social significance. More often than not, the new opportunities created by technology are minor in nature: an ingenious new way of grinding coffee beans, for instance, is unlikely to lead to the overthrow of the state. But sometimes the consequences can be profound. In the industrial revolution, the invention of power looms and stocking and spinning frames threatened to displace the jobs of skilled textile workers. Some of those workers, known as the Luddites, launched a violent rampage through the English countryside, destroying the new machines as they went. We still use the word Luddite to describe those who resist the arrival of new and disruptive technologies.

The *economic* consequences of innovation, as in the case of the Luddites, will often require a political response. But new technologies can raise *moral* challenges too. A few years from now, for instance, virtual reality systems will enable behaviour that was previously the stuff of science fiction, including (for instance) the possibility of virtual sex using platforms capable of simulating human sexual activity. This raises some interesting questions. Should it be legally permissible to have virtual sex with a digital partner designed to look like a real-life person—a celebrity, say—without that person's knowledge or consent? People are likely to have strong feelings about this. What about having virtual 'sex' with a virtual incarnation of your best friend's husband? It would seem wrong for the law to allow this, but you might argue that what you get up to in the privacy of your virtual world, without harming anybody, is no business of anyone else, let alone the law. To take a more extreme example: what about having virtual sex with an avatar of a child, in circumstances where no child is actually harmed in the creation of that experience?

These are new questions. We've not had to answer them before. Maybe you've already formed views on them. Perhaps those views differ from mine. And that's precisely the point: these are inherently *political* questions, and they ought to be answered carefully by reference to some acceptable set of moral principles. As one distinguished author has noted, new technologies can lead us to look again at our political views, just as a new dish on the menu of our favourite restaurant can lead us to challenge our taste in food.[18]

Some technologies are 'inherently political' in that they actually require 'particular kinds of political relationships' or are at least strongly compatible with them.[19] Langdon Winner, writing in 1980, cited nuclear power as an example: if you are going to adopt a system of nuclear power, you must also accept that a powerful 'techno-scientific-industrial-military elite' will be required to administer and supervise the plants and infrastructure.[20]

Then there are other technologies that are not inherently political, but which are made political by their context. In the United Kingdom, for instance, a licensed firearm is generally a technology for the hunting of wild animals. Guns are not part of mainstream culture and people are mostly happy with the idea of strict regulation. But in the United States, the right to bear arms is guaranteed by the Second Amendment. Cultural opposition to regulation is much stiffer. Same technology, different political context.

A final, more subtle, connection between technology and politics is that our inventions have a way of inveigling themselves into our political and intellectual life. A good example is the mechanical clock. As Otto Mayr explains, some ancient civilizations imagined the state as being like a human body, with individual human members of the political community forming appendages like a 'hand' or a 'foot'.[21] In the late Renaissance, the metaphor of the body was joined by others like the 'ship of state'. After Copernicus, the monarch came to be seen as a great sun around which his subjects revolved.[22] In the six-teenth to eighteenth centuries, the dominant metaphor was that of the clock, an ingenious contraption that commanded 'unprecedented veneration'.[23] After a while, thinkers came to see politics from the perspective of clockwork.[24] Harmony, temperance, and regularity became the prevailing political ideals.[25] The people yearned for an 'ever vigilant statesman–engineer' who could 'predict as well as repair all potential troubles'. Thus it was that a particular technology and a set of political values went hand in hand, reaching a climax in the seventeenth century with its 'extraordinary production of clocks' and the 'conspicuous flourishing of the authoritarian conception of order.'[26]

The twentieth-century political imagination was also coloured by technology. Even before the digital age, the prospect of all-powerful computing machines inspired a great deal of art and fiction. In *The Machine Stops* (1928) E. M. Forster portrayed a world in which humans were subordinated to The Machine, a global techno-logical system that monitored and controlled every aspect of human existence.[27]

Some writers have tried to pin down what they see as the ideology of our own time. Evgeny Morozov, for instance, has written of the 'Google Doctrine' ('the enthusiastic belief in the liberating power of technology accompanied by the irresistible urge to enlist Silicon Valley start-ups in the global fight for free-dom');[28] 'cyber-utopianism' ('a naive belief in the emancipatory nature of online communication that rests on a stubborn refusal to acknowledge its downside');[29] and 'solutionism' ('Recasting all complex social situations either as neatly defined problems with definite, computable solutions or as transparent and self-evident processes that can be easily optimized—if only the right algorithms are in place').[30]

My own view is that the technologies in question are too young for us to know what lasting imprint they will leave on our political thought. It took hundreds of years for the idea of the mechanical clock to soak into European political and intellectual life. How we choose to interpret the technologies of our time, and how they in turn shape our perception of the world, are matters that are yet to be determined.

Digital Technologies

So much for the general relationship between technology and politics. But this book is about *digital* technologies—what used to be called information and communication technologies, or ICTs. And it appears that these are not just inherently political; they are *hyper*-political. That's because they strike at the two most fundamental ingredients of political life: *communication* and *information*.

All political order is built on coordination, cooperation, or control. It's impossible to organize collective life without at least one of the three. And none of them is possible without some system for exchanging information, whether among ordinary people or between ruler and ruled.[31] This is why *language* is so important. As James Farr puts it, without language politics would not only be 'indescribable' but 'impossible':[32]

> Emerging nations could not declare independence, leaders instruct partisans, citizens protest war, or courts sentence criminals. Neither could we criticise, plead, promise, argue, exhort, demand, negotiate, bargain, compromise, counsel, brief, debrief, advise nor consent. To imagine politics without these actions would be to imagine no recognisable politics at all.

Yuval Noah Harari elegantly observes that language played a crucial role in the earliest days of human politics. A small number of sounds and signs enabled our ancestors to produce an infinite number of sentences, each with a distinct meaning. This allowed them to talk about each other, about big, complex things in the world around them, and about things lacking any physical existence like myths and stories. Such lore still holds communities together today.[33]

Writing may have begun as a way to reflect reality, 'but it gradually became a powerful way to reshape reality'.[34] The Domesday Book of 1086 was commissioned by William the Conqueror to discover the extent and value of people's property so it could be more efficiently taxed. The contents of the Book were set in stone and could not be appealed. In 1179 a commentator wrote of this 'strict and terrible last account' that:[35]

> its sentence cannot be quashed or set aside with impunity. That is why we have called the book 'the Book of Judgement'... because its decisions, like those of the Last Judgement, are unalterable.

All that mattered was what was written in the Book. Little has changed. As Harari puts it, 'Anyone who has ever dealt with the tax authorities, the educational system or any other complex bureaucracy knows that the truth hardly matters. What's written on your form is far more important.'[36]

But language is not enough. Advanced political communities also need to be able to process large amounts of *information*—from poverty and economic growth rates to fertility, unemployment, and immigration figures. If you want to govern a nation, you must first know it.[37] The eighteenth-century revolutionaries who sought to replace the decaying *ancien régimes* understood this well. Revolution in France in 1789 was followed by an intense effort of standardization and rationalization, the introduction of a unified system of weights and measures, division of territory into *départements*, and the promulgation of the Civil Code.[38] Across the Atlantic, the founding fathers of the United States enshrined the need for a decennial census (an 'Enumeration') in the first article of the Constitution. Alexander Hamilton believed that the federal government ought to be 'the center of information', because that would make it best placed to 'understand the extent and urgency of the dangers that threaten'.[39]

The connection between information and politics is fundamental, and it has left its mark on our vocabulary. The word *statistics* comes from the eighteenth-century German term *Staatwissenschaft*, the 'science of the state' taught by university professors to princelings of the Holy Roman Empire.[40] Indeed, the functional definition of statistics—'to make a priori separate things hold together, thus lending reality and consistency to larger, more complex objects'[41]— is pretty much the same as the purpose of politics, one in numerical abstraction and the other in human reality. Originally, the English word *classified* had one principal meaning: the sorting of information into taxonomies. During the nineteenth century, as the British state grew in power and its empire grew in scale, *classified* came to hold its additional modern meaning: information held under the exclusive jurisdiction of the state.[42] Even the word *control* has informational origins, deriving from the medieval Latin verb *contrarotulare*, which meant to compare something 'against the rolls', the cylinders of paper that served as official records.[43]

In *Economy and Society* (1922), the most important sociological work of the early twentieth century, Max Weber heralded the 'precision instrument' of *bureaucracy* as the most advanced way of organizing human activity up to that point.[44] Weber's view, vindicated by the next hundred years, was that a unified scheme for the systematic organization of information, premised on '[p]recision, speed, unambiguity, knowledge of the files, continuity, discretion…unity', would be the most effective system of political control.[45] This lesson was not lost on Franklin D. Roosevelt when he assumed the presidency of the United States in 1933. His famous programme of economic intervention was paired with a less glamorous but nonetheless vital effort to transform the federal government's approach to statistics. The years 1935 to 1950 saw the emergence of the 'three essential tools of social statistics and modern economics': surveys based on representative samplings, national accounts, and computing.[46] If Hamilton, Weber, and Roosevelt were around today, they'd all be interested in the gargantuan amounts of data we now produce and the increasingly capable systems we use to process it.

Stepping back, it's now possible to advance a tentative hypothesis that can guide us throughout this book: that how we gather, store, analyse, and communicate our information—in essence how we organize it—is closely related to how we organize our politics. So when a society develops strange and different technologies for information and communication, we should expect political changes as well.

It doesn't appear to be a coincidence, for instance, that the first large-scale civilizations emerged at the same time as the invention of the technology of writing. The earliest written documents, cuneiform tablets first used by the Sumerians in the walled city of Uruk in around 3,500 BC, were wholly administrative in nature, recording taxes, laws, contracts, bills, debts, ownership, and other rudimentary aspects of political life. As James Gleick observes, these tablets 'not only recorded the commerce and the bureaucracy but, in the first place, made them possible'.[47] Other ancient civilizations mushroomed on the back of script, then the most powerful known method of gathering, storing, analysing, and communicating information. In *Empire and Communications* (1950), Harold Innis explains how the monarchies of Egypt and Persia, together with the Roman empire and the city-states of that time, were all 'essentially products of writing'.[48]

Closer to our time, some of the earliest computing systems were actually developed so that governments could make more sense of the data they had gathered. Herman Hollerith, originally hired to work on the 1880 United States census, developed a system of punch-cards and a 'tabulating machine' to process census data, which he leased to the US government. Building on his earlier invention, Hollerith went on to found the International Business Machines Corporation, now better known as IBM.[49]

Turning from the past to the future, we need to ask how transformative digital technologies—technologies of information and communication—will affect *our* system of politics.

That's the question at the heart of this book.

We know that technology's effects vary from place to place. The introduction of printing technology in China and Korea, for example, did not cause the same kind of transformation that followed the introduction of the Gutenberg press in Europe, where society was better primed for religious and political upheaval.[50] Differences like these can usually be explained by economic and political circumstances. Who owns and controls the given technology, how it's received by the public, whether its possible uses are contemplated in advance, and whether it is directed toward a particular end, will all affect its impact.

This means we should be slow to assume that the development of a given technology will inevitably or inescapably lead to a given social outcome. Think of the internet: because its network structure was inherently well-suited to decentralized and non-hierarchical organization, many confidently predicted that online life would be quite different from that found in the offline world. But that's not quite how things turned out. Largely as a result of the commercial and political world into which it was born, the internet has increasingly come under the direction and control of large corporate and political entities that filter and shape our online experience.

Additionally, we can't assume that technology means progress. In *What Technology Wants* (2010), Kevin Kelly memorably shows that ours is not the first age in which the beneficial promise of technology was massively overhyped. Alfred Nobel, who invented dynamite, believed that his explosives would be a stronger deterrent to war 'than a thousand world conventions'. The inventor of the machine gun believed his creation would 'make war impossible'. In the 1890s, the early days of the telephone, AT&T's chief engineer announced, 'Someday we will build up a world telephone system ... which will join all the people of the Earth into one brotherhood.' Still optimistic in 1912, Guglielmo Marconi, inventor of the radio, announced that, 'the coming of the wireless era will make war impossible, because it will make war ridiculous.' In 1917 Orville Wright predicted that the aeroplane would 'make war impossible' and in the same year Jules Verne declared, 'The submarine may be the cause of bringing battle to a stoppage altogether, for fleets will become useless...war will become impossible.' As Kelly explains, these creations, together with the torpedo, the hot air balloon, poison gas, land mines, missiles, and laser guns, were all heralded as inventions which would lead to the end of war.[51] None did.

While Lenin described Communism as 'Soviet power plus the electrification of the whole country',[52] Trotsky saw that technological progress was no guarantor of moral progress. '[A]longside the twentieth century', he wrote, there lives 'the tenth or thirteenth':

A hundred million people use electricity and still believe in the magic power of signs and exorcism... What inexhaustible reserves they possess of darkness, ignorance and savagery!... Everything that should have been eliminated from the national organism in the... course of the unhindered development of society comes out today gushing from the throat.[53]

We cannot take any particular outcome for granted. Nor can we assume that our moral faculties will automatically develop along with our inventions. Everything is still to play for.

Overview of *Future Politics*

Despite its futuristic subject-matter, this book is structured the old-fashioned way. It's meant to be read from start to finish (although tech fanatics may prefer to flick over bits of Part I).

Part I lays the foundations. It sketches out a vision of the future with three defining features. The first is *increasingly capable systems*: machines that are equal or superior to humans in a range of tasks and activities (chapter one). The second is *increasingly integrated technology*: technology that surrounds us all the time, embedded in the physical and built environment (chapter two). The third is *increasingly quantified society*: more and more human activity (our actions, utterances, movements, emotions) captured and recorded as data, then sorted, stored, and processed by digital systems (chapter three). The term I use to describe this future is the *digital lifeworld*, a dense and teeming system that links human beings, powerful machines, and abundant data in a web of great complexity and delicacy.

Chapter four, 'Thinking Like a Theorist', surveys the political and intellectual challenges thrown up by the digital lifeworld, and the theoretical tools we have to address those challenges.

Part II turns to the future of *power*. Its central argument is that certain technologies will be a source of great power in the digital lifeworld (chapter five). Some of these technologies will exert power by applying a kind of *force* to human beings. Imagine a self-driving car that refuses to park on a yellow line, or a shopping app that won't process orders for materials that look like those needed to make a bomb (chapter six). Others will exert power through *scrutiny*, by gathering and storing intimate details about us, and even predicting our behaviour before it happens (chapter seven). A final set of technologies will exert power by controlling our *perception*. These platforms will be able to filter what we know of the wider world, set the political agenda, guide our thinking, stoke our feelings, and mobilize our prejudices to a greater extent even than the media barons of the past (chapter eight).

These three forms of power—*force*, *scrutiny*, and *perception-control*—are as old as politics itself. What's new is that digital technology will give them a potency that far exceeds any previous instruments of power known to humankind. The main consequence for politics, I suggest, will be that those who control these technologies of power will be increasingly able to control the rest of us. Two groups stand to benefit the most: political authorities and big tech firms. That's the focus of chapter nine.

This change in the nature of power will affect every aspect of political life. Part III looks at the implications for liberty. On the one hand, new inventions will allow us to act and think in entirely new ways, unleashing exciting new forms of creation, self-expression, and self-fulfilment. On the other hand, we should expect to see a radical increase in the capacity of political author-ities to enforce the law, leading to a corresponding reduction in what we are able to get away with. In short, the digital lifeworld will be home to systems of law enforcement that are arguably *too* effective for the flawed and imperfect human beings they govern (chapter ten). What's more, an increasing number of our most cher-ished freedoms—including the freedom to think, speak, travel, and assemble—will increasingly be entrusted to private tech firms, whose engineers and lawyers will design and operate the systems through which those freedoms are exercised. For freedom of speech, we'll rely on the restraint of social media and communications platforms; for freedom of thought, we'll depend on the trustworthiness of news and search algorithms; for moral autonomy, we'll rely on the judgement of those who determine what we can and can't do with their digital systems. That's chapter eleven.

Growth in the power of political and tech élites will, I suggest, urgently require parallel growth in the power of citizens to hold those élites to account. That's the focus of Part IV, which considers the future of democracy. It suggests a number of ways in which democracy might be transformed, for better or worse, by greater human participation in the form of Direct Democracy, Deliberative

Democracy, and Wiki Democracy; or by greater machine involvement in the form of Data Democracy and AI Democracy (chapters twelve and thirteen).

In Part V we turn to the future of social justice. In the digital lifeworld, I suggest, algorithms will play a central role in the *distribution* of important social goods, like jobs, loans, housing, and insurance (chapter fourteen). Algorithms will also increasingly be used to rank, rate, score, and sort us into social hierarchies of status and esteem (chapter fifteen). Who is seen and who remains unseen? Who finds popularity and who is forgotten? Who matters and who, in social terms, might as well not exist? These are important questions of *recognition*. Both distribution and recognition are essential to social justice, and previously they were left to the market, the state, and society. In the digital lifeworld, questions of social justice will depend, in large part, on the decisions taken by those who curate the relevant algorithms.

The digital lifeworld will also give rise to new and strange forms of injustice. Think of the online passport system in New Zealand that rejected the passport photograph of a man of Asian descent because it concluded his eyes were closed.[54] Or the voice-recognition systems that couldn't recognize women's voices because they had only ever 'heard' the voices of men.[55] Or the online auto-tagging systems that tagged photographs of black people as 'apes' and pictures of concentration camps as 'sport' and 'jungle gym'.[56] These are real examples, and they're just the beginning. It used to be that only humans could humiliate and degrade us. No longer. The implications for social justice are profound (chapter sixteen).

There are compelling reasons to suspect that the digital lifeworld could give rise to serious economic disparities between rich and poor, particularly as digital systems come to perform more and more tasks previously only done by humans, potentially leading to large-scale technological unemployment (chapter seventeen). Chapter eighteen addresses the concern that the future economy might favour only an élite class of 'owners' of productive technologies, while a struggling majority is left to fight over a shrinking share of the pie. I call this the Wealth Cyclone. To avert this outcome, I suggest, we may need to revisit the very notion of *property* itself.

The central danger identified in this book is that gradually, and perhaps initially without noticing, we become increasingly subjugated to digital systems that we can scarcely understand, let alone control. That would place us, in turn, at the mercy of those who control those digital systems. In chapter nineteen I suggest two ways in which such a fate might be avoided. The first is *transparency*: to make sure that those who have the power to affect our core freedoms, or to affect the democratic process, or to settle matters of social justice, are never allowed to operate in darkness. The second is what I call the *new separation of powers*: to make sure that no entity is allowed to secure control over more than one of the means of *force*, *scrutiny*, and *perception-control*, or achieve a monopoly over any of them.

The book closes with a brief foray into a time *after* the digital lifeworld, where the world is so transformed that the idea of *politics* itself loses its meaning (chapter twenty).

A few final words before we press on.

This book can only scratch the surface. Whole swathes of political life are left untouched, along with ideas from outside the western philosophical tradition. Some issues are necessarily truncated and simplified. This is both a concession to my own limitations and an act of mercy for readers: any effort to be comprehensive would have resulted in a book of biblical girth. 'May others,' as Wittgenstein says, 'come and do it better.' [57] (Except I mean it.)

I try throughout not to be dogmatic (a difficult thing for a lawyer). My aim is to offer a guide, not a manifesto. Some of the technologies and ideas may be new and unfamiliar, but our aim, at least, is as old as humankind: to be 'at once astonished at the world and yet at home in it'.[58]

PART I

The Digital Lifeworld

'It shall seem to men that they see new destructions in the sky, and the flames descending therefrom shall seem to have taken flight and to flee away in terror; they shall hear creatures of every kind speaking human language; they shall run in a moment, in person, to divers parts of the world without movement; amidst the darkness, they shall see the most radiant splendors. O marvel of mankind! What frenzy has thus impelled you!'

Leonardo Da Vinci (1452–1519)

ONE

Increasingly Capable Systems

'There is nothing that man fears more than the touch of the unknown.' **Elias Canetti, *Crowds and Power* (1984)**

We begin with a survey of where we are now and a sketch of the world to come.

In the next century, politics will be transformed by three developments: increasingly capable systems, increasingly integrated technology, and increasingly quantified society. Together these changes will give rise to a new and different type of collective life: the *digital lifeworld*. The strange word *lifeworld* is taken from the German *lebenswelt*, meaning all of the immediate experiences, activities, and contacts that make up our individual and collective worlds. When you imagine the digital lifeworld, imagine a dense and teeming system that links human beings, powerful machines, and abundant data in a web of great delicacy and complexity. In this Part, I don't presume to evaluate or critique the effects of the technologies I describe. The purpose is to identify and understand them, and then (in chapter four) to inspect the intellectual tools that will help us to think clearly about what it all means for politics.

The next three chapters contain many real-life examples, and the idea is not to hold them all in mind at once. Instead our aim is to glimpse, if only in outline, the future that stalks us. That begins with acknowledging that we will soon live alongside computing machines of extraordinary capability. These are *increasingly capable systems*,[1] and they will be the first defining feature of the digital lifeworld.

Artificial Intelligence

The field of artificial intelligence (AI), which began in earnest in 1943, is concerned with the construction of 'intelligent' digital systems. When I refer to AI here, I am describing systems that can perform tasks that in the past were believed to require the cognitive and creative processes of human beings.[2] Progress hasn't always been smooth but today it is impressive and gathering in speed. There are thousands of activities formerly done only by humans that digital systems can now do faster, more efficiently, with greater precision, and at a different order of magnitude.

AI systems are now close to surpassing humans in their ability to translate natural languages, recognize faces, and mimic human speech.[3] Self-driving vehicles using AI are widely expected to appear on our roads in the next few years (Ford is planning a mass-market model by 2021).[4] In 2016 Microsoft unveiled a speech-recognition AI system that can transcribe human conversation with the same number of errors, or fewer, as professional human transcriptionists.[5] Researchers at Oxford University developed an AI system capable of lip-reading with 93 per cent accuracy, as against a 60 per cent success rate among professional lip-readers.[6] AI systems can already write articles about sports, business, and finance.[7] In 2014, the Associated Press began using algorithms to computerize the production of hundreds of formerly handcrafted earnings reports, producing fifteen times as many as before.[8] AI systems have directed films and created movie trailers.[9] AI 'chatbots' (systems that can 'chat' to you) will soon be taking orders at restaurants.[10]

Ominously, engineers have even built an AI system capable of writing entire speeches in support of a specified political party.[11] It's bad enough that politicians frequently sound like soulless robots; now we have soulless robots that sound like politicians.

Every day, algorithms carry out countless transactions in financial markets on behalf of investors. They're trusted to pursue complex strategies based on changing market conditions. Deep Knowledge Ventures, a Hong Kong-based investor, has appointed an algorithm called VITAL (Validating Investment Tool for Advancing Life Sciences) to its board of directors.[12] In medicine, AI systems can now differentiate between lung cancers and predict survival periods better than human pathologists. Researchers believe the same is likely to be true of other cancers in due course.[13] In law, an AI system correctly guessed the outcomes of 79 per cent of hundreds of cases heard at the European Court of Human Rights.[14] Lethal autonomous weapons systems are under development. That's missiles, armed drones, and weaponized robots supported by AI. If deployed into the theatre of war, they'd have the capacity to select targets based on certain criteria before homing in and destroying them—potentially, in due course, without intervening human decision-making.[15]

Games of skill and strategy are considered a good way to gauge the increasing capability of digital systems. In short, they now beat the finest human players in almost every single one, including backgammon (1979), checkers (1994), and chess, in which IBM's Deep Blue famously defeated world champion Garry Kasparov (1997). In 2016, to general astonishment, Google DeepMind's AI system AlphaGo defeated Korean Grandmaster Lee Sedol 4–1 at the ancient game of Go, deploying dazzling and innovative tactics in a game exponentially more complex than chess. 'I . . . was able to get one single win,' said Lee Sedol rather poignantly; 'I wouldn't exchange it for anything in the world.'[16]

A year later, a version of AlphaGo called AlphaGo Master thrashed Ke Jie, the finest human player, in a 3–0 clean sweep.[17]

A radically more powerful version was developed shortly thereafter, called AlphaGo Zero. AlphaGo Zero beat AlphaGo Master 100 times in a row.[18] Now even AlphaGo Zero has been superseded—by AlphaZero.[19]

As long ago as 2011, IBM's Watson vanquished the two all-time greatest human champions at *Jeopardy!*—a TV game show in which the moderator presents general knowledge 'answers' relating to sports, science, pop culture, history, art, literature, and other fields and the contestants are required to provide the 'questions'. *Jeopardy!* demands deep and wide-ranging knowledge, the ability to process natural language (including wordplay), retrieve relevant information, and answer using an acceptable form of speech—all before the other contestants do the same.[20] The human champions were no match for Watson, whose victory marked a milestone in the development of artificial intelligence. This was a system that could answer questions 'on any topic under the sun...more accurately and quickly than the best human beings'.[21] The version of Watson used on *Jeopardy!* was said to be the size of a bedroom; by the early 2020s it's expected that the same technology, vastly improved, will sit comfortably in a smartphone-sized computer.[22] Today, what IBM calls 'Watson' no longer exists in a single physical space but is distributed across a 'cloud' of servers that can be accessed by commercial customers on their computers and smartphones.[23] As IBM is keen to stress, different versions of Watson do more than win game shows. In late 2016, one Watson platform discovered five genes linked to amyotrophic lateral sclerosis (ALS), a degenerative disease that can lead to paralysis and death. The system made its discovery after digesting all the published literature on ALS and parsing every gene in the human genome. This took Watson a matter of months; humans would have taken years.[24] In early 2017, Fukoku Mutual Life Insurance in Japan sacked thirty-four of its staff and replaced them with Watson's 'Explorer' platform, which will chomp through tens of thousands of medical records and certificates, data on hospital stays, and surgical information to calculate payouts to policyholders.

AI has spawned a multiplicity of sub-fields, each applying different methods to a wide range of problems. There is a spectrum of approach, for instance, between those who seek to recreate the neural engineering of the human brain, just as 'early designs for flying machines included flapping wings', and those who employ entirely new techniques tailored for artificial computation.[25] Some researchers seek the holy grail of an *artificial general intelligence* like the human mind, endowed with consciousness, creativity, common sense, and the ability to 'think' abstractly across different environments. One way to achieve this goal might be *whole-brain emulation*, currently being pursued in the Blue Brain project in Switzerland. This involves trying to map, simulate, and replicate the activity of the (more than) 80 billion neurons and tens of trillions of synapses in the human brain, together with the workings of the central nervous system.[26] Whole-brain emulation remains a remote prospect but is not thought to be technically impossible.[27] As Murray Shanahan argues, our own brains are proof that it's physically possible to assemble 'billions of ultra-low-power, nano-scale components into a device capable of human-level intelligence'.[28]

Most contemporary AI research, however, is not concerned with artificial general intelligence or whole-brain emulation. Rather, it is geared toward creating machines capable of performing specific, often quite narrow, tasks with an extraordinary degree of efficacy. AlphaGo, Deep Blue, and Watson did not possess 'minds' like those of a human being. Deep Blue, whose only function was to play chess, used 'brute number-crunching force' to process hundreds of millions of positions each second, generating every possible move for up to twenty or so moves.[29]

It's tempting to get hung up on the distinction between machines that have a narrow field of cognitive capacity and those able to 'think' or solve problems more generally. The latter is a juicier target than the former. But distinctions between 'narrow' and 'broad', or 'strong' and 'weak' AI, can obscure the fact that even narrowly focused AI systems will create vast new opportunities and risks worthy of careful attention in their own right. Soon computers will be able to do things that humans can do, even if they don't do them in the same way—and a lot more besides. And it doesn't matter that one system may only be able to perform a small number of tasks. At the very least, it looks like the digital lifeworld will play host to a multitude of overlapping AI systems, each engineered to perform specific functions. And for those of us on the receiving end, it may be impossible to distinguish between a system that possesses general intelligence, and one that uses fifty different sub-systems to give the *impression* of general intelligence. In most important respects, the effect will be the same.

Machine Learning

The most important sub-field of AI at present is *machine learning*. As Pedro Domingos explains in his book *The Master Algorithm* (2015), the traditional way of getting a computer to do something was 'to write down an algorithm'—a series of instructions to the computer— 'explaining…in painstaking detail' how to perform that task.[30] In contrast with an ordinary algorithm, a *machine learning algorithm* can discover on its own how to recognize patterns, create models, and perform tasks. It does this by churning through large amounts of data, identifying patterns, and drawing inferences. Machine learning algorithms can learn both knowledge ('if a thing looks like X it is a Y') and skills ('if the road curves left, turn the wheel left').[31] The idea is that after a certain point 'we don't have to program computers'; instead 'they program themselves'.[32]

Many of the AI systems I have described employ machine learning techniques. Indeed, machine learning algorithms are all around us:[33]

> Amazon's algorithm, more than any one person, determines what books are read in the world today. The NSA's algorithms decide whether you're a potential terrorist. Climate models decide what's a safe level of carbon dioxide in the atmosphere. Stock-picking models drive the economy more than most of us do.

When the time comes for you to take your first ride in a self-driving car, remember that:[34]

> no engineer wrote an algorithm instructing it, step-by-step, how to get from A to B. No one knows how to program a car to drive, and no one needs to, because a car equipped with a learning algorithm picks it up by observing what the driver does.

Machine learning, to borrow from Domingos, is the automation of automation itself.[35] It's a profound development because it liberates AI systems from the limitations of their human creators. Facebook's engineers, among others, are working on a machine learning algorithm that can build other machine learning algorithms.[36]

Machine learning algorithms generally 'learn' in one of three ways. In *supervised* learning, the human programmer sets a series of defined outcomes and provides the machine with feedback about whether it's meeting them. By contrast, in *unsupervised* learning, the machine is fed data and left to look for patterns by itself. An unsupervised machine can therefore be used to '*discover* knowledge', that is, to make connections of which its human programmers were totally unaware.[37] In *reinforcement* learning, the machine is given 'rewards' and 'punishments' telling it whether what it did was right or wrong. The machine self-improves.

Many of the advances described in this chapter, particularly those involving images, speech, and text, are the result of so-called 'deep learning' techniques that use 'neural networks' inspired by the structure of animal brains. Google launched one in 2012, integrating 1,000 large computers with more than a billion connections. This computer was presented with 10 million 'random' images from YouTube videos. It was not told what to look for, and the images were not labelled. After three days, one unit had learned to identify human faces and another had learned to respond to images of a cat's face (this is YouTube after all).[38] Engineers at Google now use 'duelling' neural networks to train *each other*: one AI system creates realistic images while a second AI system plays the role of critic, trying to work out whether they're fake or real.[39]

The rapid increase in the use of deep learning can be seen from the AI systems used in games. The version of Deep Blue that beat Garry Kasparov at chess in 1997 was programmed with many general principles of good play. What's most remarkable about AlphaGoZero and AlphaZero—the latest and most powerful incarnations of the Go-playing AI systems—however, is that they 'learned' not by playing against the very best humans or even learning from human play, but by playing against *themselves* over and over again, rapidly improving over time.[40]

Machine learning has been around for a while. Its rapid growth and success in the last couple of decades has been enabled in part by the development of new algorithms, but mostly by the explosion in processing power and the growth in available data. Data is critical to machine learning. Too little will stunt the growth of a machine learning algorithm, but given enough, 'a learning program that's only a few hundred lines long can easily generate a program with millions of lines, and it can do this again and again for different problems.'[41] This is why data has been called 'the new coal'[42] and those who gather it are called 'data miners'.

As we'll see at various points in this book, however, reliance on flawed real-world data can cause havoc with machine learning systems. Microsoft launched its AI chatbot Tay on Twitter on 23 March 2016. Tay was intended to mimic the speech of a nineteen-year-old girl and to learn from interactions with other Twitter users. Sixteen hours after its launch, Tay was removed from active duty after posting a series of racist and sexually inflammatory tweets, including one which captioned a photo of Adolf Hitler with the tag 'swag alert', and another saying 'Fuck my robot pussy daddy I'm such a naughty robot'. Tay had 'learned' to communicate this way from other users on Twitter. This example says as much about humans on social media as it does about machine learning.

A final point about machine learning: it used to be that the computing power that fuelled any particular system was physically present within the system in question. The most powerful digital devices literally contained the processors that made them run. The arrival of cloud computing in the last decade or so has meant that the computing power no longer needs to be located within the device itself. Like Apple's Siri it can be accessed over the internet. This has major implications for the integration of technology, as it means that small devices can draw on big computing resources (chapter two). But it's also important for machine learning because it means that machines don't need to 'learn' from their own separate experiences; they can learn from others' too, so that every machine in a swarm or fleet adds to the collective 'intelligence' of the whole.

Exponential Growth in Computing Power

Progress in artificial intelligence and machine learning has been underpinned by advances in mathematics, philosophy, and neuroscience. But as mentioned, it has depended above all on two developments: an explosion in the amount of available data, and an explosion in computing power.

For the last fifty years or so, computing power—the ability of computer chips to process data—has grown at an exponential rate, doubling roughly every two years. This progress is generally expected to continue. On current trends a computer in 2029 will be sixty-four times faster than it was in 2017. If the technology continued to improve at the same rate, then in 2041 it would be 4,096 times faster. After thirty years, the computer would have grown *millions* of times more powerful. Ray Kurzweil and others predict that within the next decade or so, a normal desktop machine (costing $1,000 or thereabouts) will rival and surpass the processing power of the human brain. By 2050, 'one thousand dollars of computing will exceed the processing power of all human brains on earth'.[43] If this sounds unlikely, look back to where we have come from. Just thirty years ago, it would have needed 5,000 desktop computers to rival the processing power of today's iPad Air.[44] Sixty years ago, 2010's iPad2 (now hopelessly out of date) would have cost $100 trillion, roughly twenty-five times the United States federal budget for 2015.[45] The average smartphone has more processing power than the Apollo Guidance Computer that sent Neil Armstrong to the moon.[46]

Our brains are not wired to think exponentially. We tend to think of change as happening in a straight upward line, not noticing when the underlying rate of change is itself accelerating. To put it in perspective, try (in Pedro Domingos' example) to imagine a single *E. Coli* bacterium of miniscule proportions. This bacterium divides in two and doubles in size roughly every fifteen or twenty minutes. Assuming the right conditions, after a few hours it will mushroom into a colony of bacteria, but still too small for the human eye to see. After twenty-four hours, however, that *E.coli* microbe will swell into a bacterial mass the size of planet Earth itself.[47] That's the power of exponential growth.

The theory that computer processing power doubles roughly every two years is often referred to as Moore's Law. This 'law', which is not actually a law but rather an observable pattern of development, has been called 'Silicon Valley's guiding principle, like all ten commandments wrapped into one'.[48] It originated with a 1965 article by Gordon Moore, the co-founder of Intel, who predicted that the number of components that could be crammed onto an integrated circuit would double roughly every two years. At the time, Moore predicted that this trend would continue 'for at least ten years'.[49] Others were sceptical, giving it a couple of years at best. Moore's Law has now persisted for more than five decades. As Walter Isaacson notes, it became more than a prediction: it was a 'goal for the industry, which made it partly self-fulfilling'.[50] Interestingly, processing power is not the only technology improving at an exponential rate. A host of others, including hard-disk capacity, bandwidth, magnetic data storage, pixel density, microchip density, random access memory, photonic transmission, DNA sequencing, and brain-scan resolution, are all developing in the same way.[51] If Moore's Law continues, the next few decades will witness the arrival of machines of remarkable capability. A world in which every desktop-sized machine has the processing power of *all of humanity* will be radically different from our own.

It's sometimes said that Moore's Law will grind to a halt in the next few years, mostly because it will become physically impossible to cram any more transistors into the same microchip, and because the economic efficiencies enjoyed in the last half-century are set to diminish. There is certainly some evidence of a slowdown, although Moore's Law has been given its last rites countless times in the past.[52] However, it is probably wrong to assume that the *current* computing paradigm—the integration of transistors onto 2D wafers of silicon (*the integrated circuit*)—is the final computing paradigm, and cannot itself be improved upon by some other method. History, market forces, and common sense all suggest otherwise. Before the integrated circuit, computers were built using individual transistors. Before that, in the days of Alan Turing, they relied on vacuum tubes, relays, and electromechanics. The story of computing is the story of a succession of increasingly powerful methods of processing information, each developing exponentially, reaching its physical limitations, and then being replaced by something better. Exponential growth in computing processing power stretches back to the seventeenth century and 'the mechanical devices of Pascal'.[53] Nothing is inevitable, but Moore's Law did not begin with the integrated circuit and it is unlikely to end with it either.

The interesting question is what will come next. A host of new methods are already in development, aimed at reaching the frontier of silicon-based computing and pushing beyond it. One approach has been to use non-silicon materials in chips for the first time.[54] Another possibility is to move from the current paradigm of '2D' integrated circuits—where transistors are arranged side-by-side on a wafer of silicon—to a '3D' approach where transistors are piled high.[55] Another approach might be to abandon silicon altogether in favour of carbon nanotubes as the material for building even smaller, more efficient transistors.[56] Yet another approach, currently taken by Google, would be to use more special-purpose computer chips for particular functions—chips that do fewer things but much faster.[57] Microsoft increasingly uses a new type of chip that could combine much greater speed with flexibility.[58]

In late 2019, it was widely reported that engineers at Google had achieved 'quantum supremacy', solving a particular problem in a significantly shorter time than would have been expected of a classical computer. 'Quantum computing' could offer an entirely new—and much more powerful—computing paradigm. Another possible alternative to silicon might be to use 2D graphene-like compounds and 'spintronic' materials, which compute by harnessing the spin of electrons rather than moving them around.[59] There's also the growing field of neuroelectronics, which seeks to reverse-engineer the neural networks of the human brain while potentially requiring less power than silicon.[60] In the still longer term, Quantum Dot Cellular Automata ('QDCA') technology may yield an unimaginably small semiconductor capable of doing the work of a transistor, but using much less power and wasting little energy.[61]

Many of these technologies are still in their infancy and nothing certain can be said about the future of Moore's Law. But the least likely outcome is that computer science simply grinds to a halt, with hungry young Silicon Valley engineers hanging up their circuit boards and heading for retirement. Whatever the next computing paradigm turns out to be, it's certainly not unreasonable to assume that computing power will continue to grow at the same rate as it has since Pascal's invention of the calculator 400 years ago.

TWO

Increasingly Integrated Technology

'a new world has come into existence; but it exists only in fragments.' **Lewis Mumford, *Technics and Civilization* (1934)**

In the digital lifeworld, technology will permeate our world, inseparable from our daily experience and embedded in physical structures and objects that we never regarded previously as 'technology'. Our lives will play out in a teeming network of connected people and 'smart' things, with little meaningful distinction between human and machine, online and offline, virtual and physical, or, as the author William Gibson puts it, between 'cyberspace' and 'meatspace'.[1] This is what I call *increasingly integrated technology*.

It already feels like we can't escape from digital technology. Take smartphones. It's estimated that more than 90 per cent of us keep ours within 3 feet of our bodies twenty-four hours a day.[2] Sixty-three per cent of Americans check their device every hour. Nearly 10 per cent check every five minutes.[3] It's hard to believe they've only been with us for a decade or so. Yet the *quantity* of digital technology in the world is set to grow massively in the next few decades. Tens of billions, and eventually trillions, of everyday objects, from clothing to household appliances, will be endowed with computer processing power, equipped with sensors, and connected to the internet. These 'smart' devices will be able to make their own decisions by gathering, processing, and acting on the information they absorb from the world around them.[4] As technology and design improves, we may stop noticing that digital objects are even 'technology'. David Rose describes a world of 'enchanted objects'—'ordinary things made extraordinary'.[5] This phenomenon, or variants of it, has been variously called 'the internet of things', 'ubiquitous computing', 'distributed computing', 'ambient intelligence', 'augmented things', and perhaps most elegantly, 'everyware'.[6]

There are five underlying trends. Digital technology is becoming more *pervasive*, more *connective*, more *sensitive*, more *constitutive*, and more *immersive*. Let's look at each in turn.

Pervasive

First, technology is becoming increasingly *pervasive*. Although estimates vary, it's predicted that by 2025 there will be 75 billion devices connected to the internet.[7] Almost inconceivably, the Internet Business Solutions Group at Cisco Systems estimates that 99 per cent of the physical objects in the world will eventually be connected to a network.[8] In such a world, processing power would be so ubiquitous that what we think of as 'computers' would effectively disappear.[9]

At home, refrigerators will monitor what you eat and replenish your online shopping basket; ovens and washing machines will respond to voice commands; coffee machines will brew your beverage when you stir in bed. Sensors will monitor the heat and light in your home, changing the temperature and opening the blinds accordingly. Your house might be protected by 'smart locks' that use biometric information like handprint, face, and retina scans, to control entry and exit.[10]

Although they're not to everyone's taste, wearables are grow-ing in use. Pull on a Ralph Lauren 'PoloTech' shirt and it will monitor your steps, heart rate, and breathing intensity—provid-ing you with personalized performance feedback.[11] Snapchat Spectacles and similar early accoutrements, already on the market, can capture what you see in shareable ten-second clips.[12] In the future, more sophisticated products will supersede the first generation of Nike Fuelbands, Jawbone fitness trackers, Fitbit wristbands, and Apple watches. 'Epidermal electronics'—small stretchy patches worn on the skin—will be able to record your sun exposure, heart rate, and blood oxygenation.[13] Meanwhile, when you toss a ball around the garden, the pigskin itself will record the distance, velocity, spin rate, spiral, and catch rate for post-game analysis.[14]

In public, smart waste bins will know when they are full, high-ways will know when they are cracked, and supermarket shelves will know when they are empty. Each will feed information back to the persons (or machines) responsible for fixing the problem. Smart signs, streetlamps, and traffic lights will interact with the driverless cars that pass by.[15] 'Smart cities' are expected to grow in number. Authorities in Louisville, Kentucky, have already embedded GPS trackers inside inhalers to measure which parts of their cities are hotspots of air pollution.[16]

Connective

As well as permeating the physical world, technology will continue to grow more *connective*, facilitating the exchange of information between people, between people and machines, and between machines themselves. Since the turn of the millennium, the number of people connected to the internet has grown radically, from around 400 million in 2000 to 3.5 billion in 2016.[17] That number is expected to rise to nearly 4.6 billion by 2021.[18] It looks like most of the earth's population will eventually be connected to wireless internet-based networks, not just through desktop computers but via 'smart' devices, smartphones, tablet computers, games consoles, and wearables. Facebook now boasts more than 2 billion active users.[19] Twitter has more than 330 million.[20] YouTube has at least 1 billion.[21]

Digital technologies have also changed the nature of human connectivity as well as its extent. Perhaps the most profound change is the growth of decentralized modes of producing and distributing information, culture, and knowledge. Wikipedia is the most famous example. Together, tens of thousands of contributors from around the world have produced the greatest repository of human knowledge ever assembled, working cooperatively, not for profit, outside the market system, and not under the command of the state. Similarly, file-sharing websites like Tor are increasingly popular, and in 2015 there were more than 1 billion uses of Creative Commons, a collaboration-friendly copyright system that encourages the use and adaptation of content by others without further permission by the originator. As Yochai Benkler argues in *The Wealth of Networks* (2006) and *The Penguin and the Leviathan* (2011), it's not that human nature has changed in the last twenty years to make us more cooperative. Rather, it's that this scale of cooperative behaviour would have been impossible in the past. Connective technology has made it possible.[22]

The last few years have seen the emergence of another technology with potentially far-reaching implications for connectivity and cooperation. This is 'blockchain', invented by the mysterious pioneer (or pioneers) Satoshi Nakamoto. It's best known as the system underpinning the cryptocurrency Bitcoin, launched in 2009. The workings of blockchain are technically complex, but the basic premise can be described simply. Imagine a giant digital ledger (or spreadsheet) of the kind we would previously have put on paper. This ledger contains a record of every transaction that has ever taken place between its users. Every few minutes, it is updated with a new 'block' of information containing all of the last ten minutes' transactions. Every new block refers back to the previous block, creating an unbroken chain of custody of all assets reaching back to their inception.[∝] The ledger is not stored in a single place. Instead, it is stored ('distributed') simultaneously across thousands of computers around the world. For security, it can only be added to, and not changed; it is public and can be scrutinized; and most importantly, transactions are secured by powerful 'public key' cryptography.

Blockchain's social significance is that it enables secure transactions between strangers without the help of a trusted third-party intermediary like a bank, credit card company, or the state. It, purports to solve a longstanding problem in computer science (and politics), which is how to create 'trust', or something like it, between people with no other personal connection. Digital currency is perhaps the most obvious use for blockchain technology, but in theory it can be used to record almost anything, from birth and death certificates to marriage licences.[23] It could also provide solutions to other problems of digital life, such as how to produce and retain control over secure digital 'wallets' or IDs.[24] Looking further ahead,

---

∝ This process is overseen by so-called 'miners', generally paid in cryptocurrency for their efforts, who turn the information in the latest 'block', together with some other information, into a short, seemingly random sequence of letters and numbers known as a 'hash'. Each hash is unique: if just one character in a block is changed, its hash will change completely. As well as the information in the previous block, miners also incorporate the hash of the last block. Because each block's hash incorporates the hash of the block before it, it is very hard to tamper with, because it would mean having to rewrite *previous* blocks, stretching back in time, as well as the latest one.

it's plausible to imagine 'smart' assets managing themselves by combining AI and blockchain: 'Spare bedrooms, empty apartments, or vacant conference rooms could rent themselves out...autonomous agents could manage our homes and office buildings...'[25]

Blockchain also offers a potential means of regulating more complex legal and social relations beyond simple rights of property or usage. A so-called 'smart contract', for instance, is a piece of blockchain software that executes itself automatically under pre-agreed circumstances—like a purchase agreement which automatically transfers the ownership title of a car to a customer once all loan payments have been made.[26] There are early 'Decentralised Autonomous Organisations' (DAOs) that seek to solve problems of collective action without a centralized power structure.[27] Imagine services like Uber or Airbnb, but without any formal organization at the centre pulling the strings.[28] The developers of the Ethereum blockchain, among others, have said they want to use DAOs to replace the state altogether.

Blockchain still presents serious challenges of scale, governance, and security.[29] Yet for a youthful technology it is already delivering some interesting results. The governments of Honduras, Georgia, and Sweden have trialled the use of blockchain to handle land titles,[30] and the government of Estonia is using it to record patient health records.[31] In the UK, the Department for Work and Pensions has piloted a blockchain solution for the payment of welfare benefits.[32] In the US, the Defense Advanced Research Projects Agency (DARPA) is looking into using blockchain technology to protect its military networks and communications.[33]

Increasingly connective technology is not just about people connecting with other people. It is also about increasing connectivity between people and machines—through Siri-like 'oracles' which answer your questions and 'genies' that execute commands.[34] In the future, when you leave your house, 'the same conversation you were having with your vacuum cleaner or robot pet will be carried on seamlessly with your driverless car, as if one "person" inhabited all these devices'.[35] Samsung is looking to put its AI voice assistance Bixby into household appliances, like TVs and refrigerators, to make them responsive to human voice command.[36]

Self-driving cars will communicate with one another to minimize traffic and avoid collisions. Within the home, Bluetooth Mesh technology could increasingly be used to connect 'smart' devices with one another, using every nearby device as a range booster to create a secure network connection between devices that would previously have been out of range.[37] (It's important to note, however, that one of the challenges for the 'internet of things' will be developing a unified protocol that enables devices to communicate seamlessly with one another.)[38]

Looking further ahead, developments in hardware could yield new and astonishing ways of communicating. In 2014, a person using an electroencephalogram (a recording of brain activity, known as an EEG) headset successfully sent a 'thought' to another person wearing a similar device in France, who was able to understand the message. This was the first scientific instance of 'mind-to-mind' communication, also known as telepathy.[39] You can already buy basic brainwave-reading devices, such as the Muse headband, which aims to aid meditation by providing real-time feedback on brain activity.[40] Companies such as NeuroSky sell headsets that allow you to operate apps and play games on your smartphone using only thoughts. The US army has (apparently not very well) flown a helicopter using this kind of technology.[41] Brain–computer interfaces have been the subject of a good deal of attention in Silicon Valley.[42]

Overall, increasingly connective technology appears set to deliver the vision of Tim Berners-Lee, inventor of the world wide web, of 'anything being potentially connected with anything'.[43]

Sensitive

In the future, we can expect a dramatic rise in the number of sensors in the world around us, together with a vast improvement in what they are able to detect. This is increasingly *sensitive* technology. Our handheld devices already contain microphones to measure sound, GPS chips to determine location, cameras to capture images, and several other sensors. Increasingly, the devices around us will use radar, sonar, lidar (the system used in self-driving cars to measure the distance to an object by firing a laser at it), motion sensors, bar code scanners, humidity gauges, pressure sensors, magnetometers, barometers, accelerometers, and other means of sensing, and hence interacting with, the physical world.

There are many reasons why we might want more sensors in our own homes and devices—for recovering lost or stolen items using GPS, for instance, or monitoring the security or temperature of our homes from afar.[44] Industrial entities, too, benefit from real-time feedback on their machinery, whether in relation to humidity, air pressure, electrical resistivity, or chemical presence. Transit and delivery companies can monitor the workload and stress placed on their fleets. Engineering and architectural firms can measure corrosion rates and stress. Similarly, within water systems, 'sensors can measure water quality, pressure and flow, enabling real-time management and maintenance of the pipes'.[45] Automatic meter reading technology feeds back usage data to utility providers, allowing them to detect faults, match supply with demand, and send out bills automatically—with little or no human intervention.[46]

Municipal authorities already recognize the value of a 'dense sensor network' to enable 'the monitoring of different conditions across a system or place'.[47] Automated number plate recognition technology can be used to track vehicles as they cross a city and to levy penalties for traffic violations.[48] The city of Santander in Spain has distributed 12,000 sensors in urban areas to measure 'noise, temperature, ambient light levels, carbon monoxide concentration, and the availability and location of parking spaces'.[49] Following its mission in Afghanistan, the US military left 1,500 'unattended ground sensors' to monitor Afghan and Pakistani population movement.[50] Researchers at the Senseable City Lab at the Massachusetts Institute of Technology are working on a cheap package of sensors to be put on top of street lights, which would make it possible to measure noise and pollution 'almost house by house in real time'.[51] More remarkable still are plans, presently on hold, for the development of 'PlanIT Valley' east of the city of Porto, in Portugal. This city would use an 'Urban Operating System' to gather information from more than 100 million embedded sensors, feeding the data back to applications that monitor and control the city's systems.[52]

From the macro to the micro, 'smart dust' technology involves micro-electromechanical systems measuring less than 2 millimetres by 2 millimetres, equipped with tiny sensors capable of gathering a variety of data. One pilot study called 'Underworlds' seeks to harness the 'data' that is 'flushed down the toilet.' It envisages small robots moving through sewers, collecting samples for analysis, and measuring patterns of food intake, infectious diseases, and gastric health.[53]

Sensors are also moving into sensory realms previously only experienced by living creatures. One company, for instance, is developing a mobile chemical sensor able to 'smell' and 'taste'. (One hopes that the sewer-robots described in the previous paragraph are not endowed with this ability.) Helpfully, your smartphone will be able to test your blood alcohol level, blood glucose level, or whether you have halitosis, using about 2,000 sensors to detect aromas and flavours—far more than the 400 sensors in the human nose.[54] Scientists at MIT recently developed a type of spinach—implanted with nanoparticles and carbon nanotubes—capable of detecting nitro-aromatics in the soil around it and sending live feedback to a smartphone. The result? Bomb-detecting spinach.[55] (At last, someone has found a use for spinach.)

In the field of machine vision, AI systems are increasingly able to find the most important part of an image and generate an accurate verbal caption of what they 'see' (e.g. 'people shopping in an outdoor market'),[56] and computerized face recognition is now so advanced that it is routinely used for security purposes at border crossings in Europe and Australia.[57] Less loftily, face recognition technology is used by the toilet paper dispensers in machines at Beijing's Temple of Heaven park to make sure that no individual takes more than their fair share.[58]

Increasingly sensitive technology will prompt a change in how we urge machines to do our bidding. We are currently in the era of the 'glass slab'—smartphones and tablet computers that respond mainly to our touch, and other stimuli such as voice commands.[59] Soon, machines will respond to other forms of command, such as eye movements[60] or gesture: there are already robotic toys that 'sit' in response to a specific wave of the hand. Some interfaces will be of an entirely new kind, like the temporary tattoos developed by MIT which can be used to control your smartphone,[61] or the Electrick spray-paint that turns any object into a sensor capable of reading finger presses like a touchscreen.[62] In 2015, workers at the Epicenter hub in Stockholm implanted microchips into their hands, enabling them to open secure doors and operate photocopiers by waving over a sensor.[63]

The most intimately sensitive technologies will gather data directly from our bodies. Proteus Biomedical and Novartis have developed a 'smart pill' that can tell your smartphone how your body is reacting to medication.[64] Neuroprosthetics, still at an early stage of development, interact directly with nerve tissue. A chip implanted in the motor cortex of a paralysed patient enabled him to spell out words by moving a cursor on a screen with his thoughts.[65] In a survey of 800 executives conducted for the World Economic Forum, 82 per cent expected that the first implantable smartphone would be available commercially by 2025.[66] By then, smartphones will have truly become what US Supreme Court Chief Justice John Roberts called, 'an important feature of human anatomy'.[67]

Machines are becoming sensitive in a further important sense, in that they are increasingly able to detect human emotions. This is the field of *affective computing*. By looking at a human face, such systems can tell in real time whether that person is happy, confused, surprised, or disgusted. One developer claims to have built 'the world's largest emotional data repository with nearly 4 million faces', from which the system has learned to interpret subtle emotional cues.[68] Raffi Khatchadourian writes in the *New Yorker* that:[69]

> computers can now outperform most people in distinguishing social smiles from those triggered by spontaneous joy, and in differentiating between faked pain and genuine pain. They can determine if a patient is depressed...they can register expressions so fleeting that they are unknown even to the person making them.

If Ludwig Wittgenstein is right that the face is the 'soul of the body',[70] then affective computing will mark a spiritual upheaval in the relationship between humans and machines. And the face is not the only portal into our internal life:[71]

> Emotion is communicated, for example, through...body movement that can be measured, for instance, by gyroscopic sensors; posture, detected through pressure-sensing chairs; and skin-conductance electrodes can pick up indicative changes in perspiration or in electrical resistance. It is also possible to infer emotional states from humans' blinking patterns, head tilts and velocity, nods, heart rate, muscle tension, breathing rate, and, as might be expected, by electrical activity in the brain.

Machines are well placed to detect these signals. For instance, it's possible to use the vocal pitch, rhythm, and intensity of a conversation between a woman and a child to determine whether the woman is the child's mother.[72] By bouncing ordinary WiFi signals off the human body, researchers at MIT claim to be able to determine, about 70 per cent of the time, the emotional state of a person they have never studied before. This rate improves with people known to the system.[73] Another biometric is human gait (manner of walking), which AI systems can use to identify a known person from afar, or even to recognize suspicious behaviour in strangers.[74]

As well as reading our emotions, machines can increasingly adapt and respond to them too. This is *artificial emotional intelligence*. Its uses are manifold—from ATMs able to understand if you are in a relaxed mood and therefore receptive to advertising[75] to AI 'companions' endowed with 'faces' and 'eyes' that can respond in 'seemingly emotional ways'.[76] Technologists are already working to replicate the most intimate connection of all, with artificial romantic partners capable of sexy speech and motion.[77]

Constitutive

By increasingly *constitutive* technology, I mean digital technology that makes itself felt in the hard, physical world of atoms and not just the 'cyber' world of bits. In large part this is the province of robotics. The practice of building mechanical automata dates back at least 2,000 years to Hero of Alexandria, who built a self-powered three-wheel cart.[78] The earliest reference to the idea of an autonomous humanoid machine is the Golem of Jewish lore. In the imagination of Jorge Luis Borges the Golem was 'a mannequin shaped with awkward hands', which:

raised its sleepy eyelids,

saw forms and colors that it did not understand,

and confused by our babble

made fearful movements

Modern robotics remains a challenging field, in part due to Moravec's paradox, which is that (contrary to what might be expected) 'high-level reasoning requires very little computation, but low-level sensorimotor skills require enormous computation resources.'[79] Thus it has always been easier to design problem-solving machines than to endow them with balance or athletic prowess equal to that of a human or animal. We still don't have robots we would trust to cut our hair.

Nevertheless, the world population of robots is now said to number at least 10 million,[80] of which more than 1 million perform useful work (robots, for instance, now account for 80 per cent of the work in manufacturing a car).[81] Amazon's robots, which look like roving footstools, bring goods out of storage and carry them to human employees.[82] Most crop spraying in Japan is done by unmanned drones.[83] In 2016 about 300,000 or so new industrial robots were installed,[84] and global spending on robotics is expected to be more than four times higher in 2025 than it was in 2010.[85]

We already trust robotic systems to perform complex and important tasks. Foremost among these is surgery. Using advanced robotics, a team of surgeons in the United States was able to remove the gall bladder of a woman in France, nearly 4,000 miles across the Atlantic.[86] Perhaps the most commonplace robots in the future will be self-driving cars, able to navigate the physical world safely 'without getting tired or distracted'.[87] Google's fleet of autonomous vehicles has driven more than 2 million miles with only a handful of incidents, only one of which is said to have been the fault of the vehicle itself.[88] Since human error is the 'certain' cause of at least 80 per cent of all crashes, increased safety will be one of the principal advantages.[89] We are likely to see, in the next decade, driverless trucks and boats, as well as airborne drones of varying autonomy.

Nature has been the inspiration for many recent developments in robotic locomotion. Some robots can 'break themselves up and re-assemble the parts, sometimes adopting a new shape—like a worm (able to traverse a narrow pipe), or a ball or multi-legged creature (suitable to level or rough ground respectively).'[90] At Harvard, researchers are working on RoboBees, measuring less than an inch and weighing less than one-tenth of a gram. They fly using 'artificial muscles' comprised of materials that contract when a voltage is applied. Potential applications include crop pollination, search and rescue, surveillance, and high-resolution weather and climate mapping.[91] Work is underway on robotic cockroaches[92] and 'spiders, snakes, dragonflies, and butterflies that can fly, crawl, and hop into caves, cracks, crevices, and behind enemy lines'.[93] Researchers in the field of 'soft robotics' have developed the 'Octobot', a thumb-sized autonomous mollusc made from soft silicone rubber without any rigid structures in its body.[94]

Companionship is an increasingly important function of robots. Toyota's palm-sized humanoid Kirobo Mini is designed to provoke a similar emotional response as a baby human.[95] 'Paro' is a cuddly interactive baby seal with 'charming black eyes and luxurious eyelashes'. It appears to be beneficial for elderly people and those with dementia. Future models will monitor owners' vital signs, sending alerts to human carers where necessary.[96] Zenbo, which costs the same as a smartphone, is a cute two-wheeled robot with a round 'head', equipped with cameras and sensitive to touch. It can move independently, respond to voice commands, and display emotions on its screen-face.[97]

The potential uses for robots are limited only by our creativity. In 2016, Russian authorities 'arrested' the humanoid 'Promobot', which was canvassing attendees at a rally on behalf of a candidate for the Russian parliament. After failing to handcuff the offender, the police eventually managed to escort it from the premises. Promobot reportedly put up no resistance.[98]

Nanotechnology, the field for which the 2016 Nobel Prize in Chemistry was awarded, is another burgeoning area of research. It involves the construction of devices so small that they are measured in nanometres—one-billionth of a metre. The nanoscale is 1–100 nanometres in size. A red blood cell, by comparison, is 7,000 nanometres wide.[99] The possibilities of nanotechnology are mind-boggling: nanobots can already 'swim through our bodies, relaying images, delivering targeted drugs, and attacking particular cells with a precision that makes even the finest of surgeons' blades look blunt'.[100] There are nanobots that can release drugs in response to human thought, potentially enabling them to detect and prevent an attack of epilepsy at the precise moment it occurs. Another less salubrious application of the same technology would be to 'keep you at the perfect pitch of drunkenness, activated on demand'.[101] Nanotechnology also has implications for data storage. Researchers at Delft University in the Netherlands have created an 'atomic hard drive' capable of storing 500 terabits of information in a single square inch. Put another way, it could store the entire contents of the US Library of Congress in a cube measuring 0.1 mm each way.[102]

Another constitutive technology is 3D printing, also known as additive manufacturing. It enables us to print physical things from digital designs. Some think it could herald an era of 'desktop manufacturing' in which many people have 3D printers in their home or office and can 'print' a wide range of objects.[103] Or municipal 3D printers could allow people to print what they need using open-source online digital templates.[104] So far, some of the most useful 3D-printed objects have been in medicine. Printing splints for broken

limbs is now relatively common,[105] and customized replacement
tracheae (windpipes) can now be printed in fifteen minutes.[106]
Surgeons have printed stents, prosthetics, and even bespoke replacement segments of human skull.[107] Researchers at Cornell University
have printed a human ear.[108] Human kidneys, livers, and other
organs, as well as blood vessels, are in development.[109] A 3D-printed
exoskeleton embedded with bionic technology has restored mobility
to people unable to walk.[110]

Outside medicine, 3D printers have been used to make full-sized
replica motorbikes,[111] bikinis,[112] aeroplane parts,[113] entire houses,[114]
synthetic chemical compounds (i.e. drugs),[115] and replicas of sixteenth-
century sculptures.[116] Food is an area of growth, with 3D-printed
chocolate, candy, pizza, ravioli, and chickpea nuggets all 'on the
menu'.[117] Eventually, it is predicted, a plethora of materials will be
used as ingredients for 3D-printing, including plastics, aluminium,
ceramic, stainless steel, and advanced alloys. Producing these materials used to be the work of an entire factory.[118] '4D' printing is also
in the works—intended to create materials programmed to change
shape or properties over time.[119]

Immersive

Humankind, wrote T. S. Eliot, '[c]annot bear very much reality'. In
the future, we won't have to. Technology will become radically
more immersive, as a result of developments in augmented and
virtual reality.

In the mid-twentieth century, the computer was 'a room' and if
we wanted to work with it, we had to 'walk inside'. Programming
often meant 'using a screwdriver'.[120] Later, the 'desktop' became
the primary interface between humans and computers, where

information on a screen could be manipulated by means of a keyboard and, later, a mouse.[121] As noted earlier, we're now in the era of the 'glass slab'.[122]

The term 'augmented reality' (referred to as AR) is used to describe the enhancement of our sensory experience of the physical world by computer-generated input, such as sound, graphics, or video. Smart glasses, still in their infancy, allow the wearer to experience digital images overlaid onto the physical world. They might show directions to the park, or assembly instructions for a new wardrobe. They might identify a wild bird or flower. They could even provide facts about the person the wearer is talking to—helpful for politicians expected to remember thousands of faces and names. In audial AR, Google has already developed earbuds said to be able to translate forty foreign languages in almost real time.[123] The most celebrated early AR application is Snapchat's *Lenses*, which allow selfie-takers to edit their portraits with animations and filters.

Another prominent (though slightly faddish) application is the smartphone game *Pokémon Go*, which overlays the real world with fantastical beasts to capture and train. Victory in *Pokémon Go* doesn't come from lounging at a digital terminal; the player must seek glory in the real world of physical space. In one unfortunate incident, a Koffing (a spherical beast filled with noxious gases) was found to be roaming the Holocaust Museum in Washington, DC.[124] The game has provoked some unreasonably strong feelings, with Saudi clerics declaring it 'un-Islamic' and a Cossack leader saying it 'smacks of Satanism'.[125] Democracy protesters have used *Pokémon Go* as a pretext for holding illegal meetings in Hong Kong, where, otherwise, gatherings must be legally registered and authorized.[126] Holograms are another form of AR. A controversial gangster-rapper located in California 'performed' in Indiana via hologram. (The concert was shut down.) Protesters in Spain staged a hologrammatic virtual protest in a public space from which they had been banned.[127]

In time, more advanced AR will make it almost impossible to distinguish between reality and virtuality, even when both are being experienced simultaneously. The secretive startup MagicLeap is working on a 'tiny 3D technology that can shine images on your retinas' which 'blends the real world with fantasy'.[128]

More profound still is the emergence of virtual reality (VR). When you put on a VR headset, you enter and experience a vivid three-dimensional world. Touch controllers bring your hands in with you as well.[129] 'Haptic' clothing gives you sensual feedback through tiny vibrating motors spread over your body— you *really* don't want to get stabbed or shot.[130] Once inside, you're free to see, feel, explore, and interact with a new dimension of existence. While AR technologies operate within the real world, VR technologies create an entirely new one. Technology giants including Facebook (Oculus Rift), Microsoft (HoloLens), Samsung (Gear VR), Google (Daydream), and Playstation (Playstation VR) are already in fierce competition to develop the best VR hardware.

Virtual reality feels remarkably real. After a few moments of adjustment, even resistance, users' senses begin to adapt to the new universe around them. As time passes, disbelief is suspended and sensual memory of the 'real' world, as being something separate, begins to fade. I can affirm, from first-hand experience, that even a simple racing game can stimulate real feelings of exhilaration and fear. When testing an early VR racing system, my 'car' spun off the track and hurtled toward a steel barricade. Momentarily, I believed I was about to die. (For what it's worth, my life did not flash before my eyes. In fact, the whole episode was rather less dramatic than I might have expected for a final reckoning.) Reporters have spoken of what it is like to be sexually assaulted in VR: even though the groping is not 'real' in the physical sense, it can cause lasting feelings of shock and violation.[131]

While much of today's focus on VR is on gaming, in due course VR will be used to experience a great deal of life. Workers will attend virtual meetings, shoppers will peruse virtual supermarkets, sports fans will frequent virtual stadiums, artists will create in virtual studios, political philosophers will pontificate in virtual cafés, historians will wander virtual battlefields, socialites will hang out in virtual bars, and punters will seek out virtual brothels. Importantly, the experience in each case will not be limited by the constraints of the 'real' world—VR can generate entirely new worlds where ordinary rules (whether law, norms, or even the rules of physics) do not apply. Imagine being, in VR, an astronaut ferociously battling alien craft, an antelope galloping across the Serengeti, or a squid swimming through the deep. This isn't about making the real virtual; it's about making the virtual seem real. Eventually we may live in a *mixed reality*, where VR and AR are so advanced that the digital and physical become indistinguishable. In such a world it will be hard, and possibly futile, to discern where 'technology' begins and ends.

THREE
Increasingly Quantified Society

'The first and most basic rule is *to consider social facts as things*...To treat phenomena as things is to treat them as data, and this constitutes the starting point for science'
Émile Durkheim, *The Rules of Sociological Method* (1895)

In the digital lifeworld, a growing amount of social activity will be captured and recorded as data then sorted, stored, and processed by digital systems. More and more of our actions, utterances, movements, relationships, emotions, and beliefs will leave a permanent or semi-permanent digital mark. As well as chronicling human life, data will increasingly be gathered on the natural world, the activity of machines, and the built environment. All this data, in turn, will be used for commercial purposes, to train machine learning systems, and to predict and control human behaviour. This is *increasingly quantified society*.

The twenty-first century has seen an explosion in the amount of data generated and processed by human beings and machines. In 2020 there is expected to be at least 40 zettabytes of data in the world—the equivalent of more than 3 million books for every living person.[1] Today, humans generate roughly the same amount of information every couple of hours as they did from the dawn of civilization until 2003.[2] Already we

create as much every ten minutes as the first ten thousand generations of humans combined.[3] Like computer processing power,
the speed with which we produce information is expected to
continue to grow exponentially.[4]

What is data, and where is it all coming from?

In *Big Data* (2013), Viktor Mayer-Schönberger and Kenneth
Cukier explain that data is 'a description of something that allows it
to be recorded, analyzed, and reorganized'. The process of turning a
phenomenon into data has been called datafication.[5] We have data-
fied and *digitized* (turned into binary code legible by machines)
enormous swathes of activity on earth. As late as 2000, only a quarter
or so of the world's stored information was in a digital form. Now, it
is more than 98 per cent.[6] Four factors have made this possible. First,
much more data is *gathered*, because an increasing amounts of social
activity is undertaken by and through digital systems and platforms.
Second, in the last fifty years the cost of digital *storage* has halved
every two years or so, while increasing in density '50-million fold'.[7]
Third, the explosion in computational power has given us the ability
to *process* what we store. Fourth, digital information has almost no
marginal cost of reproduction—it can be replicated millions of times
very cheaply. Together, these factors explain why the transition from
a print-based information system to a digital one has yielded such an
explosion of data.

Mayer-Schönberger and Cukier compare current developments
with the last 'information revolution': the invention of the printing
press by Johannes Gutenberg nearly 600 years ago. In the fifty years
following Gutenberg's innovation, more than 8 million books were
printed. This change was described as 'revolutionary' by the scholar
Elizabeth Eisenstein because it meant, in all likelihood, that more
books had been printed in half a century than had been handwrit-
ten by 'all the scribes in Europe' in the previous 1,200 years.[8] Yet if
it took fifty or so years for the amount of data in existence to

double in Gutenberg's day, consider that the same feat is now being achieved roughly every two years.[9]

Much of the data in the world originates with human beings. Sometimes we deliberately bring it into existence, as when we use our devices to record and communicate. Every day we send billions of emails[10] and fire off hundreds of millions of tweets.[11] Even when they don't seem rich in data, these communicative acts can capture the internal life of humans in a way that was previously impracticable. Even something as paltry as a tweet, limited to 280 characters, is deceptively rich in information. It includes thirty-three items of metadata ('information about information') which can be quite revealing in aggregate:[12]

> an analysis of 509 million tweets over two years from 2.4 million people in 84 countries showed that people's moods followed similar daily and weekly patterns across cultures around the world—something that had not been possible to spot before. Moods have been datafied.

Away from social media platforms, some people deliberately choose to monitor the data emitted by their bodies—generally for health and wellness reasons, but sometimes for fun or curiosity. For a small group, the phenomenon of *sousveillance* goes beyond breathing rate and pulse. There are plans for:[13]

> comprehensive life-logs that would create a unified, digital record of an individual's experiences...a continuous, searchable, analysable record of the past that includes every action, every event, every conversation, every location visited, every material expression of an individual's life, as well as physiological conditions inside the body and external conditions (e.g. orientation, temperature, levels of pollution).

Needless to say, much of this data is shared with the manufacturers of the devices in question. If we so choose, the deepest workings of our bodies are now quite datafiable, right down to the information

contained in our DNA. It took 'a decade of intensive work' to
decode the human genome by 2003. The same task can now be
done in a day.[14]

Even when not consciously creating or hoarding data, we leave
a 'data exhaust' just by going about our lives.[15] The trail of digital
breadcrumbs we leave is discreetly hoovered up by the devices
on or around us. Some, like our tax and phone records, are fairly
mundane. Others are less so, like applications on smartphones,
which use GPS to trace and record our location, even when location has nothing to do with the app in question. According to
Marc Goodman, 80 per cent of Android apps behave in this way.[16]
In 2012, researchers were able to use a smartphone-cellular system (another way of tracking location, in addition to GPS) to
predict, within 20 metres, where someone would be *twenty-four
hours later*.[17] Some 82 per cent of apps track your general online
activity.[18]

We submit around 60,000 search requests to Google every second—more than 3.5 billion each day.[19] Each one, together with
what Google knows about the identity of the searcher, is fed into
Google's growing silo of information. If all the data processed in
one day by Google were printed in books, and those books were
stacked on top of each other, then the pile would now reach
more than halfway from earth to the moon. That's just each day.[20]
Facebook, too, holds a remarkable amount of information about
each of its users. When Max Schrems, an Austrian privacy activist
who used Facebook occasionally over a three-year period, asked to
see the personal data about him stored by Facebook, he received
a CD-ROM containing a 1,222-page document, including phone
numbers and email addresses of his friends and family; the devices
he had used to log in; events to which he had been invited;
his 'friends' and former 'friends'; and an archive of his private
messages—including transcripts of messages he thought he had

deleted. Even this cache was probably incomplete: it excluded, for instance, facial recognition data and information about his website usage.[21] Mr Schrems was just one of (then nearly, now more than) 2 billion active users, from whom Facebook has built an extraordinary rich profile of human life.

Finally, data is increasingly generated by machines. Some are juggernauts belching out large amounts of data. When firing, the Large Hadron Collider at CERN generates 40 terabytes of data every *second*.[22] In its first few weeks of operation in 2000, the Sloan Digital Sky Survey telescope harvested more data than had been previously been gathered in the history of astronomy.[23] In the future, the largest data-contributors will be pervasive devices distributed around the planet. Mid-range cars already contain multiple microprocessors and sensors, allowing them to upload performance data to carmakers when the vehicle is serviced.[24]

Data scientists have always wrestled with the challenge of turning raw *data* into *information* (by cleaning, processing, and organizing it), then into *knowledge* (by analysing and interpreting it).[25] The arrival of big data has required some methodological innovation. As Mayer-Schönberger and Cukier explain, the benefit of analysing vast amounts of data about a topic rather than using a small representative sample has depended upon data scientists' willingness to accept 'data's real-world messiness' rather than seeking precision.[26] In the 1990s IBM launched Candide, its effort to automate language translation using ten years' worth of high-quality transcripts from the Canadian parliament. When Google began developing its translation system in 2006, it took a different approach, harvesting many more documents from across the internet. Google's scruffy dataset of around 95 billion English sentences, including translations of poor or middling quality, vastly outperformed Candide's repository of 3 million well-translated

English sentences. It's not that Google's initial algorithm was superior. What made the difference was that Google's unfiltered and imperfect dataset was tens of thousands of times larger than Candide's. The Google approach treated language as 'messy data with which to judge probabilities', an approach that proved to be considerably more effective.[27]

Data is valuable, and the more is gathered in one place, the more its value increases. When we search the web, for instance, the contents of each search are of infinitesimal value—but when searches are aggregated they offer a profound window into searchers' thoughts, beliefs, concerns, health, market activity, musical tastes, sexual preferences, and much more besides. We surrender our personal data in exchange for free services—something I call the *Data Deal* in chapter eighteen. The commercial value of Facebook lies primarily in the data that it harvests from its users, which can be used for a range of purposes, from targeted advertising to building face recognition AI systems. When Facebook went public in 2012, each person's profile was estimated as being worth $100 to the company.[28] Famously, when sales improved after book recommendations were generated by algorithms rather than people, Amazon sacked all its in-house book reviewers. This is why data has been called a 'raw material of business' and a 'factor of production', and 'the new coal'.[29] The ensuing rush has spawned a multi-billion-dollar industry 'that does nothing except buy and sell the personal data we share online'.[30]

It's not just businesses that are interested in big data. Governments are too—from municipal regimes designing smart cities to central governments using it to monitor compliance. The British tax authorities, for instance, use a fraud detection system that holds more data than the British Library (which has copies of every book ever published in the United Kingdom).[31] It is also increasingly apparent that governments use personal data for global surveillance. Two US National Security Agency (NSA) internal databases code-named HAPPYFOOT and FASCIA contain comprehensive location information of electronic devices worldwide.[32]

An increasingly quantified society is one that is more available for examination and analysis by machines and those who control them. As more and more social activity is captured in data, systems endowed with exceptional computational power will be able to build increasingly nuanced digital maps of human life—massive, incredibly detailed, and updated in real time. These schematics, abstracted from the real world but faithfully reflecting it, will be invaluable not just to those who wish to sell us stuff, but those who seek to understand and govern our collective life. And when political authorities use data not just to study or influence human behaviour, but to *predict* what will happen before we even know it—whether a convict will reoffend, whether a patient will die—the implications are profound. As explained in the Introduction, there has always been a close connection between information and control. In an increasingly quantified society, that connection assumes even greater importance.

No Way Back?

The future described in the last three chapters is not inevitable. In theory at least, we could halt the innovation already in progress, so that the digital lifeworld never comes into existence. But this is unlikely. Innovation is driven by powerful individual and shared human desires—for prosperity, security, safety, convenience, comfort, and connectivity—all of which are nurtured by a market system designed to stimulate and satisfy those desires. My view is that politics in the future will largely unfold within the parameters of the lifeworld generated by these new technologies, with debate centred on how they should be used, owned, controlled, and distributed, and not on whether we can force the genie back into the lamp. In chapter four, we consider how to think clearly and critically about what this means for politics.

FOUR

Thinking Like a Theorist

'To grasp the world of today we are using a language made for the world of yesterday. And the life of the past seems a better reflection of our nature, for the simple reason that it is a better reflection of our language.'

Antoine de Saint-Exupery, *Wind, Sand, and Stars* (1939)

In the past, advances in science and technology helped humans to strip away some of the mystery of the world. Max Weber, writing at the turn of the twentieth century, identified the central feature of modernity as *Entzauberung*, translated as *de-magification* or *disenchantment*. This was the process by which magic and superstition were replaced by rational observation as the preferred means for explaining the mysteries of life. The present generation may be the first to experience the opposite effect, the *re-magification* of the world. As time goes on we'll increasingly find ourselves surrounded by technologies of extraordinary power, subtlety, and complexity—most of which we can barely understand, let alone control. 'Any sufficiently advanced technology,' says Arthur C. Clarke, 'is indistinguishable from magic.'[1] If that's right, then the digital lifeworld holds in store a magic show quite unlike anything we've seen.

Before now, we've never had to coexist with nonhuman systems of exceptional power and autonomy. We've never lived in a world where technology is seamlessly integrated into the fabric of social life—pervasive, connective, sensitive, constitutive, and immersive all at once. We don't know what it's like for large swathes of our lives to be recorded, tracked, and processed. As we journey onward into the digital lifeworld, the main risk is that we lose track of our political and moral intuitions, unwilling or unprepared to think critically about changes that we come to take for granted. In this chapter, we examine the intellectual tools that can help us prevent this from happening.

We begin with the concept of politics itself, looking to find an appropriate definition for the twenty-first century. We then turn to political concepts more generally—the building blocks that frame and structure the way we think about politics. Next, we take a brief look at the discipline of political theory, which is concerned with the development and analysis of such concepts. Finally, we consider a bold new endeavour within that discipline, dedicated to under-standing the *future* of political ideas. Politics, language, and time: these are the themes of this chapter.[2] By the end, we will be well prepared for the analysis in Part II.

What is Politics?

You may be relieved to learn that there is no inherently right or wrong definition of 'politics'. Like all concepts, it is a construct of the human mind, invented to describe phenomena in the world. No activity comes labelled 'political' or 'non-political'.[3] Every linguistic community—that is, every group of people sharing the same expectations about the use of language—is free to say that some things are political and are others are not.

Most people have a rough idea of what they think politics is: how and why we live together, the ways we govern ourselves, and so forth.

But we can be more precise. For some, politics is synonymous with government: the process by which lawmakers decide on society's collective goals and devise laws and policies through which they can be realized. On this view, politics is something that takes place in parliaments, government buildings, and town halls, under the control of politicians and civil servants. The average citizen may participate in politics (so understood) to a greater or lesser extent, through voting or activism, depending on both the nature of the state and that citizen's own inclinations.

An alternative view is that politics can be found almost every-where,[4] not just in the public realm but in private too: between friends and colleagues, and within families; in clubs, teams, and religious establishments; in government but also in art, architecture, science, literature, and embedded in language itself. Politics is present wherever there is cooperation, conflict, or control; or wherever it is possible that some particular social relation might be ordered differently, from workplace politics to sexual politics. On this view, politics isn't something you can avoid or ignore.

Politics can therefore be understood narrowly or broadly, with a range of perspectives in between. And within every perspective there will always be a clutch of hotly contested sub-perspectives. One area of controversy, for instance, is whether a dictatorship can be called a political system. Some scholars say that reconciling differing interests through political institutions (like parliaments or congresses) is the very essence of politics. One person ruling brutally in his own interests isn't merely a bad political system: it's not a political system at all.[5] Others disagree, arguing that this perspective wrongly elevates one particular *conception* of politics—a liberal one—above all others, including dictatorship, which can qualify as political just the same. Another grey area is whether *war* is part of politics as opposed to something distinct. The Prussian general Carl von Clausewitz famously believed that war is the 'continuation of politics by other means', yet scholars such as Sir Bernard Crick and Hannah Arendt argue that war represents the *breakdown* of politics, not politics itself.[6]

You may already be clutching your cranium in despair. How are we supposed to make sense of the future of politics when we can't even agree on what politics *is*? The good news is that, as I mentioned before, there is no inherently right or wrong answer. The bad news is that it does still matter—a lot—which definition we choose to adopt. This is because a too-narrow definition can wrongly exclude certain topics from the political agenda. A classic example relates to the treatment of gender and sexuality. As Judith Squires argues, a narrow conception of politics-as-government excludes, by definition, the private sphere of domesticity and sexuality from political discourse.[7] This is significant because, as anyone who has chaired a meeting knows, the surest way to stop something being done or changed is to prevent it from being on the agenda in the first place.[8] A definition of politics that only includes the formal institutions of government means that the issue of male violence against women is never even discussed. This gives rise to a political discourse which is not merely incomplete, but prejudicial. This—brace yourself—has been called 'the politics of politics'.[9]

But hold on, it might be objected, what's wrong with saying that gendered violence should be part of the discussion, but it doesn't have to be part of *political* discussion? Just as scientists shouldn't lose sleep over whether a particular discovery can properly be called a development in biology, chemistry, or medicine, why should it matter whether a thing counts as 'politics' or not?

There are at least two answers to these questions. The first is that political discourse is closely linked to political power: if a particular topic is part of mainstream political discussion, on the lips of politicians, commentators, academics, lobbyists, and campaigners, it's likely to have more impact than one that is siphoned off into a separate conversation. The second answer, related to the first, is that politics has a special quality of seriousness and gravity. Yes, politics can be tawdry, sleazy, and frustrating; and political discourse is often shallow and crude. But to say something is a *political* issue is, implicitly, to say that it's an issue of importance, of relevance to the community as a whole. Campaigners recognize this. It's why the women's liberation movement in the 1970s fought so hard to convince the world not merely that the personal is relevant, or the personal is important, but that *the personal is political.* It's partly why, borrowing the phrase, I now say *the digital is political.*

How we choose to define politics will affect our ability to think incisively about politics in the digital lifeworld. There is a risk, in adopting crusty or timeworn definitions of politics, that we blind ourselves to new developments that would not fall under the old definition but which are plainly political in nature. For example, a popular view among scholars is that what makes politics truly distinctive is the notion of the use of *force* by the state.[10] According to this view, the state is distinguished from other forms of collective association, like golf clubs, by the fact that it alone can compel you to obey its rules by the (more or less legitimate) threat or actual use of force. If you don't pay your taxes, state officials can deprive you of your liberty by handcuffing you and hauling you off to prison. By contrast, if you fail to pay your membership fees at the golf club, the club cannot imprison you in the caddyshack. However, the 'force' definition of politics assumes that humans only govern themselves using rules set by the state, backed by the threat or use of force. While this may have been a reasonable assumption in the past, it won't be in the future. As the next Part shows, in the digital lifeworld we'll increasingly be subject to new forms of control where rules are embedded in the technology we encounter. A self-driving car that cannot drive over the speed limit because its software prevents it from doing so presents a quite different constraint on a human driver than a car which can drive at that speed but whose driver refrains from doing so in the face of potential punishment by the state. This raises fundamental questions of power and liberty—the very stuff of politics—even though no one has necessarily been forced to do anything by the state. The 'force' definition is thus inadequate because it excludes a relevant line of inquiry from the start.

I propose a broad and inclusive definition of politics that allows us to be confident that when we think about politics in the future, we don't inadvertently close our minds to new and radical social formations not yet come into being. It's as follows:

> *Politics* refers to the collective life of human beings, including why we live together, how we order and bind our collective life, and the ways in which we could or should order and bind that collective life differently.

This definition does not presuppose any particular form of political system. In fact, all it assumes is (a) that humans will continue to live collectively, (b) that our collective life is capable of being ordered and bound in more than one way. These are minimal suppositions, allowing us to proceed with an open mind.

Political Concepts

Imagine it's election season and you are watching a political advertisement. The clip shows a politician, mid-speech, standing behind a podium. He wears a dark suit and a warm smile. He's reaching the climax of his speech: 'Liberty. Justice. Democracy.' With each word of the slogan, he brings his fist emphatically into an open palm. The crowd cheers and he continues, lifting his voice over the applause, 'You know that's what I stand for. And I promise you today, that if you give me your vote, I will make these ideals a reality for you, your families, and all the people of this great nation. That is my pledge.' The oration ends, the crowd erupts, the politician waves, confetti cannon explode, music pumps, the advertisement finishes, and you switch off the television.

Has he won your support?

Liberty, justice, and democracy. Your first thought is that the country could certainly do with more of those. But after a moment you begin to wonder: if he's for liberty, justice, and democracy, what's the other candidate for? Unfreedom? Injustice? Tyranny? It can't be that simple. In fact, you begin to recall, doesn't she *also* say that she's in favour of those things? Isn't it just that her ideas of liberty, justice, and democracy mean something different?

Next, imagine that you're at a dinner party and the person next to you begins to lecture you about her political views. You listen, politely and noncommittally, as she gets more and more animated, concluding, with a flourish of her fork, 'anyway, it doesn't matter what I think. Or what you think. We don't live in a democracy any more. All the power in this country is in the hands of business élites and the mainstream media.' Walking home that evening, you wonder whether you agree with what she said. If by 'democracy' she meant the formal system by which citizens of a country elect political leaders to generate legislation, then you don't agree. Private interests may have a powerful influence on the legislative process—too powerful in your view—but that hasn't done away with the democratic system altogether. But if by 'democracy' she meant something broader, like the general principle that every person ought to have an equal say in the political decisions that govern their lives, then maybe she has a point. What hope do ordinary folk have to make their voices heard equally when corporate lobbyists can schmooze politicians in expensive hotels and resorts? And what did she mean by *power*? If she meant that business and the media literally control the country—setting the agenda, making rules, enforcing them—then that would be going too far. Perhaps what she really meant was that business, lobbyists, and the media disproportionately influence the laws of the land.

Power, liberty, democracy, justice—these are fundamental political concepts. We use them when thinking and speaking about politics.[11] Yet as the tales of the politician and the dinner party show, each concept can have more than one meaning. It's possible, for example, that I define the concept *liberty* as being freedom from government intrusion, while you define it as the ability to choose your own goals and pursue them to the best of your natural ability. Both of us are still using the *concept* of liberty acceptably. Neither of us is wrong. We say that these are different *conceptions* of the same *concept*.[12] The only limit on the way we use concepts is that each one has some part of it, some irreducible core, that cannot be taken away without the concept ceasing to mean what it means.[13] The irreducible core of *liberty*, arguably, is *the absence of restraint*—which was present in both our definitions. We just differed on the nature of the restraint in question—government intrusion or lack of autonomy.

Who determines what the irreducible core of a given concept should be? Everyone. Or at least, everyone in the same linguistic community. The meaning of a concept is determined by how it's used by the community as a whole. I can't sensibly use a word to mean one thing if everyone else thinks it means something different. That's why I can't meaningfully say that liberty means ice cream, or that justice means the colour blue: there wouldn't be enough peers in my linguistic community who recognized that as an acceptable definition. Concepts have no objectively right or wrong definition; what matters is whether a particular usage would be acceptable—or becoming acceptable—'to a significant number' of its users.[14]

The difference between linguistic communities' usage of concepts can be crucial. When a British person hears the word 'liberal', she might think of someone who supports minimal government inter- ference in the economy and civil society. An American who hears the word 'liberal' may imagine precisely the opposite—something more akin to a socialist who favours a larger welfare state. The differences are even starker when time, as well as place, is introduced as a variable. Take the concept of *property*. To us, *property* generally refers to *things*. This seems obvious, but it was not always the case. In Babylon in around 1776 BC, property could just as easily refer to *people*, as children were owned by their parents. If I killed your daughter, my own daughter would be executed in restitution.[15] Similarly, today we consider it obvious that property can be bought and sold. Yet in very early Greek and Roman law, the sale of property was pretty much forbidden, as it belonged not to individuals but to families, including dead ancestors and unborn descendants.[16] Hence Plato's Athenian in the *Laws*:[17]

> you are the owners neither of yourselves nor of this property, which belongs to your family as a whole, past and to come

There are many other examples. In seventeenth-century England, the concept of *revolution*, borrowed from astronomy, meant the restoration of a previous form of government. After the French Revolution in 1789 it assumed entirely the opposite meaning— that of abrupt and violent political change.[18] The term 'new media' is widely used today to describe online platforms like Facebook, Twitter, and Reddit. Back in the early 1940s, the US Supreme Court used the term to refer to a truck carrying a bullhorn.[19]

It's not just political concepts that change in meaning across time and place. The same has historically been true of terms in science and technology. Before Sir Isaac Newton, for instance, the words *force*, *mass*, *motion*, and *time* had no precise scientific meaning. Newton endowed them with the strict definitions that have lasted for centuries. Similarly, in the nineteenth century, natural philosophers 'mathematicized' the word *energy*, which previously meant vigour or intensity, into a central concept in the discipline of physics.[20] Today, a computer is a machine that sits on your desk. A hundred years ago, a computer was a person, usually a woman, who performed arithmetic and tabulation.[21]

Why might the same concept mean something different in a different time or place? In short, communities adapt their language to the political, social, and cultural needs of their time. As Karl Marx observes, the 'production of ideas, of conceptions, of consciousness,' is 'at first directly interwoven with the material activity and the material intercourse of men.'[22] What we call our *perspective* is drawn deeply from our cultural and social environment. For this reason, the sociologist Karl Mannheim argues in *Ideology and Utopia* (1929) that even *knowledge* can be dated according to its historical style, just like works of art or architecture. 'Strictly speaking' he says:[23]

> it is incorrect to say that the single individual thinks. Rather it is more correct to insist that he participates in thinking further what other men have thought before him.

A still more radical position, championed by Marx, holds that in every epoch 'the ruling ideas' are 'the ideas of the ruling class'.[24] On this view, a particular group of people produces and distributes ideas that serve its own selfish interests.

We do not need to go as far as Marx or even Mannheim to see the wisdom in what they say. Political concepts do not fall from the sky. Each one is dreamed up and used by persons living and thinking in a particular time and place. As the examples of *property* and *revolution* show, when times change, we adapt our old concepts to deal with the new facts of life. Old ideas can be salvaged and repurposed for new ends. As you'll see in the remainder of this book, many venerable concepts, categories, distinctions, theories, and arguments can (with a little care) continue to guide us in the future as they did in the past.

Sometimes, however, an epoch breaks so comprehensively from the past that it requires the coinage of entirely new concepts. 'In such busie, and active times,' said Thomas Sprat in 1667, 'there arise more new thoughts of men, which must be signifi'd, and varied by new expressions.'[25] In *The Age of Revolution* (1996) Eric Hobsbawm lists some of the words that were invented or gained their modern meanings in the short period from 1789 to 1848:[26]

> 'industry', 'industrialist', 'factory', 'middle class', 'capitalism', and 'socialism' . . . 'aristocracy' . . . 'railway', 'liberal' and 'conservative' as political terms, 'nationality', 'scientist', and 'engineer', 'proletariat', and (economic) 'crisis'. 'Utilitarian' and 'statistics', 'sociology' and several other names of modern sciences, 'journalism' and 'ideology' . . . 'strike' and 'pauperism'.

What new words have come about in our own time? Since 2000, the *Oxford English Dictionary* has admitted the words 'internet', 'CCTV', and 'geek', together with 'wiki', 'microchip', 'metadata', 'machine learning', 'double click', 'cyber-', 'glitch', 'genetic engineering', 'transhumanism', 'text message', 'upload', and 'web site', as well as many others. Some entries, like 'world wide web' and 'Information Superhighway', already sound dated.

Political language is supple: concepts are subject to change depending on the usage of the linguistic community at any given time. There are no eternal concepts, no eternal meanings. This also means that we cannot simply assume that there is any universal political truth, scary though that may be. In a fast-changing world, political theorists have a duty to ask how much of our current wisdom is based on the way we *currently* live, or *used* to live, and whether it would still make sense if our experience was transformed beyond recognition. Cultures that treat ideas as eternal, when in fact they are specific to a particular time, are condemned forever to describe the new in terms of the old. Their concepts, developed to explain and understand the world, instead become 'mind-forg'd manacles' that serve only to obscure it.[27] That's why it is necessary, from time to time, to test whether our political vocabulary is up to the task of interpreting and ordering the world. This, in large part, is the role of political theory. It is also one of the purposes of this book.

Political Theory

'Politics' is also an academic discipline. This book draws on the ideas and methods of one sub-discipline in particular: political theory. (I do not maintain a rigid distinction between the terms 'political theory,' 'political philosophy,' 'political ideas,' and 'political thought'. There is no consensus on what the distinction should be, it's hard to draw, unwieldy to sustain, and of no practical significance.) Political theorists develop and study the concepts we use to think and speak about politics, asking what they *mean*, where they *come from*, and whether it can be said that they are *true*, *false*, *right*, or *wrong*. Political concepts, which we've been looking at, are the substance of political theory. The next few pages are dedicated to understanding its methods. There are three: conceptual analysis, normative analysis, and contextual analysis. Don't worry if you find them unfamiliar to begin with. This book, and not just this chapter, is dedicated to the practice of thinking like a theorist.

Conceptual Analysis

Ambiguity is everywhere in politics—in speeches, conversations, manifestos, pamphlets, articles, books, and blogs. A good deal of political argument, particularly on social media, is dissatisfying because participants misunderstand, misinterpret, or don't bother to find out what the other person actually means. This is where conceptual analysis comes in. As Adam Swift explains, 'Before we know whether we agree with someone . . . we have to know what it is she is saying'.[28] Conceptual analysis is the process of trying to understand what people mean when they say things about politics.[29] It involves subjecting political utterances to careful probing and questioning, teasing out definitions and distinctions in order to find clarity, consistency, and simplicity. In academic political theory, this sort of work is done in books and articles. A large amount has been written, for instance, about the different conceptions of *power*, *liberty*, *justice*, and *democracy*. In university seminars, good professors quiz their students until all ambiguity has been squeezed from their arguments. In ordinary life, conceptual analysis begins less formally, with asking someone, 'what exactly do you mean by that?' To the politician in the story above, we might ask: Liberty to do what? Justice for whom? And how does your vision of democracy differ from the version with which we currently live?

Conceptual analysis can slice through the waffle—and, let's be honest, bullshit—that make up so much political discourse, whether in parliaments, the media, or the academy. Political speech is sometimes *designed* to deceive, shut down argument, obfuscate, and bamboozle; in Orwell's words, 'to make lies sound truthful and murder respectable, and to give an appearance of solidity to pure wind'.[30] And while such trickery might be expected of a politician, it cannot be so easily forgiven in a theorist, whose role is to clarify rather than confuse. Those who insist on saying things in a complicated way, where a simple expression would do, give the discipline a bad name. (This, by the way, is also true of writing on tech.)

Of course, not all political speech can be crystal clear. Some is unclear because the world is complicated and political ideas are not always easy to explain or understand. And some is deliberately stylized or rhetorical, designed to inspire or enrage rather than clarify or classify. Confronted with the rousing battle-cry, 'Give me liberty, or give me death!' only the dorkiest pedant would reply, 'Yes, but, pausing there, what *kind* of liberty are you referring to?' In the cut-and-thrust of real-world politics, a degree of imprecision can give politicians the wiggle-room they need to reach a compromise. This is perhaps what Winston Churchill had in mind when he spoke of the 'enormous and unquestionably helpful part that humbug plays in the social life of great peoples dwelling in a state of democratic freedom'.[31] But there is enough obfuscation in the world. For our purposes it is better to remember Wittgenstein's dictum, that 'what can be said at all can be said clearly'.[32]

Conceptual analysis also involves trying to understand the moral connotations of concepts as well as their meaning. To call something *gorgeous* is to say *both* that it has certain characteristics *and* that those characteristics are in some way desirable. To call something foolish or repulsive is to say that it has certain characteristics *and* that those characteristics are undesirable. The same is true in politics. To describe something as *corruption*, for instance, is both to describe a state of affairs *and* implicitly to ascribe a negative quality to it. Nowadays, to say a process is *democratic* is (generally) to ascribe a positive quality to it. Words like corruption and democracy are *appraisive*—they have a recognized moral connotation as well as a recognized meaning.[33] Not all political speech is appraisive—to call something *property*, for instance, is not necessarily to say something good or bad about it either way. Part of conceptual analysis is trying to understand whether a concept is being used appraisively or not.

Conceptual analysis is particularly helpful when new technologies are the focus of discussion. Technology invites the use of weird terminology which can be vague or hard to understand. Exciting new gadgets can lead even the most sober writers into foaming hyperbole. The result, too often, is analysis that is both unintelligible and hysterical. The purpose of conceptual analysis, which prizes precision above all, is to clear away the fog.

Normative Analysis

'Normative analysis' is simply about trying to work out what is right and wrong, good or bad, in the domain of politics. It is closely related to the disciplines of moral philosophy and ethics. What duties of justice do we owe to one another? Should we be obliged to participate in public life? What should we be free to do, and what should we be prevented from doing? Is it ever acceptable to disobey the law? These are normative questions. Normative analysis allows us, through reflection and argument, to identify principles to guide us in how we live together. It helps us to work out the difference between what *is* and what *ought to be*. As one scholar observes, 'You could pile up a mountain of data about the differences between, say, democracies and dictatorships, but without the normative element ... nothing would follow about which form of government ought to be implemented.'[34] The most influential political theory (John Stuart Mill's *On Liberty* (1859), for instance) tends to consist largely of normative analysis. But normative analysis is also something we commonly engage in during casual discussion. The rights and wrongs of collective life are what make politics interesting.

Normative theories are used to illuminate the moral rightness or wrongness of particular systems, policies, or principles. Sometimes we expect language to do the arguing for us—when we use appraisive terms like *corrupt* or *democratic*—but part of normative analysis is asking not just *whether* a word carries a normative connotation, but whether it *deserves* to. Although today the word *democratic* is positively appraisive, for most of human history it was used as a term of insult (chapter twelve).

Contextual Analysis

Contextual analysis, finally, is the work of finding out where concepts come from, why they came into existence, what they meant in the context in which they were conceived, and how their meaning has changed over time. It is closely linked to the discipline of history, specifically 'intellectual history' or the 'history of ideas'. One reason to study concepts in context is that, as discussed, they change in meaning from time to time and place to place. We can't assume that someone demanding *liberty* in eighteenth-century France meant the same as someone demanding *liberty* in twenty-first century England.

The three methods—conceptual, normative, contextual—can be fruitfully blended together. Conceptual analysis, for example, normally comes before normative analysis: it's helpful to understand the meaning of a political statement before arguing whether it is right or wrong. But the different methods are not always kept separate. When you say, 'Democracy means majority rule' and I say, 'No, democracy means full respect for the rights of minorities', both of us are trying to say what democracy means (a conceptual claim) *and* what it ought to mean (a normative claim). If you then say, 'In America in 2018, democracy means nothing if it does not mean full respect for the rights of minorities', you're making a point that is contextual, conceptual, and normative all at once. The idea is not to insist on the rigid separation of the three methods, but rather to be mindful about when we are thinking conceptually, when normatively, and when contextually. This will help us when we come to think about the future of politics.

The Future of Political Ideas

Political theory is sometimes criticized for not being useful in the rough-and-tumble of real-world politics. Abstract ideas might be interesting in lecture halls and debating societies, it is said, but can do little to solve actual political problems. Theory doesn't make the trains run on time. An idea never fed a hungry child. Your average citizen wants decent local schools and pothole-free roads, not a dreary seminar on Hegel's theory of the dialectic. I have some sympathy with this sort of argument. Too many works of political theory get bogged down in 'navel-gazing, intellectual masturbation'[35]— of the kind that seldom leads to satisfaction. But it doesn't have to be this way. Political theory can help to provide answers to the urgent political problems of the day. Some of our greatest theory has come in times of upheaval—revolutions, world wars, civil wars— when people desperately sought to understand the crises unfolding before them. Our most powerful writing on civil disobedience, for instance, comes from the 1960s, the decade of the civil rights movement and Vietnam.

We are living through a time of equivalent upheaval. So where's the theory?

There are three thoughts to take away from this chapter. The first is that we're all political theorists, whether we like it or not. There is no such thing as a value-free or neutral political stance; every political utterance, act (like voting), or omission (like not voting) contains within it an implicit hierarchy of priorities and values. Political theory matters so much because it brings those priorities into focus and makes them available for rational debate. Second, political theory should be firmly rooted in the facts of life. That way, theory never becomes divorced from reality. Instead of starting with abstract ideas and theories about human nature, or the good life, we should start with the world as we expect to find it—moving from the 'earth' of reality to the 'heaven' of ideas rather than the other way round.[36] Third, and most importantly, political theory should be able to give practical guidance as to how we live together.

My hope is that anyone who cares about the future of politics will contribute to the great effort of imagination that the digital lifeworld is going to require. Thinking about the future is difficult but the theoretical method, at least, ought to be simple: begin by adopting a series of predictions about the future. Then see what light our existing political ideas might shed on such a future. This could mean asking what Alexis de Tocqueville would have thought about the idea of tech-enabled direct democracy, or how Marx's theories could be applied to the ownership and control of AI systems. If it turns out that our concepts are inadequate to describe or critique the world that's emerging, then the final task is to develop a *new* political vocabulary that can better make sense of the world we're building. That's the approach I take from here on in.

Up Next: Part II

We are ready, now, to take the first steps into virgin political territory, beginning with the most fundamental political concept of all: power.

PART II

Future Power

'Where I found a living creature, there I found will to power; and even in the will of the servant I found the will to be master.'

Friedrich Nietzsche, *Thus Spake Zarathustra* **(1891)**

FIVE

Code is Power

'Nearly all men can stand adversity, but if you want to test
a man's character, give him power.'   **Abraham Lincoln**

How do the strong dominate the weak? By exerting *power*: the bedrock of political order and the godfather of political concepts. Any serious effort to understand the future of politics must involve an investigation of what power is, what forms it will take, and who will wield it. That's the work of the next five chapters.

I suggest that power in the future will take three forms. The first is *force* (chapter six). The second is *scrutiny* (chapter seven). The third is *perception-control* (chapter eight). Digital technology will increasingly be the main source of all three. This means that those who control the technologies of force, scrutiny, and perception-control will be powerful; and those who don't will be correspondingly powerless. My contention is that, over time, power will grow more concentrated in the hands of the state and large tech companies.

In this chapter our aim is to clarify what we mean by power and take a closer look at the relationship between power and digital technology. The vision that unfolds may seem a little heartless in places, focused on the brute fact of power rather than its legitimacy. Don't worry: the rights and wrongs come later.

What is Power?

Power is a difficult concept to define, and it's tempting to plump for the definition once given by a judge to the term 'hardcore pornography': I know it when I see it. But while this sort of laxity might pass muster in a US court of law, political theorists try to be a little more precise, particularly given that power is not always something you can *see*. We begin, instead, with the simple distinction between having *power over* someone and having *power to* do something.[1] *Power over* is the power of a boss who tells her subordinate to perform a task, or a schoolteacher who orders her pupils to sit quietly. It's about a ruler getting the ruled to comply. *Power to*, by contrast, does not imply a system of rulers and ruled. It refers to the ability, facility, or capacity of a person to do something—to walk down the street, for instance, or lift a heavy barbell. It is a broad definition. With *power to*, the fact that one person or group is powerful does not require that someone else is correspondingly powerless.

Both of these conceptions of power are linguistically acceptable. Neither is wrong. But at this stage of our inquiry, *power over* is more relevant than *power to* for two reasons. To begin with, if we want to understand the future of politics then we need to know who will be powerful over whom, what forms that power will take, and to what ends it might be directed. Our concern is with the power relations *between* people rather than the ability of each individual to do as he or she pleases. Second, *power to* is similar enough to the distinct concept of *liberty* that we can more sensibly tackle it under that heading in Part III.

Adopting *power over* as our primary focus, then, what does it mean to say that person A has power over person B? There are several possible answers to this question, but it's worth starting with the intuitive one offered by the great political scientist Robert Dahl: 'A has power over B to the extent that he can get B to do something that B would not otherwise do.'[2] This definition covers a variety of situations, from the power of the state to lock up its citizens to the power of a mother to order her child to bed. Implicitly, it also covers situations where A *stops* B from doing something that B would otherwise do. But Dahl's definition is not without its problems. If, through my erratic driving, I inadvertently force you to change lane on the freeway (which you would not otherwise have done), have I exercised power over you? Or would we say that I only exert power when I *intend* to? What about if, in a fit of rage, I demand that my girlfriend returns the birthday gift that she kindly bought for me—even though I like the gift and my welfare would be better served by keeping it? Do I still exert power when it's not in my *interest* to do so?[3] These are questions on which people may legitimately disagree and I don't propose to get bogged down in them. We should, however, ask whether power is always a matter of getting someone to do something they would *not otherwise do*. I have no intention of stealing your wallet, but does that mean that the criminal law has no power over me because it only prevents me from doing something I would never have done anyway? This seems strange. For this reason, it would seem sensible to loosen Dahl's definition to say that A has power over B to the extent that he has the *capacity* to get B to do something that B would not otherwise do.

So what makes a person or entity *powerful*? That depends on three factors. First, if their power has a wide *range*, in terms of geography and the number of people and issues it affects. Thus a national government, whose power extends over all the people within a territory across a wide range of matters, is more powerful than a headteacher whose power is confined to the boundaries of the school and issues relating to its administration. Second, a powerful person or entity is one whose power pertains to matters of *significance*. The power of a judge to determine the liberty or rights of others is greater than the power of the cafeteria server who determines how much potato mash you receive at lunchtime (important though that may seem to you in the hunger of the moment). Finally, the *stability* of power is important. Your power to pin me to the floor until I wriggle free is less stable than the power you would obtain by chaining me to the wall in a dungeon.[4] Pulling together the threads:

> A person or entity is powerful to the extent that it has a stable and wide-ranging capacity to get others to do things of significance that they would not otherwise do, or not to do things they might otherwise have done.

This definition is important. We'll refer to it throughout the book. Together with the understanding of politics set out in chapter four ('the collective life of human beings, including why we live together, how we order and bind our collective life, and the ways in which we could and/or should order and bind that collective life differently') it's broad enough to include new forms of power that might not traditionally have been thought of as 'political'.

The Different Faces of Power

Power comes in a number of forms. To illustrate this, imagine that Matt intends to shoot and kill his friend Larry, but a third person, Kim, stops him. In one scenario, Kim wrestles Matt to the ground and confiscates the firearm, allowing Larry to escape unharmed.

This form of power we call *force*: where Kim strips Matt of the choice between compliance and noncompliance. In another scenario, Kim tells Matt, 'If you murder Larry, I will burn down your house.' Matt decides not to shoot Larry because he does not want his house burned down. In this instance, Kim's power takes the form of *coercion*: Matt complies because he fears punishment or deprivation. Another alternative is for Kim to persuade Matt that shooting Larry would be morally wrong. If she does so successfully, then Kim has exerted *influence*: securing Matt's compliance without resorting to a threat of deprivation. Next, imagine that Kim is a religious leader who sees Matt on the verge of slaying Larry and cries, 'Don't shoot!' Matt complies, not because he fears deprivation or even because he believes that slaying Larry would be wrong, but because he respects Rabbi Kim's *authority* as a source of moral commands. Finally, imagine that Kim stops Matt from murdering Larry by falsely promising him a million dollars if he refrains. In this situation, power takes the form of *manipulation*—where Matt complies because he has not been shown the true basis of the request being made of him.[5]

Force, *coercion*, *influence*, *authority*, and *manipulation* are helpful ways of thinking about different forms of power. But we shouldn't think them inherently right or wrong—that influence and authority, for instance, are morally preferable to force or manipulation. All forms of power may be used for good or ill. Force can be legitimate (a police officer who handcuffs an escaped convict), influence can be malign (ISIS propaganda aimed at young Muslim men), and authority can be abused (paedophile priests). Even manipulation can be morally justifiable: think of an undercover agent who deceives to infiltrate a criminal enterprise.

We turn now to the relationship between digital technology and power.

Digital Technology and Power

Code and Algorithms

All digital systems run on software, also called code. Code consists of a series of instructions to hardware—the physical stuff of technology—telling it what to do. It is written in programming language, not natural language of the kind understood by humans. But the word 'language' is apt to describe it, because it also has its own formal rules of grammar, punctuation, and syntax.[6] One important difference between code and natural language is that code is meant to be much more precise. It seeks to generate unambiguous commands with no grey areas or room for interpretation.

The key thing to remember about digital technology is that it can only operate according to its code. If you ask a calculator what $5 + 5$ is, it can only answer 10. This will be true regardless of how much you might want it to give another answer. The wishes of human users are irrelevant to the functioning of code unless and until it is programmed to take those wishes into account. For the same reason, you can't use a calculator to draft an email.[7] Asking a particular technology to do something it's not programmed to do is like walking into a closet and asking it to take you to the fifth floor.

Code and *algorithms* are often referred to interchangeably, but they're not strictly the same thing. The word *algorithm* can be traced back to the ninth-century Persian mathematician Abd'Abdallah Muhammad ibn Mūsā Al-Khwārizmī. The translation of his name, *algorismus*, came to describe any mathematical approach to reasoning, calculating, and manipulating data.[8] Today, the word *algorithm* describes a set of instructions for performing a task or solving a problem. It need not be written in computer code. A set of driving directions is one form of algorithm, specifying what to do under various conditions: 'Go down the street, turn right at the post office and then left at the lights. If the traffic is bad, take a right at the garage and take the second exit at the roundabout...' Relationship advice often takes algorithmic form: 'If he continues to lie, break up with him. But if he apologizes, cool off for a bit, see how you feel, then try to talk things through...'

When we talk about digital technology, the algorithm is the formula and the code is the expression of that formula in programming language. A great deal of code contains algorithms, whether for making decisions, learning skills, finding patterns, sorting data, or predicting events.

Code is Power

What's this got to do with power? Well, when we interact with digital technologies we *also* necessarily submit to the dictates of their code.[9] To take a simple example, you can't access a password-protected document unless you enter the correct password: the machine has no choice or discretion in the matter, and neither do you.[10] It makes no difference that the document contains vital medical information that could save your life. It also doesn't matter that the only reason you don't know the password is because you forgot it. The code won't allow you to do what you would otherwise do.

We regularly come up against the constraints of code. Think about digital content such as music, films, and e-books. In 2009, President Obama gave the visiting UK Prime Minister Gordon Brown a gift of twenty-five classic US films. Returning to London, however, the most powerful man in the land found that the movies wouldn't play on his UK DVD player.[11] Why? To the untrained eye, it would seem like a bug or glitch. But the opposite was true.

The prohibition had been *coded* into the DVDs by their manufacturers and distributors, to protect their commercial interests and
enforce the law of copyright. This is usually referred to as Digital
Rights Management (DRM).

Because of code's ability to direct our conduct in a finely honed
way, many distinguished thinkers, following the pioneering work
of Harvard professor Lawrence Lessig, have argued that *code is
law* (or at least that code is *like* law).[12] Thinking about the DVD
example, it's clear that code exerted a kind of *force*: Mr Brown was
entirely deprived of the choice whether or not to play the DVD.
But it could just as well have been *coercion*. Imagine that you send
your sister an email containing a digital music file that you downloaded from iTunes. That music may well refuse to play on her
computer, because of DRM rules prohibiting the duplication of
copyrighted content. Again, a kind of *force* would prevent your
sister from doing what she wanted. Now imagine that iTunes did
not actually *prevent* her from accessing the content, but instead that
it *punished* her for trying to do so, by locking her (or your) account
for 24 hours thereafter. If you knew this was going to happen, you
might think twice before downloading and sending the music, and
your sister wouldn't try to listen to it if you did. In this example,
you both behave a certain way because you fear punishment. That's
coercion rather than *force*. In this instance, code certainly shares some
features with law.

But to say that code is law, or even like law, is no longer enough.

Code can also be used to *influence* and *manipulate* people in
ways that don't resemble the workings of the law. As we'll see,
this happens when tech is used to scrutinize people and when it is
used to control their perception of the world. A more helpful
formulation than *code is law* is that *code is power*: it can get us to do
things we wouldn't otherwise do by means of force, coercion,

influence, and manipulation. And it can do so in a stable and wide-ranging way. That's why code will play such a crucial part in the future of politics.

Code's Empire

In 2006 Lessig wrote that 'Cyberspace is a place. People live there . . . and then at some point in the day they jack out and are only here . . . They have returned.'[13] Typically prescient at the time, this way of thinking is now outdated. In the digital lifeworld 'cyber-space' will be a less helpful way to talk about our interaction with technology. As we've seen, our lives will play out in a field of connected people and 'smart' things with less distinction between human and machine, online and offline, virtual and physical. Whether we *want* to interact with digital technologies, or even whether we are *aware* of them, will matter less and less.

I suspect that the distinction between 'cyber' and 'real' is already losing most of its psychological resonance for younger readers, for whom chatting on WhatsApp is just chatting, not 'cyber-chatting', and buying stuff on Amazon is simply shopping, not 'e-commerce'. (We long ago stopped speaking of Amazon.com.) In the long run, the distinction between cyberspace and real space will lose its explanatory value. There's nothing 'virtual' about the digital pace-maker keeping you alive. No one would say that a passenger in a self-driving car was travelling in 'cyberspace'. While in 2000 you might have been able to leave code's empire by logging off and shutting down, in the digital lifeworld that will be much rarer. Trying to escape the reach of technology will be like trying to escape the law: possible in theory by fleeing to the wilderness, but highly impractical if you want to live a normal life. And as virtual reality (VR) systems grow in popularity, some of us will spend time in universes entirely constituted by code—where code is not only power, but nature and physics too.

The Future of Code

Much of the code that surrounds us in the digital lifeworld will be able to reprogram itself and change over time as it learns to recognize patterns, create models, and perform tasks. As we saw in chapter one, machine learning and AI systems more generally will become more autonomous and 'intelligent' in ways that do not necessarily mimic human intelligence or the intentions of their original programmers. They will also be much more common. This will profoundly affect the way that code exerts power. The academic literature has traditionally treated code as a kind of stable architecture, setting inflexible side-limits on what we can and can't do. This is reflected in the structural metaphors we use to talk about code: 'platforms, architectures, objects, portals, gateways'.[14] In the future, code will be a much more dynamic, sensitive, and adaptable referee of our conduct—capable of changing the rules as well as enforcing them. Sometimes it will seem ingenious or bewildering. Sometimes it might seem irrational or unfair. But that's power.

The Next Four Chapters

Digital technologies exert power on us when we interact with them, by defining what we can and can't do, by scrutinizing us, and by controlling what we perceive of the world. In the digital lifeworld, such technologies will be everywhere. The code animating them will be highly adaptable and 'intelligent', capable of constraining us in a flexible and focused way. Therefore certain digital technologies will provide a means of exerting great power in the digital lifeworld, and those who control these technologies will exert great power through them. Even if no particular person or group can be said to be 'in charge' at a given time, humans will be constantly subject to power from many different directions, constraining and guiding their behaviour. That's the essence of power in the future, and it's the subject of the next four chapters.

SIX

Force

'Moses summoned the elders of the people and put before
them all that the Lord had commanded him. And the people
answered as one: "All that the Lord has spoken we will do!" '

Exodus: 19:7–19:8

Introduction

This chapter is about the use of *force*—where one party secures
the compliance of another by removing the latter's option to
comply or not. Force is the purest and most direct form of power.
Its legitimate use has historically been reserved to the state and
officials of law enforcement. As we'll see, in the digital lifeworld
the use of force will be subject to three important changes in years
to come. The first is what I call the *digitization* of force: a shift from
written law to *digital law*. The second is what, borrowing from
Lawrence Lessig, we can call the *privatization* of force, eroding the
state's long-held monopoly on its use. The final change is the *automation* of force, with the emergence of autonomous digital systems
that can exert force against humans without immediate human
oversight and control.

Each of these transformations will be of profound political
significance.

The Digitization of Force

We are accustomed to thinking of laws as rules written in prose, unalterable except by the legislature, with sanctions applied by human officials when the rules are breached. The digital lifeworld raises the prospect of a new kind of law. It differs in four important ways from what we live with now. First, it will be enforced by digital systems rather than humans. Second, it will *force* us not to break the law rather than simply punishing us afterwards. Third, it will be adaptable, changing according to the situation. Finally, it will be possible to draft it in code rather than natural language. This is *digital law*, a wholly new way for others to force us to do things or refrain from doing them.

Self-Enforcing Law

The first and simplest change made possible in the digital lifeworld is a shift from law enforced by *humans* to law enforced by *digital systems*.

In the past, the work of enforcing the law mostly fell to people. A few hundred years ago, it would have been private or quasi-private authorities such as feudal landlords, parish guilds, charity associations, local constables, and roving magistrates—all of whom were tasked with keeping order within their domain. For a long time, the most important official was the executioner, memorably described by Joseph de Maistre as the 'cog between the prince and the people'.[1]

The last few centuries saw the arrival of a new class of law-enforcement professionals whose full-time role was to implement the law systematically and methodically. Foremost among them were police forces, for the first time trained and uniformed like armies.[2] This revolution in policing came with a revolution in punishing. Whereas the gaols of old had rarely housed criminals for longer than a few nights, new carceral institutions such as prisons and madhouses grew to hold thousands of inmates on a semi-permanent and permanent basis. They were staffed by trained wardens and guards. The judicial system, tasked with prosecuting and sentencing criminals, relied on judges, jurors, and probation officers.[3]

In the future, digital technology will be able to do much of the work of law enforcement previously done by human officials. We saw in the last chapter how code embedded in DVDs and music files basically takes law-enforcement officials out of the picture. The rules enforce themselves: there's no need for police, investigators, prosecutors, judges, or juries (at least until the DRM technology is hacked, which is a crime). We already have automated speed cameras on roads; it's not a great leap from there to a 'smart' enforcement system that automatically removes cash from your digital wallet for every ten seconds you drive over a speed limit, or fines you for every minute you spend parked illegally.[4] Increasingly, digital technology could be used to perform roles traditionally associated with the judiciary as well. Ebay's online dispute resolution system, without courts or lawyers, is said to resolve some 60 million commercial disagreements a year—more than three times the total number of lawsuits filed in the entire US court system.[5]

As more human activity falls to be undertaken by digital systems, digital enforcement will become a matter of necessity, not just convenience. Think about financial trading. In time, fewer trades will be executed by screaming stockbrokers on trading floors, and more by intelligent trading algorithms that respond to market events at lightning speed. There used to be 600 US cash equities traders at Goldman Sachs' headquarters in New York. Now there are just two. One-third of the bank's staff are computer engineers.[6] The introduction of automated trading systems has made financial regulation a nightmare for human officials. The algorithms are often too fast, too complex, and too adaptable to be subject to human oversight.

Increasingly, the most practical method of enforcing the law is through the use of *other algorithms* programmed to detect and prohibit errant behaviour, or at least flag it for human attention.[7] One of the digital systems being used to keep tabs on financial behaviour is a version of IBM's Watson.[8]

The idea of digital law enforcement may seem weird but it doesn't really disrupt our notion of what the law *is*. Indeed, it sits comfortably with the definition of law set out in the early twentieth century by Hans Kelsen, the Austrian jurist whose mighty work, *Pure Theory of Law* (1934), has propped open the doors of law students for generations. Kelsen argues that the law does not actually prohibit things; it simply requires that if certain conduct is performed, an official should apply some sanction.[9] On this theory, in H. L. A. Hart's rendering, there is no law prohibiting murder: 'there is only a law directing officials to apply certain sanctions in certain circumstances to those who do murder.'[10] In the digital lifeworld, this could just as well read, 'There is no law prohibiting wrongdoing: there is only law, in the form of code, directing digital systems to apply certain sanctions in certain circumstances to those who perpetrate that wrongdoing.'

Enforcement through Force not Coercion

Imagine a self-driving car which, instead of using the threat of a fine for the time it spends over the speed limit, is coded so that it cannot drive over the speed limit at all. You would never be fined for parking illegally because the car itself wouldn't let you. You wouldn't be able to trespass on property that its GPS systems knew to be private or secure. You couldn't use your car to run down pedestrians in a terrorist attack. Manufacturers have already indicated that their vehicles will recognize emergency vehicles such as ambulances and fire trucks, and pull over to let them past. Your car could also be programmed to stop for the police—irrespective of whether you wish to or not. If you've ever commandeered a golf buggy, you may be familiar with this experience: the vehicle won't drive over a certain speed, it slows when approaching main roads, and comes to a complete stop near greens and water features. This, of course, takes all the fun out of it.

This is the second major shift: from law enforced by *coercion* (where people refrain from unlawful conduct through fear of punishment) to law enforced by *force* (where people refrain because they have no choice but to refrain). This is another of Lessig's seminal insights.

By and large, we are free today to break the law but deterred from doing so by the threat of later being caught and punished. Officials can rarely *force* us not to break the law. They don't know when we're going to commit a crime and are unlikely to be there when it happens. Yes, we might be caught red-handed and stopped, but this is the exception rather than the rule. We all take steps to prevent and deter crime—locking the windows at night and setting the burglar alarm—but enforcement of the law by the state generally only takes place *after* the deed has already been done.

Historically, this was not a very effective way of making people obey. In early modern Europe, smuggling, looting, tax evasion, debauchery, theft, and all manner of other crimes routinely went unpunished. In his masterwork *Discipline and Punish* (1975), the French philosopher and turtleneck connoisseur Michel Foucault describes 'massive general non-observance, which meant that for decades, sometimes for centuries, ordinances could be published and constantly renewed without ever being implemented'.[11] Many highways and hinterlands, infested with roving brigands, languished outside the law entirely. Querulous populations would rise in revolt against their rulers. This semi-anarchy was made possible, in part, by the technological conditions of the time. It was impractical for officials to keep a close eye on communities in distant provinces.

As a result, when people broke the law they did not generally expect to be caught.

If people didn't expect to be caught, how did the state keep order at all? By making the retribution so horrific that it wasn't worth the risk. This meant gruesome public punishment: chains, gallows, wheels, gibbets, pillories, scaffolds, and stocks. In full view of townsfolk, criminals and suspects would be humiliated, lashed, tortured, dismembered, decapitated, amputated, burnt, mutilated, starved, broken, ripped, and quartered.[12] The average serf would think twice before thieving a loaf of bread if he thought there was even a small chance that he might be disembowelled in punishment. These days, as Foucault observes, we no longer fear physical punishment but we do see it as highly *likely* that the state will catch and punish us in some way. 'Inevitability' of punishment has replaced 'intensity' as the reason many do not commit serious crimes.[13]

The shift from law enforced by *people* to law enforced by *technology* means that power will increasingly lie in *force* rather than *coercion*, with self-enforcing laws that cannot be broken because they are encoded into the world around us. To recap—A *coerces* B by threatening deprivation, while A *forces* B by removing the choice whether or not to comply. Lessig uses the analogy of a locked door: 'A locked door is not a command "do not enter" backed up with the threat of punishment by the state. A locked door is a physical constraint on the liberty of someone to enter some space.'[14] In the future, the law will more often represent a locked door than a command not to enter.

Think again about DRM technology. When we refrain from sharing music downloads, it's not because we fear later punishment, but because the code prevents us from breaching copyright law in the first place. In finance, we literally *can't* withdraw money from a bank account after it is 'frozen'. The banking software prevents it. It's not hard to imagine that in the future, criminals placed under house arrest or curfew might be made subject to 'smart lock' technology that physically prevents them from accessing or leaving certain properties. The technology already exists for 'smart guns' operated by biometric sensors, which only shoot when held by the hand of the legitimate owner. If stolen or traded illegally, these weapons are useless.[15] Airborne drones may be programmed so as to be unable to fly within certain radii, whether for reasons of security (near an airfield) or privacy (over GPS-delineated private property). Looking further ahead, robotic systems may refuse to obey commands that they know to be illegal. Software platforms may refuse to complete certain transactions: imagine Amazon refusing to process an order for materials that look like those needed to make a bomb.

Looking at it from another angle, recall that the law doesn't just command us not to do things; it also confers powers, rights, and duties that enable us to make and uphold wills, contracts, and marriages.[16] Nowadays, if you and I wish enter into a contract, there is nothing *forcing* you to perform your end of the bargain. If you breach the contract I can ask the state to force you to do what you promised to do, or at least to compensate me for what I have lost by your failure. But as with the criminal law, force generally only follows the breach. A digitally enabled 'smart contract' operates in a different way. Rather than a written instrument, it is a piece of code embodying the terms of an agreement, possibly solemnized on a blockchain to ensure that its terms cannot be changed or tampered with (chapter two). The code itself executes the bargain between the parties. Say, for instance, that you buy a car using third-party financing. After you pay your final loan instalment to the lender, a smart contract would automatically transfer title of the vehicle from the lender to you. There would be no need for further human intervention, and no chance that the lender would renege on its side of the deal.[17] Smart contracts differ from normal contracts in that they are based on code rather than on the trust-worthiness or prudence of the parties. For this reason they are much harder to breach. As time goes on, more transactions may be administered through ironclad agreements enforced in code. Governments are already exploring the idea of administering welfare benefit payments using smart contracts to reduce fraud, error, and delay.[18]

Adaptive Law

We don't normally expect laws to change without some kind of formal process, like the passage of new legislation or the interven-tion of the courts. In the digital lifeworld, however, it will be possible to implement *adaptive* laws capable of responding quickly and precisely to changing circumstances.[19]

Imagine that instead of a static speed limit on the freeway, limits were set in real time depending on traffic and weather conditions as monitored by digital systems.[20] At quieter times or in better weather conditions your car (autonomous or not) would be permitted to go faster. Or, because I am a better driver than you, with fewer crashes on my record (or perhaps with higher ratings from other drivers) I am entitled to drive my motorcycle at 70 miles per hour while yours will not go faster than 40. Because I have a clean crim-inal record, I am entitled to send $10,000 a day to other bank accounts, while the software limits you (with your previous conviction for fraud) to only $500. These kinds of adaptations would not be difficult for an advanced digital system to administer and enforce, chiefly because they are *rule*-based: applying a speed limit does not require discretion or judgement.[21]

The alternative to *rules* are *standards*. A rule says: do not drive more than 50 miles per hour. A standard says: drive with appropriate care in the circumstances. Standards, like whether something is *appropriate*, *reasonable*, or *excessive,* have considerable grey areas. Today the law contains both rules and standards, but it has generally been thought that only rules are appropriate for digital enforcement, because systems cannot or should not be entrusted with discretion. In the digital lifeworld, I suggest, this may no longer be the case.

Imagine a digital system that is able to predict, with a high degree of accuracy, the outcome of particular medical decisions. With the right information, it can estimate the likelihood that an operation will successfully excise a tumour, or that a particular antibiotic will cure an infection. Call this system Robot MD. Robot MD can make such predictions not because it understands medicine and thinks like a doctor, but because it has a huge amount of processing power, machine learning capabilities, and access to data concerning thousands of previous cases. It can identify patterns and correlations that even human doctors might not see. (We already have AI systems that differentiate between lung cancers and predict survival periods better than human pathologists. Others can identify the signs of Alzheimer's with 80 per cent accuracy a decade before the symptoms show themselves.[22] And AI systems like MedEthEx can generate advice about the ethically appropriate way for a doctor to proceed in a given situation.)[23]

Now consider a second AI system, Robot JD, which 'knows' nothing about medicine but a lot about law. If a doctor gives Robot JD the facts relating to a particular patient and explains what course of action he proposes to take, Robot JD is able to predict with a high degree of accuracy the likelihood that a judge would find that course of action to be negligent. Again, this is not because Robot JD can think either like a doctor or a lawyer, but rather because it has munched through thousands of previous cases and can predict the court's decisions in the presence of particular facts. (An early system of this kind has already correctly guessed the outcomes of 79 per cent of hundreds of cases heard at the

European Court of Human Rights.[24] Another was better able to predict the rulings of the US Supreme Court than a team of eighty-three expert human jurists, nearly half of whom had themselves clerked for Justices of the Court. The experts predicted correctly just under 60 per cent of the time. The algorithm was right 75 per cent of the time.)[25]

How might Robot MD and Robot JD contribute to the development of adaptive law? Drawing on the remarkable research of Anthony Casey and Anthony Niblett, we might predict as follows.[26] In the first place, doctors could just use Robot MD and Robot JD in an advisory way, asking their opinion in order to help inform the decision-making process. But as usage of Robot MD and Robot JD became more commonplace, it would be increasingly odd *not* to consult them before taking a particular course of action, just like today it would be odd not to consult an MRI scan before operating on someone's brain. If you have a machine that can accurately predict whether a procedure might go terribly wrong, or be negligent under the law, why *wouldn't* you consult it beforehand? In time, legislators or the courts may decide that *not* following the course of action prescribed by Robot MD or Robot JD is itself negligent, just like not consulting an MRI would be today. In this world, the dictates of the AI systems themselves have, in a sense, *become* law—and that law will adapt depending on the situation. And so we arrive at adaptive law based on *standards*, not just rules.[27]

Using systems to predict the law, and thereby determine it, is not as heretical as you may think. In a famous article in 1897, Oliver Wendell Holmes, who became a leading figure in the school of American Legal Realism, claimed that the law *is* '[t]he prophecies of what the courts will do in fact, and nothing more pretentious'.[28] This was a respectable notion even before we had machines like Robot JD that could predict the outcomes of legal decisions.

Now those machines are a real possibility. This means that legal standards can be stated upfront (and if necessary, enforced) for any situation in which the facts are known, without having to wait for a court to decide retrospectively whether a particular standard was breached.

Code-ified Law

If the law is increasingly computed by AI systems applying general standards to specific situations, it means that our rules will increasingly be 'drafted' in code itself. This is perhaps the most radical possible change of all, and has been called the 'code-ification' of law.[29] In a world of code-ified law, the authoritative statement of the law would lie not in a written statute or judgment. It would lie in the code that served both to describe and enforce it. In its weak form, code-ified law could be used for formalities such as the registration of land and marriages. Today, if your marriage is not registered in the civil registry, you are simply not married. In the future, the status of a marriage could depend on what has been entered into the correct digital registry. (What code hath joined together, no man may put asunder.) But code-ified law could also take a stronger form, as would be the case if the dictates of Robot JD and Robot MD came to govern the law of negligence, and the law is simply what the program states it to be at any given time. This would make sense in a world where code is increasingly used both to determine what the law *is* and to *enforce* it.

At first, the idea of replacing written law with code might appear astonishing. We think of laws as being written in prose and recorded in print. We expect them to be capable of being understood and interpreted, if not by every literate person then at least by trained lawyers. We don't bother memorizing their exact contents because they are precise, relatively stable, and permanently recorded. And if we want to know the law on a particular issue, we look it up. But it's important to remember that while code-ified law seems bizarre to us, *written* laws would have made no sense to our forebears. What we understand to be 'law' is a distinctly literate conception. Things were very different in cultures 'with no knowledge whatsoever of writing or even of the possibility of writing.' In these times, no one ever looked up a law. In fact, as Walter Ong observes, no one ever looked up anything:[30]

> In a primary oral culture, the expression 'to look up something' is an empty phrase: it would have no conceivable meaning. Without writing, words as such have no visual presence, even when the objects they represent are visual. They are sounds. You might 'call' them back— 'recall' them. But there is nowhere to 'look' for them. They have no focus and no trace (a visual metaphor, showing dependency on writing), not even a trajectory. They are occurrences, events.

Think for a moment about how different law was before the invention of writing. Instead of precisely codified rules, communities lived according to customs, folkways, and habits. Acting lawfully meant behaving in a way that was generally understood to be decorous, seemly, and appropriate. There was no detailed regulation. Rules were rehearsed in the form of maxims, sayings, and poems. The general principles were well-known to most, but mastery was reserved to a small class of learned elders.[31] Oftentimes, legal principles were not expressed in general terms but indirectly through folk tales and parables. Homer's oral poems, for instance, contain no abstract concept of *justice* (which would have been nearly impossible in a culture without writing). Instead they contain *justices* (*dikai*) in the form of small narrative episodes with implicit lessons about how life ought to be lived.[32] Sometimes, in preliterate societies, a judgment or decision on a specific matter would be issued by a governmental authority. Today, we would say such a decision is *handed down*, but in an oral society it would literally have been *proclaimed*. And instead of being carefully recorded for later consultation, such proclamations remained in force only as long as they were remembered.[33]

It was only fairly late in human history, with the invention of script and other pictorial representations, that humans were able to record the law with precision. This was revolutionary. It led to a massive proliferation in the quantity and complexity of the law. It enabled rules to become stable, certain, and durable over time. And it helped to nurture the doctrine of precedent, according to which judicial decisions are taken on the basis of similar decisions made in the past.[34] Even then, however, we probably idealize the virtues of written law more than is deserved. In truth, little has changed since 1835 when Alexis de Tocqueville wrote that 'only lawyers' are familiar with the law: 'The nation at large hardly knows about them; people see them in action only in particular cases, have difficulty in appreciating their implications, and submit to them unthinkingly.'[35] The difference with code-ified law is that we'd be less likely to accidentally break a rule of which we were unaware—because the code enforces itself.

My point is this: the idea that law will increasingly take the form of code rather than writing may sound bizarre, but perhaps no more wacky than the original idea that law might take the form of abstract written prose rather than narrative oral poems. Whether this is desirable, of course, is another question.

Digital Law in Action

So what does a self-enforcing, adaptive, code-ified law look like? Consider the following scenario, again based on the research of Casey and Niblett.[36]

You attend a hospital in today's world with a sore leg. When you arrive at the emergency room, you are sent for an X-ray and sit down to wait for the results. After a while you are attended by a human doctor, Dr Smith, who inspects your leg and tells you that it needs to be operated upon immediately. Concerned by this response, you ask whether that course of action is consistent with the X-ray results. Dr Smith replies that he has not looked at the X-ray results; he does not need to. He's an experienced physician and has seen this a hundred times before—the leg needs to be operated on immediately. The surgery takes place, and it turns out that Dr Smith was wrong: there was no need to operate after all. This would have been clearly visible from the X-ray results. Unfortunately, because of complications arising from the surgery, you suffer long-term damage to the limb. You sue the hospital. In court, unsurprisingly, the judge finds that Dr Smith acted negligently by not looking at the scans— not because the law contains a *rule* specifying that doctors must always look at X-ray results, but because the law lays down a *standard* that doctors must exercise *reasonable care* or they will be negligent. It was negligent, the court finds, to refuse even to look at the scans.[37]

Now imagine what might have happened in the digital lifeworld. Dr Smith would have been *required* to consult Robots MD and JD before operating. Not to do so would itself be negligent. In fact, it might not have been possible to register a surgical procedure on the system without having consulted Robots MD and JD first. When consulted, these AI systems would have warned Dr Smith that he ought to wait for the X-ray; operating without doing so would be likely to cause harm and be seen by a court as negligent. Most likely Dr Smith would have felt compelled to wait for the X-ray. The outcome for your leg would have been different. If Dr Smith sought to persist with the operation in the face of legal warnings by Robots MD and JD, then the medical machinery required to perform the surgery might have been automatically switched off, or Dr Smith's credentials for accessing it temporarily revoked, so that the procedure could not have been performed at all.[38]

I do not suggest that all law in the future will be digital law, or that all digital law will have all of the features described in this chapter. Yet if even some of the changes outlined here become reality, then it will mark a significant departure from the past. We shall have to decide when digital law is an appropriate way of governing human affairs and when it should be resisted. This is a new political question. We revisit it in Parts III and IV.

The Privatization of Force

One of the first ideas taught to students of politics is that, with some exceptions, the state is the only body entitled to use force against people, or permit it to be used by others. This idea is commonly cited in connection with the German theorist Max Weber:[39]

> 'Every state is founded on force,' as Trotsky once said at Brest-Litovsk. That is indeed correct... a state is that human community which (successfully) lays claim to the monopoly of the legitimate physical violence within a certain territory.

Theorists have long argued about *why* we allow the state, and the state alone, to use force against us. The most famous answer came from the seventeenth-century English thinker Thomas Hobbes, who lived through a period of bloody civil war and took a gloomy view of human nature. Without a 'common Power' to keep us 'all in awe', he argued, we would languish in a permanent 'warre' of every man against every man.[40] The Swiss thinker Jean-Jacques Rousseau took a cheerier approach, arguing that the state was the product of a great and voluntary social contract between every citizen. By submitting to the state, we lose little and gain a lot. Since every one of us 'gives himself to all, he gives himself to no one'.[41] Rousseau's contemporary David Hume (with whom he fell out spectacularly) scoffed at such notions. Like Hobbes, Hume saw submission to the state as a matter of necessity rather than consent. The state alone may use force 'because society could not otherwise subsist' and because it's impossible to escape it anyway:[42]

> Can we seriously say, that a poor peasant or artisan has a free choice to leave his country, when he knows no foreign language or manners, and lives, from day to day, by the small wages which he acquires? We may as well assert that a man, by remaining in a vessel, freely consents to the dominion of the master; though he was carried on board while asleep, and must leap into the ocean and perish, the moment he leaves her.

Whichever justification you prefer, it's a staple of political theory that the state alone is entitled to force us, on a consistent basis, to do things we wouldn't otherwise do, and not to do things we might otherwise do. But in modern systems that power is subject to the *rule of law*. The law strictly defines and limits the state's power to use force. It also provides the means through which the state imposes order and discipline on its citizens, by setting out (a) prohibited and required conduct, (b) the formalities needed to do things of a legal nature like making contracts, getting married, and leaving wills, and (c) the *sanctions* to be applied when the rules and formalities in (a) and (b) are broken. Function (c) is the business of *enforcement*—and, as we've seen, it's when the state's entitlement to use force comes into play. If you break the criminal law, the state can *force* you to surrender your property in the form of a fine or requisition, or even your personal liberty in the form of a prison sentence. A court of law can order you to uphold your obligations under a contract or a will. To disobey a court order is a crime, again sanctionable by the use of force.

So far in this chapter we have assumed that the state, and the state alone, may legitimately use force (and the threat of force) within its territorial domain. That's consistent with mainstream political theory. But if you think about it, many of the technologies already described in this book are developed, manufactured, and distributed by private corporations. The most advanced AI, for instance, is generally thought to be found in the research facilities of private companies like Google and Facebook, not national governments (or even universities). We have assumed that so long as these corporations reside in states and are subject to their laws that they may be required to use their technologies as a means through which national law can be enforced. So when we speak of a self-enforcing speed limit on roads, what we are really saying is that the state might require Google, Ford, Tesla, or Uber to code such a limit into their self-driving vehicles.

But—and here's the important point—in the digital lifeworld there will also be plenty of times when we are at the mercy of code that is private in origin and nothing to do with the national law. Whenever we use a digital platform, whether a market like Amazon or a forum like Facebook, we're entirely subject to the rules of that platform as dictated by its code.[43] In the absence of any law to the contrary, if the code requires us to give personal details as a means of authentication, we have no choice but to comply if we wish to participate. If it prescribes that all our messages will be saved forever, there is nothing we can do about that. And if we are expelled from the network, for whatever reason, the code may not permit us to rejoin.

The same analysis could apply to physical objects with code embedded into them. Rather than being 'dumb' objects, our household appliances, vehicles, utilities, buildings, robots, and drones will increasingly be animated by code that prescribes what we can and can't do with them. We'll be subject to the rules, values, and priorities that their engineers have coded into the devices. (Think, for example, of 'the traitorous coffee maker' sold by Keurig that refused to brew coffee from non-Keurig brand beans.)[44] Each individual limitation induced by these technologies may constitute only a small exertion of force, but the cumulative effect will be that we're subject to a good deal of power flowing from whoever controls those technologies. The implications for freedom are discussed in the next Part.

Take the famous example of the 'trolley problem'.[45] You are motoring down the freeway in a self-driving car, and a small child steps into the path of your vehicle. If you had control of the car, you would swerve to avoid the child. You know that this would cause a collision with the truck in the adjacent lane, probably killing both you and the trucker—but to preserve the life of the child, that's a sacrifice you are willing to make. Your car, however, has different ideas. Whether by intentional design or through 'learning' from others, it considers that saving two lives is better than saving one. And so you have no choice: you are forced to plough into the child, killing her instantly. In the aftermath, you find yourself asking: should the decision whether to kill two adults or one child really have been left to the car manufacturer? Shouldn't that sort of question be a matter of individual conscience, so that you can reprogram your vehicle according to your own ethics and accept the consequences? Or, because it potentially affects everyone, could the question require a *collective* decision as to what cars should do in that kind of situation? Lessig had it right, in a different context, when he said 'this is all the stuff of politics':[46]

> Code codifies values, and yet, oddly, most people speak as if code were just a question of engineering. Or as if code is best left to the market. Or best left unaddressed by government.

The issue goes beyond values to the question of checks and balances—because with great power, technology brings the capacity to do terrible things. Consider the nightmare scenario in which a rogue engineer at a car manufacturer remotely locks the door of your self-driving vehicle while you are inside and reprograms it to drive off the nearest cliff. Or even that the entire fleet of vehicles is reprogrammed to drive at high speed into the nearest wall. These are dramatic thoughts but they require serious consideration. The law places checks and balances on the powers of the executive branch of government. Theorists have debated for centuries what form these checks and balances should take when it comes to government. Now we need to ask what controls are necessary for digital technology, growing in power and remarkably open to abuse. It would seem naïve just to leave matters to the manufacturers themselves.

Pushing back, you might ask whether, in all these examples, anyone is really *forcing* you to do anything. You don't *need* to join a social network; and even if you want to, perhaps you don't *need* to join one whose terms and conditions are repugnant to you. You don't *need* to ride in a self-driving car, or in that particular model. Why not just move to a different one if you don't like the code?

This is a respectable objection, and it's dealt with at more length in the next Part. For now, it suffices to say that the objection depends on the existence of a marketplace of alternative options, which isn't something we can take for granted. Moreover, it's worth emphasizing that while opting out of one brand or technology may be possible, it will not be possible to opt out of the digital life-world altogether. Just by going about our daily business, we will be subject to the power of technology, whether we like it or not. The person who tries to opt out of code's empire will be like the man Hume describes as trying to opt out of the state itself; carried aboard a ship while asleep, his only means of escape to jump over-board. Or, to borrow a phrase used by Elizabeth Anderson in a different context, to say that tech firms don't exert power over us because we can opt out is a bit like saying that Mussolini was not a dictator because Italians could emigrate.[47] The reality is that we will increasingly be subject to power from various new directions. Whether the regulation of that power should be a matter of collective control, or left to the market, or subject to some other form of accountability, will be one of the defining political issues of the digital lifeworld.

The Automation of Force

We come, finally, to the most radical change of all: the possibility of autonomous digital systems capable of exerting force outside of human control. Such systems, to operate stably, would need three characteristics. First, they would have to be *self-directing* in the sense of being sufficiently coded to discharge their functions without any further intervention by human beings. This would either mean being engineered to deal with every possible situation or capable of 'learning' how to deal with new situations on the job. (This kind of self-direction does not, however, require artificial general intelligence or even a sense of morality. Aeroplane autopilot systems have a high degree of self-direction but no moral or cognitive capacity, and yet we trust their ability to keep us safe in the sky. Like an aeroplane, which is neither a moral agent nor even conscious of its own existence, a system could exert power without being aware that it is doing so.)[48]

Second, such systems would have to be *self-sustaining,* in the sense of being able to survive for a decent period without assistance from humans. This might require a permanent source of energy, such as solar power. But it might also involve the capacity to harvest its energy and maintenance needs from the commercial market-place, just like a human being or corporation. Digital systems are already able to transact with each other, like the Samsung washing machine that orders new detergent from the store when it runs out.[49] Blockchain-enabled smart contracts could enable more sophisticated commercial operations, with robots buying their sundries from other humans or machines.

Finally, such systems would have to be functionally *independent*, in the sense that they could not easily be taken over by human masters seeking to regain control. Blockchain technology could again play a part: if the operating code was distributed across multiple machines worldwide, it would be very hard to shut it down. 'Even in the case of a catastrophic event,' says the blockchain expert Primavera De Filippi, such a system could continue to operate.[50]

We are left, then, with the prospect of systems like the hypothetical example described by De Filippi:[51]

> an autonomous AI-powered robot, designed to operate as a personal assistant. This robot offers its services to the elderly and competes with other humans or machines...on both the price and quality of the services it provides. The seniors who benefit from these services can pay the robot in digital currency, which is stored in the robot's account. The robot can use the collected money in various ways: to purchase the energy needed to operate, to repair itself whenever something breaks, or to upgrade its software or hardware, as necessary...If this robot relied on more advanced artificial intelligence...there might be increased interest from the public in emancipating this robot...from centralized control...

Personal assistant droids aside, there are other technologies whose purpose might best be served by functional autonomy. It's not hard, for instance, to imagine a fleet of small autonomous agricultural robots whose purpose is to pollinate or treat plants in a delicate area of rainforest. More radically, consider Wendell Wallach's example of an airport security system that is (a) self-directing, in that it is capable of identifying suspicious individuals or known terrorists without human assistance, (b) self-sufficient, in that it does not rely on active intervention by humans to stay online, and (c) functionally independent, so it cannot be subjected easily to human override. Such a system might automatically lock down a terminal when it detects dangerous activity—without further human intervention that could cause fatal delay.[52] In the long run, the digital lifeworld may become home to a wide variety of autonomous systems like these.

It's one thing, of course, to use autonomous robots to care for plants; and quite another to use them to secure our public spaces. But to what extent should human lives be subject to force that's not under immediate human control?

Force: Implications

In the Introduction to this book I proposed a tentative hypothesis: that how we gather, store, analyse, and communicate our information, in essence how we organize it, is closely related to how we organize our collective life—and when a society develops strange and different technologies for information and communication, we might expect political changes as well. In this chapter we have started to test this hypothesis as it relates to the idea of force. Digital law, privatized force, autonomous systems of power—these all pose a pretty fundamental challenge to how we have historically thought about the use of force. In the next two chapters we look at some of the subtler but equally important implications for power, and how technology can be used to *influence* and *manipulate*.

SEVEN

Scrutiny

'A Party member lives from birth to death under the eye of the Thought Police. Even when he is alone he can never be sure that he is alone. Wherever he may be, asleep or awake, working or resting, in his bath or in bed, he can be inspected without warning and without knowing that he is being inspected. Nothing that he does is indifferent. His friendships, his relaxations, his behaviour towards his wife and children, the expression of his face when he is alone, the words he mutters in sleep, even the characteristic movements of his body are all jealously scrutinised. Not only any actual misdemeanour, but any eccentricity, however small, any change of habits, any nervous mannerism that could possibly be the symptom of an inner struggle, is certain to be detected.'

George Orwell, *Nineteen Eighty-Four* (1949)

Power does not need to be violent or threatening. It can be gentle, even tender. Some power is so subtle that it's almost invisible. Often the weak are not even aware that they are being dominated by the strong. Sometimes they *do* know they are being dominated, but they *welcome* it. The most skilful political leaders know that to force someone is to gain power over their body; while to influence or manipulate them is to gain power over their mind. This can be the deepest and richest form of power of all.

In the next two chapters we look at how, in the future, some will be able to get others to do things they wouldn't otherwise do, without using or threatening force. The focus is on two simple but remarkable instruments of power. First, *scrutiny*: the ability to gather, store, and process information about others. Second, *perception-control*: the ability to control what others know, what they think, and what they are prepared to say about the world. We start with scrutiny.

What is Scrutiny?

There is power in the ability to see without being seen; to know others while yourself remaining unknowable. Throughout history, the weak have fought to hide themselves from scrutiny while trying to impose transparency on the strong. In the French Revolution, inspired by Jean-Jacques Rousseau, some rebels dreamed of a world in which the rich and powerful had nowhere to hide:[1]

> a transparent society, visible and legible in each of its parts, the dream of there no longer existing any zones of darkness, zones established by the privileges of royal power or the prerogatives of some corporation.

With the partial exception of the totalitarian regimes of the twentieth century, most human life has been lived outside the direct gaze of political authorities. Historically, even the most efficient surveillance undertaken by government officials—police, spies, informants—was necessarily sketchy and incomplete, leaving considerable zones of privacy in which people could do as they pleased. Technologies of surveillance, until the twentieth century, were largely ineffectual.

The digital lifeworld will be different. The dream of the French revolutionaries—of a society without zones of darkness—will be made real but with a bitter twist: the lives of others will be visible only to those who control the means of scrutiny. For the rest of us, life will be 'uncovered and laid bare' before those 'to whom we must give account' (Hebrews 4:13). This marks a profound change.

By *scrutiny* I mean not just visual observance, but any way of gathering, storing, and processing information about someone else. In the future, being scrutinized will become the norm, with machines, not humans, doing most of the legwork. Less and less of what we do will escape the attention of technology. And those who gain control of the *means of scrutiny*—the technologies capable of gathering and processing information—will enjoy a great increase in their power over the rest of us, power of the kind temptingly fore-told by Satan in John Milton's *Paradise Lost* (1667):

> your eyes that seem so clear,
> Yet are but dim, shall perfectly be then
> Opened and cleared, and ye shall be as Gods.[2]

The Power of Scrutiny

Scrutiny is often spoken of in terms of its effects on dignity and privacy—and these are important—but its most profound implica-tions are for power. The relationship between scrutiny and power is close and twofold. First, scrutiny helps to gather information useful in the deployment of power. This is its *auxiliary* function. Second, scrutiny can *in and of itself* make people do things they might not otherwise do, or stop them from doing things they might otherwise do. This is its *disciplinary* function. Let's consider each in turn.

Auxiliary

The *auxiliary* function of scrutiny is straightforward: the more you know about someone, the easier it is to subject them to power, whether in the form of force, coercion, influence, authority, or manipulation. Thus if Alexandra wants to get Daniel to do something that he might not otherwise do, her task will be easier if she knows a few things about him, such as where he can be found, what he does to earn money, what property he owns, the people with whom he associates, what he likes, dislikes, and fears, whether he has a family, and so forth. With this information, Alexandra can best gauge which carrots (rewards, bribes, inducements) or sticks (threats, sanctions, deprivations) would be most likely to secure Daniel's compliance. What's more, by keeping Daniel under scrutiny Alexandra can see whether her chosen methods of exerting power are having the desired effect, or whether she needs to change approach. Knowledge is not itself power, but it is conducive to it. The auxiliary function of scrutiny is to gather that knowledge.

Disciplinary

Scrutiny's second function is *disciplinary*. Simply knowing that we are being observed makes us behave differently. It makes us less likely to do things perceived as shameful, sinful, or wrong. The child who knows her mother is watching is less likely to reach for the forbidden cookie jar. The renegade who knows that her seditious letter will be read by government spies thinks twice before sending it. The person who knows that Googling 'child porn' will get him automatically reported to law enforcement officials may be slow to undertake the search in the first place. The would-be terrorist who knows that the NSA taps a billion phone calls a day is less likely to use his smartphone to plot an attack.[3]

The disciplinary function of scrutiny can also be more diffuse. Scholars such as Sandra Bartky, for instance, have noted that while few would dream of forcing women explicitly to conform to a certain appearance, the scrutiny of others causes women to 'discipline' themselves:[4]

> The woman who checks her make up half a dozen times a day to see if her foundation has caked or her mascara has run, who worries that the wind or rain may spoil her hairdo, who looks frequently to see if her stockings have bagged at the ankle, or who, feeling fat, monitors everything she eats, has become...a self-policing subject, a self committed to a relentless self-surveillance.

That scrutiny leads to self-discipline is an idea generally credited to Michel Foucault. A rock star of twentieth-century political thought, Foucault rejected the notion that power was limited to the state forcing and coercing the masses. Instead he developed the idea of *disciplinary power*, a power achieved through constant scrutiny, which he believed could be even more effective than the use or threat of force. Such power, unlike brute force, 'reaches into the very grain of individuals, touches their bodies and inserts itself into their actions and attitudes, their discourses, learning processes and everyday lives'.[5] Writing before the internet, Foucault compared modern society to Jeremy Bentham's Panopticon, a prison in which each person is constantly watched in an 'apparatus of total and circulating mistrust'.[6] In a *Panoptic* society:[7]

> There is no need for arms, physical violence, material constraints. Just a gaze. An inspecting gaze, a gaze which each individual under its weight will end by interiorising to the point that he is his own overseer, each individual thus exercising this surveillance over, and against himself.

Foucault exaggerates when he suggests that there is *no* need for force in the maintenance of political order but his deeper point is significant: scrutiny can itself be seen as a type of coercion. It doesn't *force* you to do anything, but it does strongly encourage you to self-regulate for fear of undesirable consequences, real or imagined. That's why the 'perfect disciplinary apparatus' would be a 'single gaze' that could 'see everything constantly'.[8]

Scrutiny in the Digital Lifeworld

The digital lifeworld will bring about a transformation in the capacity of humans to scrutinize each other. In five respects, it will be different from the past. First, whole swathes of human life that were either unrecordable or too complex to be grasped in their entirety will be seen and understood by those with the means of scrutiny. Society will become vastly more *scrutable* as a result. Second, this scrutiny will be increasingly *intimate*, taking place in spaces we previously thought of as 'private'. Third, the information gathered through scrutiny may increasingly be *imperishable*, outlasting our memories and even our lives. Fourth, our behaviour will become much more *predictable* to machines tasked with looking into the future. Finally, our lives will be increasingly *rateable*, subject to scores, ratings, and rankings that pit us against each other for access to social goods. The cumulative result will be scrutiny beyond anything we've experienced in the past, and a corresponding increase in the power to which we are subject.

Scrutable

What is true of individuals is true of societies: the more you know about them, the easier it is to subject them to power.

In *Seeing Like a State* (1998), James C. Scott, a professor of political science at Yale University, argues that any significant effort to change the structure of human society—what he calls 'large-scale social engineering'—depends on society being 'legible' to its rulers.[9] (I prefer the term *scrutable* to legible.) 'If we imagine a state,' writes Scott, 'that has no reliable means of enumerating and locating its population, gauging its wealth, and mapping its land, resources, and settlements we are imagining a state whose interventions in that society are necessarily crude.'[10] Conversely, a state with powerful means of enumerating and locating its population, gauging wealth, and mapping its land, resources, and settlements is one whose interventions in that society can be extensive and profound.

How scrutable a society is, how capable of being understood through observation, will depend on (a) how easy it is to gather useful information about it, and (b) how easy it is to make sense of that information. Point (b) is important: no mere official can grasp the full complexity of human society, and even mighty élites need ways of distilling the world into a digestible format. One way to do this is through heuristics like maps, graphs, and statistics, which present a simplified version of reality. Another is to *reorganize society itself* to make it more readily comprehensible. Benjamin Constant, one of the intellectual luminaries of the French Revolution, knew this well:[11]

> The conquerors of our days, people or princes, want their empire to possess a unified surface over which the superb eye of power can wander without encountering any inequality which hurts or limits its view. The same code of law, the same measures, the same rules, and if we could gradually get there, the same language; that is what is proclaimed as the perfection of the social organisation . . . The great slogan of the day is *uniformity*.

In the digital lifeworld, human life will be uniquely scrutable and therefore uniquely susceptible to power. As explained in chapter three, we are moving into an *increasingly quantified society* in which a growing amount of our social activity will be captured and recorded as information then sorted, stored, and processed by machines. More and more of our actions, utterances, movements, relationships, emotions, and beliefs will be recorded in permanent or semi-permanent form. In the last few years there has already been an explosion in the amount of data gathered about us, mostly because more human activity is being mediated through digital technologies. For the last five decades, the cost of digital storage has halved every two years or so while increasing in density fifty-million-fold. Meanwhile, digital information can be replicated millions of times with barely any expense and no loss of quality, and *increasingly capable systems* are better enabling us to process what we store.

The result, in due course, will be a radical increase in the amount of scrutiny to which we are subject. Where we go, what we do, what we buy, what we write, what we eat, what we read, when and where we sleep, what we say, what we know, who we know, what we like, how we work, our plans and ambitions, will all be the subject of scrutiny, captured and processed by machines. This will not be the first time that society has grown radically more scrutable and therefore more susceptible to control, but this time the difference in scale will be unprecedented. Some examples from the past, taken from Scott's *Seeing Like a State*, help to put this future in context.

Think first about *cities*. If you look at a map of a settlement built in the middle ages you will likely see a tangled jumble of streets and alleys with no fixed system of street naming or numbering. The layout seems to have sprouted organically, not according to any fixed plan. The cityscape is confusing. It would be difficult for an outside élite, such as a tax collector, army recruiter, or police investigator, to exercise control over such a city. To find people and addresses would need help from willing locals. Urbanites wanting to evade the reach of power would find it easy to hide or disappear.

Such a settlement is *inscrutable* because it cannot easily be grasped, summarized, or simplified by external élites. Contrast this kind of design with the modern cities of Philadelphia, Chicago, and New York. Built to a single design, with long, straight streets arranged in grids and intersecting at right angles, these settlements are immediately scrutable to an overarching authority—even one unfamiliar with the area. In Manhattan, to make orientation easier still, the streets are consecutively numbered. Importantly, these cities are most scrutable from *above*: the bird's-eye view of the supervising official as he gazes at his map. Settlements organized in this way can be easily navigated and searched, their populations identified, taxed, conscripted, arrested, and otherwise processed.[12]

Though significant, the evolution from the higgledy-piggledy city of the middle ages to the planned metropolis of modernity will be dwarfed by the next phase of urban evolution: 'smart cities' embedded with dense networks of sensors that allow authorities to keep track of a dazzling array of variables, from noise, temperature, light, air toxicity, and utility usage, to the movement, location, and activity of vehicles and people. In the future, technology will allow authorities to monitor events and incidents not on paper maps but using surveillance and feedback systems that detail the activity of the city's inhabitants. Even compared to the grid-cities of the last few centuries, smart cities will be scrutable to an almost unimaginable degree. Very little public activity will be untraceable.

Consider now the scrutability of the *individuals* inhabiting these cities. It's difficult to govern people if you don't know who they are. Most of us take for granted the fact that we can be identified by our surname, but as Scott observes, until at least the fourteenth century most ordinary people did not have permanent second names. One name would generally suffice to be identified in the local area, and if more were needed, a person might add their occupation (Smith, Baker, Taylor), location (Hill, Wood), personal characteristic (Strong, Short, Small), or father's name (Robertson, literally 'son of Robert').[13] Scott invites us to imagine the pain of working as a tax collector and arriving in a village where 90 per cent of the male population shared the same six names: John, William, Robert, Richard, Henry, and Thomas.[14]

The introduction of surnames was, almost universally, something required by the state, beginning as early as the fourth century BC when the Qin dynasty imposed them on its people. The idea was to make *individuals* more scrutable, in place of an undifferentiated mass, so that officials could better track the ownership of property, regulate inheritance, prevent crime, collect tax, force labour, and conscript soldiers.[15] Hundreds of years later, we still bear our surnames as instruments of scrutability. An identity document without that information would be bizarre.[16] Interestingly, though, names were never the pinnacle of human scrutability: the Nazis replaced the names of inmates in Auschwitz-Birkenau with five-digit numbers tattooed onto the forearms of each inmate. Each number initially corresponded to an IBM Hollerith punch-card number.[17]

In the digital lifeworld, individuals will be even more scrutable than before. Authentication will not be a matter of external designations like names or even numbers, but will be taken directly from our living bodies, each one uniquely identifiable by the biometric details of its face, fingerprints, retinas, irises, or gait. Where we can't be identified by our biometrics, our identity will be easily discoverable from the digital exhaust left by our movements, purchases, and associates. Imagine, for instance, that a genocidal government sought to round up all the people of a particular religion or ethnicity in a given area. Their data would give them away immediately: purchase histories, social media posts, known associates, smartphone movements, and call records would all reveal them to be members of the hunted social group.

As well as being rich in data, the digital lifeworld will contain machines capable of crunching that data with unprecedented efficacy. Using only people's 'Likes' on Facebook, researchers can already guess a person's sexuality 88 per cent of the time, their race 95 per cent of the time, and their political party affiliation 85 per cent of the time.[18] That's from just a tiny fraction of the data available, and using today's technology. Imagine what will be done with the data generated by all the life activities of each person, and radically more capable systems.

It's important to understand, however, that individual scrutability in the future will not be limited to identifying us or even finding our whereabouts. It'll be about understanding our lives and desires, our plans and purposes. For that reason, digital systems won't just be interested in identifying us as flesh-and-blood individuals. Often, what will matter more is what our data says about us. And for that purpose there's no stable, defined 'me' or 'you'. As John Cheney-Lippold observes in *We Are Data* (2017), when it comes to processing and categorizing the lives of humans, what algorithms see is a shifting and fluid collection of data, constantly being deconstructed, recompiled, and re-analysed for whatever question they want to ask.[19] A predictive policing algorithm will look at my data and ask: is this man likely to commit a crime? A mortgage-lending algorithm will probe my data from a different angle, asking: is this someone who pays his debts? An NSA algorithm might ask whether my data corresponds to what they'd expect of a 'terrorist'. A marketing algorithm might note that I'm a male millennial in my twenties, and send advertisements for avocado and beard oil my way accordingly.[20] Algorithms see what they want to see depending on their purpose.

Those who scrutinize in the digital lifeworld won't care about finding a fixed stable identity, or what we might think of as our 'true' selves. On the contrary, we'll be 'made a thousand times over in the course of just one day' by 'hundreds of different companies and agencies identifying us in thousands of competing ways'.[21] Recall Yuval Noah Harari's observation that while *writing* began as a way of *reflecting* reality, it eventually became a powerful way to *reshape* reality: for bureaucratic purposes, what's written on your form matters more than the 'truth'. In the digital lifeworld, for practical purposes we will be who the algorithms say we are, whether we like it or not. Every time code makes a determination about us, it's a 'freshly minted algorithmic truth that cares little about being authentic but cares a lot about being an effective metric for classification.'[22] And most of the time we'll have no idea what's being concluded about us until (at best) after the event, just like we're taken aback when a 'targeted' ads pops up online and we wonder 'why would they possibly think I'd be interested in *that*?' What algorithms see in us will sometimes bear no resemblance to what we see in ourselves.

In the past, the inscrutability of human life was seen as a bulwark against overweening power. No one could fully understand the complexity of human society, it was said, and therefore no one could fully control it. This theme was emphasized by the Austrian economist and theorist Friedrich von Hayek, who disdained efforts by centralized authorities to make decisions on behalf of the people they governed. Such authorities, Hayek argued in 1945, would always lack knowledge of 'the particular circumstances of time and place' and even if they could secure such knowledge they would have to grossly oversimplify matters, impeding the quality of their decision-making.[23] Hayek's argument, taken up by the United States and its allies in the Cold War, was echoed fifty years later by Scott in *Seeing Like a State*. For Scott, the simplification of real life could only come at the expense of understanding it:[24]

> [A] city map that aspired to represent every traffic light, every pothole, every building, and every bush and tree in every park would threaten

to become as large and complex as the city that it depicted. And it certainly would defeat the purpose of mapping, which is to abstract and summarise.

But Hayek and Scott were writing for the twentieth century, not the twenty-first. Scott's example of a physical map, drawn on paper, is already antiquated. Administrators in the digital lifeworld—human or machine—may be able to scrutinize countless aspects of society at any given moment.

Intimate

In November 2015, Victor Collins was found dead in the hot tub at the home of James Bates in Bentonville, Arkansas. Bates was accused of murdering Collins. There were no human witnesses, however, and the death was said to have taken place in the privacy of Bates' home. But investigators took a close interest in his 'smart home' devices. His smart water meter, for instance, apparently showed that 140 gallons of water were used between 1 am and 3 am on the night of Collins' death. Detectives inferred that this unusually large amount of water may have been used to wash away evidence from the patio. More intriguingly, Bates owns an Amazon Echo, a personal assistant system that hears and executes simple spoken orders from its human masters, like 'order a pizza' or 'switch off the light'. When activated, the Echo listens to and records everything that is said to it. The police—and, eventually, the nation—were curious: perhaps the device had 'heard' the alleged murder take place. Detectives confiscated the Echo in an attempt to extract data from the night in question. For many months it looked like the Echo could prove to be the decisive witness in the murder trial of James Bates.[25] The charges, however, were eventually dropped.

A month after the death of Victor Collins, Connie Dabate was shot dead in the basement of her home in suburban Connecticut.

When the police arrived, her husband Richard told them that a masked assailant had entered the house, attacked him, tied him to a chair, and cut him with a knife before shooting his wife. But Mrs Dabate's Fitbit—an exercise device that gathers data for health purposes—told a different story. Apparently she had moved 1,217 feet around the house at the very time Mr Dabate claimed she was shot. Richard Dabate will stand trial for the murder of his wife.[26]

A marathon medallist was busted when her Garmin exercise device revealed that she hadn't run the full 13.1 miles.[27] A man whose house burnt down was charged with arson and insurance fraud when data collected by his pacemaker revealed cardiac activity inconsistent with a frantic rush to safety.[28]

Our devices snitch on us more than we know.

The sanctity of the private home has been part of English law for centuries. In 1604 the English judge Sir Edward Coke held that a person's home is his 'castle and fortress'.[29] In the digital life-world, however, even the home will not necessarily provide a refuge from scrutiny. Like the Amazon Echo in James Bates' home, domestic technologies will be able to gather information about their human owners in their most intimate spaces. And the portable technologies silently recording our lives—smartphones, wearables, implantables—are not interested in where we are at any given moment. The cameras roll just the same when we are at home as when we are out and about. If VR and AR technology replace the 'glass slab' computing paradigm,[30] device makers will literally be able to 'see the world through your eyes.'[31] In due course, we may grow to think less of the distinction between private and public, at least in so far as that distinction can be mapped spatially onto the home and outside it. In 2017 a vibrator manufacturer agreed to pay customers compensation when it turned out that their 'smart vibrator' was tracking owners' use without their knowledge and sending details of the devices' temperature and vibration intensity back to the manufacturer. It doesn't get much more intimate than that. (Not to mention that the application used to control the vibrator was 'barely secured', meaning 'anyone within Bluetooth range' could 'seize control' of it.)[32]

Law enforcement officials have made no secret of their interest in the internet of things as a means of gathering information. Says the US Director of National Intelligence:[33]

> In the future, intelligence services might use the [internet of things] for identification, surveillance, monitoring, location tracking, and targeting for recruitment, or to gain access to networks or user credentials.

Imperishable

We tend to think of forgetting as a vice. We curse our poor memories when we lose our keys or forget to call our mother on her birthday (a mistake the wise son only makes once). In Plato's *Phaedrus*, Socrates criticizes the invention of writing for its inevitable effect on men's memories:

> For your invention will create forgetfulness in the souls of those who have learned it, through lack of practice at using their memory, as through reliance on writing they are reminded from outside by alien marks, not from within.[34]

Yet forgetting actually plays an important social function.[35] Sometimes, in order to change and move on, we need to be able to put aside our old failures, regrets, embarrassments, and prejudices. We try to let go of the wrongs perpetrated against us by others, and hope that others in turn will forget our own trespasses. We hate to be reminded of long-ago failures that we have forced from our memory. And we always fear that the past might be resurrected in new and unexpected contexts. Remember that awful photograph of you dressed as a 'sexy armadillo' taken at your birthday party ten years ago? It's still lurking in the recesses of Facebook. Let's hope that it hasn't been seen by your new boss.

In reality, as Viktor Mayer-Schönberger and Kenneth Cukier argue, for most of human history *forgetting* has been the norm and *remembering* the exception.[36] The vast majority of what has been known, spoken, done, and thought by humans was never recorded and has been lost to time. In the digital lifeworld, however, the opposite will be true: much will be remembered and little will be forgotten. Information in digital form is (at an increasing rate) easier and cheaper to store and reproduce than any previous form of information, and the incentives are there to keep it. For all the reasons given in chapter three, data is a valuable commodity and will only grow in value as time goes on. In 2007 Google acknowledged that it had saved *every single search query ever entered* by its users, together with every search result those users subsequently clicked on. Every time you asked a dumb question, looked up an embarrassing medical symptom, or idly entered the name of your attractive colleague, Google stored the details of your search. And it continues to do so, albeit that after nine months the queries are anonymized.[37] Just think of the extraordinary means of scrutiny afforded to Google by its bank of *all* past and present searches. More than a billion people input search queries to Google more than three and a half billion times each day (not to mention the data it gets from more than one billion Gmail accounts). Google has a clearer window into the inner lives of human beings than any ruler or spiritual leader of the past.

Now consider all of the other information gathered about you in the digital lifeworld, and imagine that it may be kept in permanent or semi-permanent form, available to be indexed, retrieved, searched, and analysed for decades into the future. It's early days, but companies already harvest and sell dossiers about millions of ordinary people with more than a thousand individual data points per person—social security, health, education, court, criminal, and property details, together with whatever else can be gathered from your digital exhaust.[38] To paraphrase William Faulkner, in the future, 'the past is never dead. It's not even past.'

That's why some politicians (perhaps with an eye on their own misspent youths) have backed a right for under-18s to delete embarrassing material on their social media platforms.[39] It's also why the European Union has strong and restrictive laws about what may be done with personal data—laws which are set to be tightened further with the introduction of the General Data Protection Regulation in 2018. In 2014 the Court of Justice of the European Union ordered Google to remove the 'inadequate, irrelevant, or no longer relevant, or excessive' search results that came up for Mario Costeja González, a decision that has led (in Europe) to a so-called 'right to be forgotten'.[40] In her book *Ctrl+Z* (2016), Meg Leta Jones argues for a form of 'digital redemption or digital reinvention' that's more like a right to be *forgiven*.[41]

Data is sometimes lost. Not everything will be remembered. But for the disciplinary function of scrutiny this doesn't really matter: we can't easily *know* what data will be lost or destroyed, and so may increasingly act as if nothing will be forgotten.

Predictable

While most of us understand that our past and present actions may be scrutinized by others, we obviously don't expect our *future* conduct to receive the same treatment. After all, no one knows what the future holds. And yet in the digital lifeworld, increasingly capable systems will be able to predict aspects of our lives with remarkable range and precision. Already, insurance companies use algorithms to predict whether and when we will crash our cars, become ill, or die; employers predict if we might join a competitor; financial institutions predict whether we will default on our loans; officials predict which prison inmates will kill or be killed.[42] Police use digital systems to predict future crimes, offenders, and victims.[43] Clinical researchers can predict whether we will cheat on our spouses or divorce; smartphone companies can predict our location with eerie accuracy; service providers can predict whether we are going to take our custom elsewhere. The IRS predicts whether we will commit tax fraud. (Predictively ranking tax returns in terms of their risk has apparently enabled the IRS 'to find 25 times more tax evasion, without increasing the number of investigations'.)[44]

The crystal ball of tech-enabled prediction, mostly using machine learning algorithms, is a formidable new means of scrutiny. It allows the watcher to 'see' things that haven't even happened yet. That's its auxiliary function. And the disciplinary implications are perhaps even more imposing: one can imagine humans not just refraining from crime, but from perfectly legal things like going to the house of a friend who happens to have a criminal record, which might lead a system to predict, by association, that they will go on to commit crimes themselves.

Rateable

In the last few years, you may have used an online platform that 'crowdsources' ratings for certain goods—movies, Uber drivers, hotels, restaurants, dry cleaners, and so forth. You may not have suspected that, in due course, *you* might be the subject of other people's ratings.

We're already ranked and scored in various ways—from credit scores that determine whether we can secure finance to 'health scores' compiled from information gathered about us online.[45]

In the digital lifeworld it will be possible for each person to bear a holistic personal rating compiled of scores awarded by friends, colleagues, businesses, and acquaintances on whatever measures are considered socially valuable: trustworthiness, reliability, attractiveness, charm, intelligence, strength, fitness, and so forth. In turn, people's ratings could determine their access to things of social value, such as housing or jobs. This is a pretty grim prospect, and systems like it have been the subject of some fine fiction, including an episode of the sci-fi television series *Black Mirror* in 2016.

But something like it is already happening in China, where more than three dozen local governments are starting to compile digital records of social and financial behaviour in order to rate citizens' creditworthiness. According to reports, Chinese citizens may receive negative marks for a wide array of transgressions, from dodging fares on public transport to jaywalking or violating family planning rules. In time, authorities in Beijing hope to draw on wider datasets including people's online activity. Once data is gathered, algorithms will calculate citizens' ratings, which will then be used to determine who gets loans, who receives faster responses from government officials, who is allowed to access fancy hotels, and other things of social value. Chillingly, the Chinese government says that the system will 'allow the trustworthy to roam everywhere under heaven while making it hard for the discredited to take a single step.'[46]

Putting aside the Chinese example for a moment, what's clear is that an increasingly quantified society is one in which our access to things of social value could easily be determined by data about us over which we might have little control. Critically, it doesn't need to be the *state* exerting this power. Service providers of all stripes could insist on a certain rating, leading to an increase in their power and a corresponding decrease in that of the rated.[47]

Scrutiny: Implications

The auxiliary and disciplinary effects of scrutiny in the digital lifeworld are hard to conceive in their entirety. If you knew that every journey, purchase, personal message, political utterance, or act of rebellion might be captured and saved forever, and then used to determine your life chances, would you behave the same way? Or would you take a little extra care? Control of the means of scrutiny—the ability to see without being seen—will thus be an extraordinary source of power in the future. In turn, the means of scrutiny will increasingly be confronted by technologies of resistance, designed to enhance privacy and protect anonymity (chapter ten).

EIGHT

Perception-Control

'For you created my inmost being'   **Psalms 139:13**

The final way to exert power over people, without subjecting them to force or scrutiny, is to control what they know, what they think, and what they are prepared to say about the world. A good way to get someone to refrain from doing something is to prevent them from desiring it in the first place, or to convince them that their desire is wrong, illegitimate, shameful, or even insane.[1] I refer to this as perception-control.

If it's possible to create an environment that is deeply hostile to the airing of particular political grievances, then people are unlikely to bring those grievances to the fore.[2] If you know that criticizing a particular politician will bring a menacing silence to the bar where you're sitting, or cause a Twitterstorm of personal abuse to erupt against you on social media, then you might think twice about voicing that criticism in the first place. Those who seek power do well to mobilize this kind of bias against their opponents.[3] There is power in the ability to keep certain issues off the political agenda altogether.[4] What better way to maintain the status quo than to create an environment in which mere *criticism* of it is unacceptable?

Another way to exert power through perception-control is to prevent people from *having* certain grievances in the first place.[5]

This may simply be a matter of persuasion. But it can also be the fruit of skulduggery, such as censorship, which prevents issues from coming to the attention of the population at all. People can't get angry about things they don't know about.

More subtly, if the powerful can manufacture a widespread belief or conventional wisdom that what is in their interests is in everyone else's interests too, then there will be no need for force or scrutiny to ensure compliance.

To know about matters beyond our immediate experience, we rely on others to (a) find and gather information, (b) choose what is worthy of being reported or documented, (c) decide how much context and detail is necessary, and (d) feed it back to us in a digestible form. This is the work of *filtering*. How we perceive the wider world depends heavily on the filters that make it available and comprehensible to us. We know that filtered content will usually only be a fraction of the whole picture, but we hope and trust that the information we receive is true and that the most important aspects have been prioritized.

Filtering is an incredibly powerful means of perception-control. If you control the flow of information in a society, you can influence its shared sense of right and wrong, fair and unfair, clean and unclean, seemly and unseemly, real and fake, true and false, known and unknown. You can tell people what's out there, what matters, and what they should think and feel about it. You can signal how they ought to judge the conduct of others. You can rouse enthusiasm and fear, depression and despair. You can shape the norms and customs that define what is permitted and what is forbidden, which manners are acceptable, what behaviour is considered proper or improper, and how shared social rituals like greeting, courtship, ceremonies, discourse, and protest ought to be performed; what may be said and what should be considered unspeakable; and the accepted boundaries of political and social conduct. If I breach a norm or custom, I may not face a sanction in the form of force, as I would with a law, but the consequences could feel much worse: ridicule, shame, ostracism, isolation, even excommunication. 'The way we feel and think,' says Manuel Castells, 'determines the way we act.'[6] Those who control the means of perception will increasingly determine the way we feel and think, and therefore the way we act.

Understanding the role of perception-control can help to explain one of the most enduring issues in politics: why the 'scrofulous, overworked, and consumptive starvelings' of the world have so seldom risen up against their rich and powerful overlords.[7] The answer given by Karl Marx and his apostles, notably the Italian thinker Antonio Gramsci, is that ordinary people have been psychologically conditioned to accept their fate passively.[8] They are held in place by illusions and delusions that make change seem impossible or even undesirable. A host of theories with cool-sounding names have been advanced to explain this phenomenon, including hegemony, ideology, false consciousness, and fetishism (not the naughty kind). Assuming the right historical conditions, Marx believed that once the realities of the world were revealed to the industrial working class (the 'proletariat') they would rise up and overthrow capitalism. The role of intellectuals, Marx believed, was to help ordinary folk shed their misconceptions: 'As philosophy finds its *material* weapons in the proletariat, so the proletariat finds its *spiritual* weapons in philosophy.'[9] (We look at *post-truth politics* and *fake news* in more depth in chapter thirteen.)

Perception-Control in the Twentieth Century

In the last century, the daily business of filtering was mostly done by the mass media: substantial corporations broadcasting information to millions, even billions, of consumers through print, radio, and television. The power entrusted to mighty media corporations and their mogul owners was frequently the subject of concern, with fears that they might manipulate or brainwash consumers with information geared toward their own interests and prejudices. Nevertheless, the political culture in the English-speaking world has strongly favoured self-regulation of the media other than in times of extreme national peril.

The invention and widespread adoption of the internet signalled the end of the traditional mass-media monopoly over the means of filtering. Alongside the old system emerged a 'networked information economy' in which social media and digital news platforms enabled people to be *producers* and *critics* of content as well as its *consumers*.[10] In the noughties, this development was generally treated with enthusiasm. Harvard Professor Yochai Benkler predicted in *The Wealth of Networks* (2006) that the result of the networked information environment would be 'space for much more expression, from diverse sources and of diverse qualities':[11]

> the freedom to speak, but also to be free from manipulation and to be cognisant of many and diverse opinions . . . radically greater diversity of information, knowledge, and culture.

To some extent this prediction has been vindicated, mainly because the structure and culture of the pre-mobile internet leant itself to a free and pluralistic flow of information.[12] But as we'll see in chapter thirteen, the new information environment has also brought its own plagues and problems, with serious difficulties for rational deliberation.

It's important to see that the internet has also been used for *more* precise and extensive control over the information we transmit and receive, and thus over how we perceive the world. Sometimes this has simply been a matter of controlling the physical infrastructure— the transmission towers, routers, and switches—through which information travels. A repressive state that wants to censor the content available on the internet within its jurisdiction, or even to create a separate walled-off network, can use this infrastructure to do so.[13] So it is with the Great Firewall of China. And as Benkler acknowledges, the infrastructure of the later wireless internet, used on smartphones, is engineered to enable *more* control over content by manufacturers and service providers.[14]

Perception-Control in the Digital Lifeworld

In the future, how we perceive the world will be determined more and more by what is revealed to us by digital systems—and by what is concealed. When we only experience a tiny fraction of the world, *which* fraction we are presented with will make a big difference. It will determine what we know, what we feel, what we want, and therefore what we do. To control these is the essence of politics.

News

The first means of perception to come under the control of digital technology will be news. Already we are increasingly reliant on social media to sort and present our news. The main difference from today will be that the human reporters, writers, editors, and moderators who have traditionally done the work of filtering will gradually be replaced by automated systems programmed to do that work in their stead. Algorithmic filters will be able to give you certain content, transmitted in your favourite format: spoken aloud to you while you shower, played in short hologrammic clips, brought to life in AR or VR, or even laid out in good old prose. You'll be able to request just the right amount of detail and context. The main promise of algorithmic filtering is an information environment uniquely tailored to each of our needs. If you trust the code.

The process of news automation is already underway, with auto-mated article-generation, automated comment moderation, and automated editors making decisions about what news you see and what you don't. Just as Amazon and Netflix recommend the books and television you should consume, until recently Facebook's news platform was said to determine the news you saw on its platform by balancing roughly 100,000 factors, including clicks, likes, shares, comments, the popularity of the poster, your particular relationship with the poster, what subjects interested you generally, and how relevant and trustworthy the item appeared to be.[15]

Search

Second, filters kick in when we go out searching for information. Search engines decide which results or answers should be prioritized in response to our queries. The precise details of Google's page ranking methods are not known, but it's generally thought that it ranks sites according to their relevance and importance to the particular query, determined in part by how often a web page is linked to and clicked on by other searchers looking for the same information. It's difficult to overstate the commercial and political importance of ranking highly on Google searches. Ninety per cent of clicks are on the first page of search results. If you or your business rank too low, then in informational terms you might as well not exist.[16] In the future, search systems will be better able to parse questions asked in natural language, so that when you 'look up' something, it will feel less like scanning a massive database and more like consulting an all-knowing personal assistant.[17] These oracular systems will decide what you need to know and what should remain unknown. As with news, there's no guarantee that the information you receive will be the same as what I receive; it could well be tailored according to what the systems perceive to be most relevant for us.

Communication

Third, whenever we communicate using digital means—which in the digital lifeworld will be very often indeed—we open ourselves up to filtering in that realm too. To take a basic example, our email messaging systems already use algorithms to determine what is spam and what isn't. (It's always vaguely upsetting to learn that someone's email system decided that your message was 'junk'.) Alarmingly, users of the Chinese version of Skype are literally prohibited by the application's code from sending certain terms to each other, including 'campus upheaval' and 'Amnesty International'.[18] This reflects a broader trend in which communications technologies are made subject to real-time censorship based on prohibited terms. The Chinese system WeChat, the fourth largest chat application in the world with more than a billion users, is censored according to key words. If you send a message that includes a prohibited word, the remote server through which it passes will simply refuse to send the message (and you are not told that this has happened).[19] Human officials in the past would only have dreamed of being able to censor conversations in real time. Our communication can be shaped and moulded in more subtle ways too. For example, Apple's decision to remove the gun emoji from the messaging applications on its devices was an interesting effort to police people's speech and thereby their behaviour.[20]

Emotions

Fourth, digital technologies will increasingly affect how we *feel* as well as what we know. A recent study conducted by Facebook showed that it could influence the emotions of its users by filtering their news content to show either 'positive' or 'negative' stories. (Controversially, this study took place without the knowledge or consent of its subjects.)[21] And this is just the beginning: increasingly *sensitive* technologies will be able to sense and adapt to our emotions with great efficacy. Surrounded by AI 'companions' with 'faces' and 'eyes' capable of responding sensitively to our needs, and perhaps provoking our sentiments in other ways, our means of perception will be even more in the grip of technology. We'll also grow increasingly subject to technogenic norms and customs that encourage us to perform in a certain way. Writers already feel pressured to create content that is likely to generate traffic or go viral. For younger people, there's social reward in revealing one's inner thoughts on Twitter, exposing one's life and body on Instagram, and revealing one's likes and dislikes on Facebook.

Immediate Sensory Experience

Finally, in the past, external filters only really came into play when we sought to know things beyond our gaze, but in the future we'll increasingly submit our immediate sensory experiences to filters as well. As I explained in chapter one, augmented reality technology (referred to as AR) supplements our experience of the physical world with computer-generated input such as sound, graphics, or video. Smart glasses (and eventually retina-based AR) may provide a visual overlay for what we see; earbuds for what we hear. As the technology grows more advanced, it will become hard (or fruitless) to distinguish between reality and virtuality, even when both are being experienced simultaneously. If VR or AR systems do come to replace the 'glass slab' computing paradigm, this type of filtering will rise in importance.

What we see and what is blocked out, which emotions are stimulated and which are left unmoved—we will entrust these decisions to the devices that filter the world for us. A world in which the homeless are literally removed from view is one in which the political importance of homelessness is low.[22] The outcome of a drinks date in which your smart retinas feed you real-time information about your companion based on the signals emitted by her body—whether her laugh is genuine, how nervous she is, whether she is attracted to you—is likely to be quite different because of the presence of that technology. The power to control our perceptions will be an extraordinary addition to the arsenal of those who seek to control.

Perception-Control: Implications

It's sometimes said that 'to govern is to choose'. The opposite is also true: to choose is to govern. Every time an algorithm chooses which news story to tell or which search results to prioritize, it necessarily leaves some information out. The choice to prioritize news or search results based on clicks and popularity, for instance, necessarily excludes and marginalizes less mainstream perspectives. It encourages sensationalism. Failure to deal with *fake news* is a choice that allows those who purvey it to wield influence (chapter thirteen). As we see throughout this book, in the context of our liberty, our democracy, and the demands of social justice, what seem to be technical decisions are often in fact political ones.

The other side of the coin, of course, is that social media and social networking platforms have also provided ordinary people with a means to have their voices heard. Political movements like the Arab Spring and Occupy relied heavily on such technologies to mobilize and organize. But these examples only emphasize the deeper point that when we use social media to communicate, we are at the mercy of those who control those platforms. We communicate with their permission and on their terms. Their code, their rules.

What are the deeper implications of living in a world where perception-control is increasingly delegated to digital technology and those who control it? There are certainly problems of societal fragmentation when we all see the world differently because of how it is filtered. Some of these problems are discussed in Part IV.

There is also the question of legitimacy, which ripples throughout this book. We seem to trust tech firms to filter the world in a fair and unbiased way—but what if they don't? There are already some reasons for disquiet and the digital lifeworld is only in its infancy. Apple, for instance, has blocked or refused to support applications critical of the working conditions of its manufacturers.[23] With search engines, it can be hard to tell whether search results are the result of corporations paying for the space or otherwise gaming the algorithms.[24]

One of the problems with giving others the capacity to control our perceptions is the risk of extreme outcomes. In 2009, after a dispute with a publisher, Amazon reached into every Kindle device and, without permission, deleted copies of a particular book—a feat made possible because the e-readers used cloud-based storage. The title of the book in question was apt: George Orwell's *Nineteen Eighty-Four*.[25] Needless to say, it would not have been possible for a print bookseller to recall, instantaneously, thousands of books it had sold to customers in 1995. It's not hard to imagine an insecure President seeking to prevent people from accessing particular information about his past dealings.

Now imagine that your news and search filters, as well as your AR devices, were all run by a technology company—call it Delphi. One day a politician decides that Delphi has become too rich and powerful, and stands for election on the basis that the corporation should be broken up and taxed at a higher rate. Delphi's executives see this politician's proposals as posing a threat to its survival and decide to take radical action to protect their position. Over time, the politician in question is blotted out of the public consciousness. People receive almost no news of her campaign events, and when they do, the news is unflattering. When they ask their Delphi search-oracles about 'election candidates' her candidacy is either overlooked, mentioned as an afterthought, or garnished with an unpleasant factoid. Those who hear her speak in person find that their AR filters make her sound unattractive. In due course, the politician loses and the status quo remains unchanged. This stylized example of what Harvard professor Jonathan Zittrain calls 'digital gerrymandering' combines a number of concerns about the power of the means of perception in the digital lifeworld.[26] It demonstrates that although power may be used for positive ends, like keeping us well-informed, it may also be used to create an environment hostile to particular ideas, to remove or downgrade certain issues from public consciousness, and to foster norms and customs that are not in the public interest.

If an algorithm skewed our perception of the world in a significant way, causing us to hold beliefs we might not otherwise have held, or have feelings we might not otherwise have felt, or do things we might not otherwise have done, it might be hard even to realize the nature of the power being exerted over us. We don't know what we don't know. Filters cloud the very perspective needed to keep an eye on the powerful.

NINE

Public and Private Power

'One does not make a child powerful by placing a stick of dynamite in his hands: one only adds to the dangers of his irresponsibility.'

Lewis Mumford, *Technics and Civilization* (1934)

Who? Whom?

Lenin is said to have distilled politics into two words: 'Who? Whom?'[1] If the previous four chapters are close to the mark, then we'll need to think hard about the *who* and *whom* of power in the future. That's the purpose of this short chapter.

Turning first to *whom*, it seems clear that most of us will become subject to technology's power in two main ways. The first is when we engage technology for a particular purpose. That might be when we use a social media, communications, or shopping platform, or ride in a self-driving car. Almost everything we do will be mediated or facilitated by digital platforms and systems of one sort or another. Most of the time we won't have a choice in the matter: in a fully cashless economy, for example, we'll have no option but to use the digital payment platform or platforms. The second is as *passive subjects*—when, for instance, surveillance cameras track our progress down a street. Just going about our lives we'll necessarily, and often unconsciously, engage with technology. Even when we try to avoid it by switching off our personal devices, then technology integrated into the world around us will often act on us in the background.

The question of *who* is a little more vexed. In modern times we've typically seen the distinction between the state and everyone else as the central cleavage in political life. This was the result of four assumptions. First, that only the state could *force* you to do things. Second, that it was the state (and not private firms) that did most of the work of *scrutiny*. Third, that the media (and not the state) properly enjoyed the power of *perception-control*. Fourth, that the power of perception–control was ultimately a less potent form of power than *force* and *scrutiny*. As we've seen, none of these assumptions are likely to hold in the digital lifeworld.

In *Spheres of Justice* (1983) the political philosopher Michael Walzer argues that '[d]omination is always mediated by some set of social goods' known as 'dominant goods'.[2] In a capitalist society, he explains, capital is the dominant good because it can be 'readily converted' into other desirable things like power, prestige, and privilege.[3] In the digital lifeworld, I suggest, the dominant good will be digital technology, because for those who control it, it won't just bring convenience, amusement, or even wealth: it'll bring power. Note that power will lie with those who *control* the technology, not necessarily those who own it. Your personal computer, your smartphone, your 'smart' thermostat, locks, and meters, your self-driving car and your robotic assistant—you may well own these things in the future, but if today's system is anything to go by, you'll very rarely control the code inside them. Tech firms have control over the initial design of their products, determining their 'formal and technical' properties[4] as well as their 'range of possibilities of utilisation'.[5] And they'll obviously retain control over platforms—like social media applications— that remain under their direct ownership. But they'll also control the code *in devices they sell*.[6] That means that technology we buy for one purpose can be reprogrammed without our consent or even our knowledge.

For tech firms, code is power.

But the state will muscle in too. Its ability to use *force* against us, for instance, would be greatly enhanced if it also had access to broad means of scrutiny. That's why although the state doesn't *own* the technologies that gather data about us, it's already tried to establish *control* over them—sometimes with the blessing of tech firms, sometimes against their will, and sometimes without their knowledge. To take a couple of examples, law-enforcement authorities don't need to scan the emails of Gmail users for evidence of child pornography because Google does it for them and reports suspicious activity.[7] Similarly, the state doesn't need to compile public and private records of all the data collected about individuals (in the US the Constitution partly prevents it from doing so) but is perfectly able to *purchase* that information from data brokers who have undertaken the scrutiny themselves.[8] Big Brother, it's said, has been replaced by a swarm of corporate 'Little Brothers'.[9] In 2011 Google received more than 10,000 government requests for information and complied with 93 per cent of them.[10] Tech firms comply with the government for various reasons: sometimes because they agree with the government's aims, sometimes because they're well-paid, sometimes because they want to collaborate on cutting-edge technologies, and sometimes because it makes business sense to stay on the state's good side.[11] In this context, Philip Howard, professor of Internet Studies at the University of Oxford, has identified what he calls a 'pact' between big tech firms and government: 'a political, economic, and cultural arrangement' of mutual benefit to both sides.[12]

As well as asking permission, the state will sometimes use the law to help it gain control over the means of scrutiny. Many European countries and the US have enacted laws requiring Internet Service Providers (ISPs) to adapt their networks to make it possible for them to be wiretapped.[13] Sometimes, however, tech companies push back, as when Apple refused to accommodate the FBI's demands that it unlock the iPhone of one of the San Bernadino terrorists.[14]

But where the state wants information that it can't buy, legislate for, or demand—it still has the illicit option of *hacking* the databases of those who hold it. One of the revelations made by Edward Snowden was that the National Security Agency (NSA) project MUSCULAR had compromised the cloud storage facilities of both Google and Yahoo, harvesting a vast trove of emails, text messages, video, and audio for its own purposes.[15]

As we saw in chapter eight, in jurisdictions such as China the state has gained control over not only the means of force and scrutiny, but also perception-control, in its ability to censor the news people receive, what they can find when they search for information, and even what they are able to say to each other using digital platforms. In the western hemisphere, too, the state has tried to muscle in on the means of perception-control, albeit in a more indirect way (people are wary of anything that looks like state control of the media). Think, for instance, of Google's agreement to adjust its algorithm to demote sites that infringe copyright. This reduces the need for scrutiny or force by the state.[16]

As well as the state and tech firms, less stable forms of power will also lie with *hackers* who temporarily assume control of given technologies. That could mean foreign governments, organized criminals, angry neighbours, naughty schoolkids, and industrial spies. More on this in chapter ten.

The Politics of Tech Firms

Looking past the old assumptions enables us to see clearly how much power could accrue to corporations in the digital lifeworld. Firms that enjoy both the means of scrutiny and the means of perception-control, for instance, will be able to monitor and manipulate human behaviour in a way that would have been envied by political rulers in the past. Imagine them being able to control our perceptions *and* scrutinize our responses in real time, reacting to our behaviour in a constant interactive cycle. They'll have the power to target each of us individually, promoting certain behaviours and enforcing them with a gaze, or even with force.

These tech firms won't be like corporations of the past. They'll possess real power: a stable and wide-ranging capacity to get others to do things of significance that they would not otherwise do, or not to do things they might otherwise have done. This is a new political development—so new, in fact, that our vocabulary isn't rich enough to describe it. Some commentators compare mega-companies like Google to a *government* or a *state*, but this is conceptually sloppy. A tech firm is a private entity that operates in a market system in pursuit of limited economic interests. It answers not to the public at large but to its owners and stakeholders. It has interests of its own, separate from those of its users. The state, by contrast, is not supposed to have interests of its own. In theory at least, it exists for the sake of the public. What Google and the state have in common, of course, is that they exert power. But the nature and scope of that power is different. Our use of language should be able to accommodate that difference.

A more satisfactory analogy is that the most important tech firms are increasingly like *public utilities*, that is, like the organizations that maintain the infrastructure for amenities like electricity, gas, sewage, and water.[17] When privately owned, utility companies are generally subject to state regulation requiring them to act in the public interest (as are other public service providers like healthcare and education companies). The physical infrastructure underpinning the internet is already seen as a kind of public utility. That's why there has long been such passionate support for the idea of *network neutrality*, that is, that private network providers should not be able to speed up or slow down connectivity depending on the user—or for that matter, block content they didn't like. As I write, the principle of net neutrality, supported by successive US governments, is being revisited by the Trump administration.[18]

The utility analogy is apt for technologies that are set to become vital public goods: municipal fleets of self-driving cars, drone-based national postal services, cloud-computing services essential to the economic life of the country, and so forth. But the analogy is imperfect. Our relationship with utilities tends to be one of *reliance* rather than *power*: we need them, to be sure, but they don't often get us to do things we wouldn't otherwise do. And unlike public utilities, most of the digital technologies we encounter in the digital lifeworld won't exist to serve a collective need. The means of scrutiny, for instance—the technologies that gather data from us in public and private—exist chiefly for the benefit of those who control them.

In essence, when we talk about powerful tech firms, we're talking about *economic entities that are politically powerful*. Not all tech firms, however, will be equal in power. They'll only be truly powerful if their power is *stable* and *wide-ranging*, touching on matters of *significance*. So a platform that provides an important forum for political debate, for instance, will be more powerful than one that offers a funky way to swap and edit images of fluffy cats. The most powerful firms will be those that control the technologies that affect our core freedoms, like the ability to think, speak, travel, and assemble (chapters ten and eleven); those that affect the functioning of the democratic process (chapters twelve and thirteen); those with the power to settle matters of social justice (chapters fourteen, fifteen, and sixteen); and those that attain a position of market dominance in any one of these areas (chapter eighteen).

This isn't the first time that economic entities have grown politically powerful. In the feudal system, the economic ownership of land also gave landowners political control over the people who lived and worked on it. The powerful could tax their serfs, conscript them into their private militias, put them to work, discipline and punish them, and prevent them from leaving.[19] The medieval guilds were another type of economic entity that exerted political power. They issued precise regulations relating to the pricing, quality, and trade of goods and their private judicial systems fined and imprisoned members who refused to comply.[20] Even the Anglican Church, an economic as well as spiritual powerhouse, exercised considerable political power. It taxed and tithed its parishioners. It punished them for breaking ecclesiastical rules. It was entitled to censor publications it saw as 'heretical or blasphemous'.[21] One of the defining features of modernity, for better or worse, was the emergence of the state as the supreme political body, a sovereign distinct from and 'above' the market, society, and the church. In our time it's often said, correctly, that the separation between money and politics is not as clear as it should be. But whereas today's corporations mostly acquire political power through lobbying, networking, PR, and campaign finance, in the future they'll enjoy their own kind of power—the kind that comes with controlling digital technology.

My claim is not that tech firms will rival states in the nature or extent of their power. Indeed, far from predicting the death of the state, I've argued that much of the power of digital technology could be co-opted by the state, supercharging its power. But we must shed the dangerous view that a body has to be as powerful as a state before we start taking it seriously as an authentic political entity. The rise of tech firms with great political power is itself a development of major significance. It deserves its own theory.

Powering On

In the last five chapters we've looked at the future of power—and already we can see that it will be very different from the past. In the realm of *force*: a shift from written law to *digital law*, a rise in the power of private entities able to use force against us, and the emergence of autonomous digital systems without human oversight and control. In the realm of *scrutiny*: a dramatic increase in the scrutiny to which we are (or may be) made subject, in the intimacy of what may be seen by others, in the capacity of third parties to rate and predict our behaviour and then remember everything about us for a long time. In the realm of *perception*: the capacity to control with increasing precision what we know, what we feel, what we want, and therefore what we do.

In the future, mighty entities—public and private—will try to wrest control of these new instruments of force, scrutiny, and perception-control. Anyone who seeks great power will dream of holding the full house of all three. The digital lifeworld will be thick with power and drenched with politics.

Can we bring these powerful and complex new technologies to heel? What hope is there for ordinary people to have a share in the powers that govern them? These questions are the subject of the next two Parts of this book, on the future of liberty and democracy. A famous author once wrote of an impressive new form of power in human affairs, one 'unlike anything that ever before existed in the world':[22]

> It covers the whole of social life with a network of petty complicated rules that are both minute and uniform, through which even men of the greatest originality and the most vigorous temperament cannot force their heads above the crowd. It does not break men's will, but softens, bends, and guides it; it seldom enjoins, but often inhibits, action ... it is not at all tyrannical, but it hinders, restrains, enervates, stifles, and stultifies so much that in the end each nation is no more than a flock of timid and hardworking animals with the government as its shepherd.

This passage could be about the power of technology in the digital lifeworld. In fact, it was written in 1837, nearly two hundred years ago, by the young French nobleman Alexis de Tocqueville. His subject? Democracy in America.

PART III

Future Liberty

'Every individual lets them put the collar on, for he sees
that it is not a person, or a class of persons, but society
itself which holds the end of his chain.'

Alexis de Tocqueville, *Democracy in America* **(1835)**

TEN

Freedom and the Supercharged State

'We can hardly touch a single political issue without, implicitly or explicitly, touching upon an issue of man's liberty. For freedom... is actually the reason why men live together in political organization at all; without it, political life as such would be meaningless.'

Hannah Arendt, 'Freedom and Politics' (1960)

Humans have fought about freedom for centuries. Countless screeds have been published, orations delivered, constitutions enacted, revolutions fought, and wars waged—all in the name of some freedom or other. But the rhetoric of freedom conceals a more ambiguous reality. The twentieth century taught us that certain types of freedom are not inconsistent with ignorance, war, and disease. 'Liberty does not mean all good things' writes Friedrich Hayek: 'to be free may mean freedom to starve'.[1] The great philosopher and psychoanalyst Erich Fromm, who fled Germany in the 1930s, wrote in *The Fear of Freedom* (1942) that freedom can bring unbearable anxiety, isolation, and insecurity. In his native land, he wrote, millions 'were as eager to surrender their freedom as their fathers were to fight for it'.[2]

Visions of the future of freedom tend to take two forms. Optimists say that technology will set us free, liberating our bodies and minds from the shackles of the old ways. Pessimists predict that technology will become yet another way for the strong to trample on the freedoms of the weak. Which vision is right? Why should we care? My own view is that there are causes for both optimism and pessimism, but what the future requires above all is *vigilance*, if we are to ensure what John F. Kennedy called the 'survival and success of liberty'.[3] This chapter looks at the relationship between individuals and the state; chapter eleven looks at the relationship between individuals and big, powerful tech firms. At the end of Part III, I offer a set of new concepts designed to help us think clearly about freedom's future: Digital Libertarianism, Digital Liberalism, Digital Confederalism, Digital Paternalism, Digital Moralism, and Digital Republicanism. But for now, don't worry about these monstrous terms. We begin simply by trying to understand what people mean when they talk about freedom.

Three Types of Liberty

For career political theorists, liberty is a gift that never stops giving. It is a concept of breathtaking elasticity with a dazzling array of acceptable meanings. Entire working lives have been spent happily trying to define it. Heretically, I suggest that freedom can be distilled into three categories: freedom of action, freedom of thought, and freedom of community (the last also known as the republican conception of freedom).

Freedom of Action

Freedom of action means the ability to act without interference: to gather with others, to travel the land, to march and demonstrate, to engage in sexual activity, to study, write, and speak unmolested. It's a physical conception of freedom, concerned with actions, activities, motions, and movement rather than the inner workings of the mind.

Freedom of Thought

A different type of freedom lies within.

The ability to think, believe, and desire in an authentic way, to develop one's own character, to generate a life plan, and to cultivate personal tastes are all fundamental aspects of freedom of thought. It's closely associated with another concept, *autonomy*, derived from the Greek words *autos* (self) and *nomos* (rule or law): mastery over the inner self.

For some thinkers, it's not enough for people to be guided by their own desires. A truly free mind, they say, is one that can 'reflect upon' those desires and change them in light of 'higher-order preferences and values'.[4] The fact that a person craves a stiff drink means little if she's unable to reflect on *why* she so craves it—and perhaps adapt her desire accordingly. Jean-Jacques Rousseau argued that 'to be governed by appetite alone is slavery'[5] and contemporary philosophers like Tim Scanlon say that true autonomy requires us to weigh 'competing reasons for action'.[6] Freedom of thought, for Scanlon, really means freedom to think in a particular way: consciously, clearly, rationally.

When we think about freedom of mind and body, it always pays to remember the myth of Odysseus. Odysseus ordered his sailors to lash him to the mast of his ship and refuse all his subsequent orders to be let loose. He did this to protect his men from the deadly Sirens, who he knew would try to lure the vessel onto the rocks with their song. Coming within earshot of the Sirens, Odysseus writhed in his bondage and demanded to be set free. But the loyal men, who had plugged their ears with beeswax, refused to obey. The ship sailed on unharmed. While tied to the masthead, Odysseus was certainly unfree in one sense. His immediate conscious 'self' was prevented from doing what it wanted, as was his body. But this unfreedom was the result of a free choice made by his earlier, rational self. The myth demonstrates something we already know: that sometimes we act intuitively, instinctively, and spontaneously, while other times we act rationally, carefully, and consciously. Sometimes we act consistently with our moral convictions, while other times we violate our own deepest ethics. It's as if within each individual there's not one unitary self but rather an assortment of competing selves that struggle and vie for supremacy. In the moment that the addict injects himself, his conscious 'self' (in the realm of action) is surely free, but his deeper, rational, 'self', which longs to kick the habit, is shackled by the ravages of addiction. Three points follow. The first, as Berlin put it, is that 'conceptions of freedom directly derive from views of what constitutes a self, a person, a man.'[7] To say 'I am free' is therefore to prompt the question: '*which* me?' Second, it's possible to be free and unfree simultaneously, depending on which 'self' is calling the shots. Third, it's possible to will ourselves into partial or temporary unfreedom. The same is true when a person voluntarily submits to the martial regimen of the armed forces or asks to be tied up by a lover. I think of Odysseus when I see people using the *Freedom* app on their smartphones. At the user's request, this app temporarily blocks access to the internet and social media, allowing them to work without distraction.[8] Like Odysseus, users can't break their digital chains until the specified time.

Before turning to republican freedom, it's worth noting that free thought and free action (typically aspects of a *liberal* view of freedom) are complementary to each other. There's not much point in being able to do loads of things if we only want to do them because someone else told us to. The 'right to express our thoughts,' Dr Fromm reminds us, 'means something only if we are able to have thoughts of our own.'[9] Hence the power of perception-control, discussed in chapter eight. Likewise, it's all very well being able to think for ourselves, but without the ability to put those thoughts into action they'd remain forever imprisoned in the mind.

Republican Freedom

While free action and free thought go hand in hand, republican freedom is a rather different beast.

While actions and thoughts are generally attributed to individual persons, there is a broader way of thinking about liberty, which holds that to be free is to be an active member of a free community. This has been called the *republican* view of freedom. Its leading modern proponents are Philip Pettit and Quentin Skinner.[10] There are three dimensions to it.

First, a free community is one that governs itself by pursuing the will of its citizens without external interference. The poet and polemicist John Milton (1608–1674), author of *Paradise Lost*, called this 'common liberty'.[11] It's a very old idea. Indeed, the word *autonomy* was used to describe city-states long before it came to describe people.[12]

Second, true freedom comes from participation in politics and the development of what Cicero called *virtus*, 'which the Italian theorists later rendered as *virtù*, and which the English republicans translated as civic virtue or public-spiritedness'.[13]

Third, *actual* freedom matters more than *arbitrary* freedom, a crucial distinction that underpinned both the English and American Revolutions. In the first half of the seventeenth century, the 'democratical gentlemen' of the English Parliament began to complain about the power of King Charles I. It wasn't just that the king was behaving badly (although he was). The bigger problem was that the king's constitutional supremacy meant that *he could behave badly any time he wanted to*. His power, as Thomas

Paine put it, was *arbitrary*, meaning that his subjects relied on his whim and favour for the survival of their freedom.[14] If the king became insane, or vindictive, or despotic, he could snatch away the people's freedom with impunity. In time, enough of the king's subjects decided that 'mere awareness of living under an arbitrary power' was itself an unacceptable constraint on their liberty.[15] The king was removed from the throne and his head was removed from his shoulders.

The English republic did not survive, but late in the next century the fledgling American Congress also won its independence from the Crown. That revolution, too, was justified by the belief that 'if you depend on the goodwill of anyone else for the upholding of your rights' then 'even if your rights are in fact upheld—you will be living in servitude.'[16]

The idea is simple: a freedom that depends on the restraint of the powerful is no kind of freedom at all.

The republican ideal can ultimately be traced back to Roman thinkers—Cicero in particular—who saw the Roman Republic as the paradigm of a free community. Unlike freedom of thought and action, the republican conception of freedom places less emphasis on individuals doing as they please and more on the freedom of the community as a whole. Paradoxically, this means that in the name of freedom, individuals might sometimes be forced to act in a more public-spirited way than if the choice had been left to them.

Digital Liberation

The challenges to freedom in the future will be many and severe. The future state, armed with digital technologies, will be able to monitor and control our behaviour much more closely than in the past. But before the doom and gloom, let's consider some of the extraordinary new possibilities for freedom that the digital lifeworld might hold.

New Affordances

It's useful to think of technologies in terms of the *affordances* they offer. The term describes the actions made possible by the relationship between a person and an object.[17] For an adult, a chair offers a place to sit. For a child, it might offer a climbing frame, a hiding place, or a diving board. Same object, different affordances. There's no doubt that technologies in the digital lifeworld will offer many new affordances. Thanks to digital technology, we already enjoy ways of working, travelling, shopping, creating art, playing games, protecting our families, expressing ourselves, staying in touch, meeting strangers, coordinating action, keeping fit, and self-improvement that would have been unimaginable to our predecessors. Still new forms of creation, self-expression, and self-fulfilment should all be possible in the future. And lots of the dross that consumes our time—cleaning, shopping, admin—will increasingly be automated, freeing up our days for other pursuits. These aren't trivial gains for freedom.

Taking one specific set of affordances, there are many people who are currently considered disabled whose freedom of action will be significantly enlarged in the digital lifeworld. Voice-controlled robots will do the bidding of people with limited mobility. Self-driving vehicles will make it easier to get around. Those unable to speak or hear will be able to use gloves that can turn sign language into writing.[18] Speech recognition software embedded in 'smart' eyewear could allow all sounds—speech, alarms, sirens—to be captioned and read by the wearer.[19] Brain interfaces will allow people with communication difficulties to 'type' messages to other people using only their thoughts.[20]

As for freedom of thought, in the short time we've lived with digital technology we've already witnessed an explosion in the creation and communication of information. This ought to pose a threat to the enemies of free thought: ignorance, small-mindedness, and monoculture. We can imagine a future in which more and more people are able to access the great works of human culture and civilization with increasing ease. I call this something to imagine, however, because it isn't inevitable. As I argue throughout this book, how we choose to structure the flow of information in the digital lifeworld will itself be a political question of the first rank.

Some futurists predict that people in the future will be able to augment their intellectual faculties through biological or digital means.[21] This would obviously represent a transformation in freedom of thought, rolling back the frontiers of the mind itself. But wisely deployed, technology can free our minds in more subtle ways. Little prods, alerts, prompts, reminders, bulletins, warnings, bits of advice—together these digital offerings could combine to make us much more organized, informed, sentient, thoughtful, and self-aware.

Of course, the line between technology that *influences* us and technology that *manipulates* us is not always clear. In the future it may be difficult to tell whether a particular form of scrutiny or perception-control is making us more autonomous or whether it's actually exerting control in a way that's too subtle to see. How far can a person's thoughts be subject to outside influences before they cease to be 'free'? The eccentric philosopher Auguste Comte (1798–1857) believed that the key to good thinking was insulation from the ideas of others. He called this 'cerebral hygiene'.[22] (Students: next time you haven't done the required reading for class, blame 'cerebral hygiene'.) I prefer the view of Helen Nissenbaum, that to be 'utterly impervious to all outside influences' is not to be autonomous: but to be a 'fool'.[23] Harvard professor Cass Sunstein has done some interesting thinking on this subject, most recently in

The Ethics of Influence (2016). For Sunstein, reminders, warnings, disclosures of factual information, simplification, and frameworks of 'active choosing' are *nudges* that influence people but preserve their freedom of choice. Influence only becomes manipulation 'to the extent that it does not sufficiently engage or appeal' to people's 'capacity for reflection and deliberation'.[24] Or as Gerald Dworkin puts it in *The Theory and Practice of Autonomy* (1989), we need to distinguish methods that 'promote and improve' people's 'reflective and critical faculties' from those that really 'subvert' them.[25]

The Supercharged State

You've heard the case for expanded freedoms in the digital life-world. Now let's hear the other side. As we've seen, technology will make it easier for political authorities to enforce the law. This is no small claim, because humans have never known a more effective, wide-ranging, or stable system of control than the one we currently live with. The modern state already enjoys the formidable power to enforce laws using the threat or application of force. And yet its powers are weedy in comparison to those that will be enjoyed by states in the digital lifeworld. This could well mean a lessening of liberty. Here I identify four areas of possible concern that could flow from the arrival of the *supercharged state*.

Expansion of Law's Empire

Have you ever grabbed a shopping bag at the supermarket without paying for it? Or paid someone cash-in-hand knowing they won't declare it for tax? Perhaps you've streamed an episode of *Game of Thrones* without paying, dodged a bus fare, nabbed an illicit refill from the soda dispenser, 'tasted' one too many grapes at the fruit stand, lied about your child's age to get a better deal, or paid less tax on a takeaway meal then eaten it in the restaurant. These are all illegal acts. But according to a poll, as many as 74 per cent of British people confess to having done them.[26] That's hardly a surprise, and not because the British are scoundrels. People aren't perfect. It would be absurd to suggest that the world would be better if each of these indiscretions was routinely detected and punished.

All civilized legal systems offer a slender stretch of leeway in which people are sometimes able to get away with breaking the law without being punished. A surprising amount of liberty lives in this space. Its existence is a pragmatic concession to the fact that *most* of us can be trusted to obey the law *most* of the time. One of the dangers in the digital lifeworld is that the law will further colonize this precious hinterland of naughtiness. This would be the natural consequence of pervasive and intimate scrutiny of our conduct; and digital law that enforces itself and adapts to different situations. There's a big difference between the world we live in now and one in which DRM makes it impossible to stream *Game of Thrones*; the bus fare is automatically deducted from your smart wallet when you board; the soda dispenser recognizes your face and refuses additional service; the 'smart' grape stand screeches when its load is lightened by theft; your child's age is verified by an instantaneous retina scan; and so forth. The difference is one of day-to-day freedom.

The Politics of Prediction

A more serious concern is that using technology to predict and prevent crime will have serious implications for freedom. We saw in chapter seven that machine learning systems are increasingly being used to predict crimes, criminals, and victims on the basis of largely circumstantial evidence. As Blaise Agüera y Arcas and others explain, there's nothing new in trying to predict deviant behaviour. In the nineteenth century criminality was associated with physical traits. The bandits tormenting southern Italy, for instance, were said to be 'a primitive type of people'. A dimple at the back of their skull and an 'asymmetric face' were said to demonstrate their innate predisposition toward crime.[27] Criminologists and a fair few quacks since have sought to find more 'scientific' ways of predicting crimes before they happen. For a while the dominant sub-field in criminology was psychiatry. It was later joined by sociology.[28] In the second half of the twentieth century, trying to predict crime became an 'actuarial' endeavour, involving the use of statistical methods to determine patterns and probabilities.[29]

A recent paper claims that machine learning systems can predict the likelihood that a person is a convicted criminal with nearly 90 per cent accuracy using a single photograph of the person's face.[30] This paper has received a great deal of criticism, but it's not exactly an outlier. An Israeli startup with the unforgivable name of Faception has claimed to be developing a system that can profile people and 'reveal their personality based only on their facial image'. That includes potentially categorizing them as 'High IQ', 'White-Collar Offender', 'Paedophile', and 'Terrorist'—based on their face.[31] This too may turn out to be far-fetched, but we already know that AI systems can learn a huge amount about what John Stuart Mill called 'the inward domain of consciousness'[32] just by observing our expression, our gait, our heartbeat, and the many other clues we give away just by being alive (chapter two). The ability to detect deception through mere observation would fundamentally alter the balance in the relationship between authorities and the people.

The upshot is that the predictive powers of digital systems are growing significantly stronger, and they're increasingly being used to enforce the law. Why does this matter for freedom? Well, if you're going to restrict people's liberty on the basis of a prediction about their future behaviour, then you should be able to *explain* that prediction. But often we can't. Predictive Policing systems frequently tell police to direct their resources to a particular area for reasons that are quite unclear.[33] But in a free society you should probably be able to ask the policeman (or police droid) standing outside your house why he (or it) is there, and get a better answer than 'because the system predicted and directed that this is where I should stand'. The same is true when it comes to the use of predictive sentencing. We already use algorithms to predict how likely an offender is to commit another crime. If the system suggests a high likelihood, then the judge may choose to impose a longer sentence. As with Predictive Policing, however, it is often unclear why a particular offender has scored well or badly. In the US case of *Wisconsin v Loomis*, the defendant's answers to a series of questions were fed into Compas, a predictive system used by the Wisconsin Department of Corrections. The Compas system labelled him 'high risk' and he went down for a long time. But *he didn't know why* and was not legally permitted to examine the workings of the Compas algorithm itself.[34] Obviously when the defendant was incarcerated, his freedom of action was curtailed. But we can also think about this case from the perspective of republican freedom: the idea that a person's liberty could be restricted on the basis of an algorithm *whose workings are utterly opaque* is the very antithesis of that ideal.∝

There's also something philosophically problematic about restricting people's freedom on the basis of predictions about their future conduct. To see why, it helps to place predictive criminology within

∝ Side point: the independent site ProPublica obtained more than 7,000 risk scores and compared the system's predictions of recidivism to actual outcomes. They disproportionately predicted recidivism among black offenders at almost twice the rate of white defendants.[35] See chapter sixteen on algorithmic injustice.

a broader intellectual tradition. For centuries, political thinkers have tried to discover general *formulae*, *laws*, and *social forces* that can explain the unfolding of human affairs. 'Race, colour, church, nation, class; climate irrigation, technology, geo-political situation; civilisation, social structure, the Human Spirit, the Collective Unconscious'— all have been said, at one time or another, to be the forces that ultimately govern and explain human activity.[36] The holy grail for thinkers in this tradition was to *predict* the future of politics perfectly:[37]

> with knowledge of all the relevant laws, and of a sufficient range of relevant facts, it will be possible to tell not merely what happens, but also why; for, if the laws have been correctly established, to describe something is, in effect, to assert that it cannot happen otherwise.

Probably the most famous exponent of this way of thinking was August Comte (the one who also advocated cerebral hygiene). Comte believed that all human behaviour was pre-determined by 'a law which is as necessary as that of gravity'.[38] Politics, therefore, needed to be raised 'to the rank of sciences of observation' and there could be no place for moral reflection. Just as astronomers, physicists, chemists, and physiologists 'neither admire nor criticise their respective phenomena' so too the role of the social scientist was merely to 'observe' the laws governing human conduct and obey them purposefully.[39] Comte, who invented the term 'sociology', would have been fascinated by machine learning systems that could predict human behaviour. I suspect that his approach to politics might still find some sympathy among the more mathematically minded today.

But here's the problem. To predict the likelihood of future offending is to accept that human conduct is, to a significant extent, governed by general laws and patterns that are nothing to do with the free choices of the individual. If a machine can take personal, sociological, and historical facts about a person and make accurate predictions about their future activity, then that suggests that humans have less freedom of thought and action than the criminal justice system might like to assume. And if people are less free, doesn't that mean they're less *morally* responsible too? This is the paradox at the heart of predictive sentencing. In many penal theories, *punishment* is premised on the notion that individuals can be held morally responsible for their choices. But *prediction* is premised on the notion that individuals' choices are often determined by factors outwith their immediate control. Instead of incarcerating people, therefore, shouldn't our response to predictive policing technology be to work harder at understanding *why* people commit crimes, and, as far as possible, to tackle those causal factors instead?

Automation of Morality

It's said that the act of freely obeying the law teaches us to behave ethically. The point can be traced back at least to Aristotle, who taught that the state exists 'for the sake of noble actions'.[40] In the *Nichomachean Ethics* he argued that virtue of character results from habit... 'Correct habituation distinguishes a good political system from a bad one.'[41]

There are many reasons why you or I might obey the law: habit, morality, convenience, prudence, fear of punishment. But whatever the dominant reason, the conscious act of obedience teaches us to behave ethically. Part of becoming a good citizen is learning to think about the rights and wrongs of what we do. That is liberty's gift. In a world where many moral decisions are made for us, however, because many options simply aren't available in the face of self-enforcing laws, or because the inevitability of punishment means it's not worth the risk, we won't be called upon to hone our character in the same way. In such a world, as Roger Brownsword puts it, the question will usually be what is *practical* or *possible* rather than what is *moral*.[42] Children born into a society in which misdeeds are made impossible or extremely difficult will not learn what it feels like to choose not to break the law.

In 1911 the mathematician Alfred North Whitehead wrote that 'Civilization advances by extending the number of important operations which we can perform without thinking about them.'[43] Again, Silicon Valley Comteans might applaud this kind of dictum but it demands philosophical interrogation. Is the state's business to prevent crime at all costs? Or would the automation of morality not diminish us in some way, stripping away an important part of our humanity—the freedom to do bad things if we wish, but also the freedom *to choose not to*?

Tidings of Totalitarianism

So far in this chapter the tacit assumption has been that we're dealing with a state that governs broadly in the interests of its citizens and doesn't seek actively to oppress them. It's also been assumed that the state has not taken complete control over the means of force, scrutiny, and perception-control. But we must now consider the threat to liberty that would be posed by an authoritarian regime that was able to harness the technologies of power for the purposes of repression. To illustrate what this kind of society might look like, imagine an individual—call him Joseph—who plans to attend a forbidden protest.

> Joseph was never meant to hear about the protest at all. An email from a comrade containing details of the venue was censored *en route* and never arrived. The AI systems that reported, filtered, and presented Joseph's news didn't mention it in their bulletins. Online searches yielded no results. Somehow, against the odds, he knows about it.
>
> What Joseph doesn't know is that the regime's crime-prediction systems are already interested in him. Various snippets of personal data

have suggested that he might be engaged in subversive activity. Three months ago he foolishly used the word 'demonstration' in a message to a friend. Picked up by surveillance systems, this prompted an automated review of his historic social media activity, which revealed that, ten years ago, he used to associate with two of the known protest ringleaders. To worsen matters, just two weeks previously he downloaded the design for a balaclava and manufactured one at home using his 3D-printer. This pretty much *guaranteed* him the wrong kind of attention.

On the morning of the protest, Joseph's digital assistant emphasizes that there's going to be awful rain and traffic all day: better to stay indoors and work from home. Joseph, however, is aware that the device is unencrypted. Has it been hacked by the state authorities? Is this a subtle form of repression?

Disquieted but undeterred, Joseph sets off to the protest site, a plaza in the centre of town. He notes with cold amusement that there is no rain and little traffic. But at his first attempt he's unable to reach the vicinity of the protest. The public transport system, a municipal fleet of self-driving cars, refuses to drop passengers within a hundred yards of the plaza where the protest is taking place. (In our time, New York's Metropolitan Transportation Authority has been known to force trains to pass through stations near to protests, 'to keep people from assembling'.)[44]

Joseph instead proceeds on foot.

Conscious of the facial-recognition cameras now ubiquitous in the city centre, he pulls the balaclava over his face. Also aware that cameras can identify him by his normal gait, he wears shoes that are a size too small in order to change the way he walks. But he can't escape attention. His posture gives him away. The way he holds his body is suggestive of furtive behaviour. Several miles above his head, unseen, a drone quietly records his progress. (The US Army and Defence Advanced Research Projects Agency (DARPA) has developed a 'surveillance platform that can resolve details as small as six inches from an altitude of 20,000 feet'.)[45]

As he approaches the plaza, Joseph clocks with concern that his comrades are nowhere to be seen. He won't know until later that they've been placed under house arrest, remotely confined to their homes by state-controlled 'smart locks' installed on their apartment doors and windows.

Reaching the outskirts of the protest area, Joseph sees that a dispiritingly small group of protesters has been penned into the middle of the plaza by a phalanx of robotic bollards. A swarm of airborne riot-control drones circles overhead, blaring out orders and occasionally spraying paintballs into the throng. (Something like this, the Skunk Riot Control Copter, has already been developed in South Africa. It's an airborne drone armed with strobe lights, cameras, lasers, speakers, and four paintball guns that can that fire, every second, 20 paintballs made of dye, pepper spray, or solid plastic.)[46]

Suddenly, and without any introduction, a holographic image appears vividly in the air in front of Joseph's eyes. He freezes. The image is of him. It shows him knelt in a stress position in a prison cell, weeping and in agony, his hands cuffed behind his back. This image, Joseph realizes, has been telegraphed directly onto his smart contact lenses.[47] It's a warning. *We see you. Go home or face the consequences.*

The blood drains from Joseph's face. He turns and runs.

What this little vignette shows is the imposing range of powers that digital technology could lend to an authoritarian government in the future. What's striking is that only a few of the means of power involve physical brutality; the rest are softer and less obtrusive. Most of them could readily be automated. That's what makes them so dangerous.

Digital Dissent

The tale of Joseph raises the question of the future of dissent. Even where the state purports to act in the best interests of its citizens, no legal system is perfect. It's why many great thinkers have argued that obedience to one's conscience is more important than obedience to the law, and that breaking the law can be justified as a means of encouraging a change in it. 'It is not desirable to cultivate a respect for the law,' writes Henry David Thoreau, 'so much as for the right.'[48] The doctrine of civil disobedience holds that laws may be deliberately broken in a public, non-violent, and conscientious way, with the aim of bringing about change.[49] As Martin Luther King Jr wrote from Birmingham City Jail in 1963, where he had been imprisoned for marching for civil rights:[50]

> There are *just* and there are *unjust* laws...I submit that an individual who breaks a law that conscience tells him is unjust, and willingly accepts the penalty by staying in jail to arouse the conscience of the community over its injustice, is in reality expressing the very highest respect for law.

Civil disobedience of the kind practiced by Dr King requires a certain minimum of liberty in order to be possible, but like all forms of lawbreaking, it will become increasingly difficult in the digital lifeworld. New forms of civil disobedience are likely to come to the fore. If power is to be exerted through digital technology, then resistance will increasingly take a digital form too. *Hacking* is likely to be foremost among them.

Political Hacking

Hacker culture has existed for a while and has a number of definitions. At its broadest it refers to a 'playful' and 'pranking' attitude among programmers and coders, although governments and corporations engage in it as well. We are concerned here with situations where a person gains unauthorized access to a digital system *for political ends*. Such hacking, to borrow Gabriella Coleman's artful phrase, will usually be 'either in legally dubious waters or at the cusp of new legal meaning'.[51] I call it *political hacking*. Its purpose might be to access information, to expose the functioning of a system, or even to alter or disable a particular system—perhaps for the sake of liberty.

A lot of hacking already has a political flavour, albeit one marked by differences in approach. According to Coleman, hackers from Europe and North and South America have typically been more 'antiauthoritarian' than their counterparts elsewhere, while hackers in southern Europe have typically been more 'leftist, anarchist' than those from the north. Chinese hackers are 'quite nationalistic' in their work.[52]

If power subsists in the control of certain digital systems, then hacking diminishes that power by reducing the efficacy of those systems. It's a serious business. As hacking grows in political importance, it would be quite naïve to entrust it entirely to the judgment of hackers themselves, particularly (as appears to be the case) if the work of hacking itself is becoming increasingly automated.[53] As with all important political activity, we'll need to develop an acceptable ethical framework by which the work of political hackers can be judged. Drawing on the ideas of John Rawls, for instance, we might say that:[54]

- Hacks can only be justified in the public interest, not the interests of the hacker or any other party.

- Hacks affecting policies that were decided democratically should be limited to cases of 'substantial and clear injustice'.

- Hacks should be proportionate to the injustice they seek to remedy, going no further than necessary to achieve their aim.

- Hacks should never cause physical harm to persons.

- As far as possible, the consequences of hacks should be public and visible rather than 'covert or secretive'.

- Hacks should be used sparingly, even when justified.

- In a democracy, hacks should be a last resort after efforts made in 'good faith' through the proper procedural channels.

- In a democracy, hackers must accept peacefully the consequences
  of their conduct, including the possibility of arrest and punishment. (And the punishment for hacking should be proportionate to the crime and take into account its political
  function.)

These are just some possible principles. There will be others.

Cryptography

As well as hacking, an increasingly important form of digital
resistance will be *cryptography*, 'the use of secret codes and ciphers'.[55]
Encryption is already an indispensable part of digital systems. It
protects the integrity of our online transactions, the security of our
databases, and the privacy of our communications. It defends our
public utilities, communications networks, and weapons systems.
Its most obvious function is to repel *scrutiny* from those who seek
to gather data about us. But it has a bearing on the means of *force*
too because it prevents platforms and devices from being hijacked
or reprogrammed from afar.

In fact, encryption is the most important defence against *malevolent* hacking, where a person gains unauthorized access to a digital
system for reasons that are not in the public interest. Some of the
hacks we hear about today are reasonably funny, like when a 'smart'
toilet was reprogrammed to fire jets of water onto the backside of
its unfortunate user.[56] Others, however, are more sinister, like the
'smart' doll that could be reprogrammed to listen and speak to the
toddler playing with it.[57] Still others are deeply troubling: in 2016,
'ransomware' held hostage people's medical records until insurance
companies paid $20 million.[58] The scale of the problem is serious.
A study of 'critical infrastructure companies' in 2014 revealed that
in the previous year nearly 70 per cent of them had suffered at least
one security breach leading to the loss of confidential information

or disruption of operations. One divulged that it had been the 'target of more than 10,000 attempted cyber attacks each month'.[59] In 2014 about 70 per cent of the devices connected to the internet of things were found to be vulnerable to hacking.[60] With so many more connected devices, the possibilities for malevolent hacking in the digital lifeworld will be radically greater than today. If weapons or large machines were hacked by criminals or hostile foreign powers, they could be devastating to our liberty. Imagine a terrorist in Syria remotely hijacking a weaponized drone in New York City, unleashing hell on its inhabitants; or a geopolitical adversary accessing a country's missile systems. 'Increasingly,' as William Mitchell put it, 'we can do unto others at a distance, and they can do unto us.'[61] Cryptography offers some protection.

The political function of cryptographic methods will be one of the most important political issues of the digital lifeworld. Interestingly, the recent trend has been toward *greater* encryption of digital platforms. To take a well-known example, the messages you send on WhatsApp are now 'end-to-end' encrypted, meaning they can't easily be intercepted or inspected either by WhatsApp or any other third party. Increased encryption by social media and news platforms has made it harder for authoritarian regimes to selectively filter the flow of information. Previously, they could remove individual 'accounts, web pages, and stories'.[62] Increasingly, encryption has forced them to choose between blocking the entire platform or none of it. Some regimes have taken the more open option: the whole of Wikipedia is now accessible in Iran, as is Twitter in Saudi Arabia. Both were previously the subject of selective blocking. On the other hand, in 2017 Turkey's government blocked Wikipedia in its entirety and Egypt did the same to the online publishing platforms *Huffington Post* and *Medium* (and many others).[63]

Encryption isn't always the friend of freedom. For every plucky dissident or journalist who uses cryptography as a shield against tyranny, there's a terrorist organization, human-trafficking syndicate, drug cartel, or fraudster who uses it to conceal criminality. End-to-end encryption naturally causes concern in the intelligence community because it makes it harder for state agencies to detect terrorist plots. It's plainly not in the interests of liberty for dangerous groups to flourish unmolested. And there's a risk that private encryption will encourage states to develop their own 'home-grown platforms' that can be more easily controlled. Iran has developed its own version of YouTube, and Turkey is making its own Turkish search engine and email platform.[64]

A recent study by Harvard's Berkman Klein Center for Internet and Society predicted that encryption probably won't become a ubiquitous feature of technology in the future, mainly because businesses themselves will want to be able to retain easy access to the platforms that we use, partly for commercial reasons (that is, data harvesting) and partly because excessive encryption can make it harder to detect and correct problems.[65]

Even the powerful can't agree on the politics of encryption. In late 2017, the British government indicated that it might crack down on end-to-end encryption, while, in complete contrast, the European Parliament was mulling a prohibition on member states securing 'backdoor access' to encrypted technologies.[66] To preserve our liberty there will have to be a balance: individuals, corporations, and governments all have competing and overlapping priorities, but at their heart should be freedom of thought, freedom of action, and freedom of community.

Wise Restraints

In the stairwell to the main library of Harvard Law School there's a plaque that reads: 'You are ready to aid in the shaping and application of those wise restraints that make men free.' I passed this plaque many times during my year as a Fellow at Harvard's Berkman Klein Center for Internet and Society, and it intrigued me every time. The same phrase—more gender-balanced—is still used to pronounce individuals as graduates of Harvard Law School.

Restraints that make us *free.* It sounds paradoxical, but it's not.

A well-designed legal system, like any good system of rules, can enhance the overall sum of human freedom even as it places restrictions on people. It does this, on the liberal conception, by guaranteeing each person the space in which to flourish unharmed by others. This is what Hobbes meant when he said that a 'common Power' was needed to keep people 'all in awe'.[67] On a republican conception of freedom, the constraints of the law can make us freer by shaping our lives in a more purposeful and civic direction. Whether your tastes are liberal or republican, the future of freedom won't just depend on the new affordances that technology will offer. It'll depend on whether we can together shape and apply the wise restraints that make us free. Those restraints will often be in code.

There are at least four points that can be made in defence of the supercharged state. The first is that if the system is one of wise restraints, then, self-evidently, enforcing those restraints should work in the interests of liberty. It's no bad thing to enforce good laws. (An unenforced law is just a collection of words on a piece of paper. A law that's both unwritten *and* unenforced—to borrow an old political theory joke—is not worth the paper it's not written on. Political theorists are not known for being hilarious.) The question becomes what those good laws should be. We might, for example, think that in a world in which the powers of enforcement are greater, the rules should be correspondingly weaker or fewer. Assuming that the state will use all the means at its disposal to enforce the law, we'll need to tailor our laws for the digital lifeworld, not for the past.

Second, as digital technologies grow in speed, complexity, and significance, we'll *need* to use digital methods of enforcement just to keep them in check. Recall the example from the financial sector in chapter six: trading algorithms are now best regulated by *other algorithms*.

Third, entrusting more power to digital systems will also mean relying less on human officials. Yes, such officials can be kind and sympathetic. But they can also be selfish, myopic, greedy, vain, cruel, even evil. The German philosopher Immanuel Kant, not known for his jollity, believed that a 'complete solution' to human self-government was 'impossible' since 'from such crooked timbers as man is made of, nothing perfectly straight can be built.'[68] Silicon, however, is not crooked. It is predictable and consistent in its operation. At least in theory, digital systems can execute code, and therefore apply the law, in an even-handed way with less room for human prejudice or fallibility. (Some of the challenges to this argument are discussed in chapter sixteen, which looks at algo-rithmic injustice.)

Finally, although digital law may control us more finely than written law, it will often do so in a less obtrusive way. Compare the airport experience of walking through a contactless body scanner, as against being physically patted down by a stranger in surgical gloves. Likewise, a discreet biometric authentication system may be an improvement on the security guard who gruffly demands that you show him your documents.

Freedom's Future

Are we devising systems of power that are too complete, too effective, for the people they govern? Although we'll enjoy many new affordances, this chapter has showed the risks that flow from the supercharged state. A system of precise and perfect law enforcement may not be well-suited to the governance of flawed, imperfect, damaged human beings. Perhaps some system of 'wise restraints' can help us to maintain a satisfactory degree of overall freedom in the digital lifeworld, but there is no room for complacency. And this is only half the story. We turn now to freedom's fate in circumstances where it is private firms, and not the state, setting the limits of our liberty.

ELEVEN

Freedom and the Tech Firm

'Let every nation know, whether it wishes us well or ill, that we shall pay any price, bear any burden, meet any hardship, support any friend, oppose any foe to assure the survival and the success of liberty.'

John F. Kennedy, Inaugural Address (1961)

Niccolò Machiavelli wrote in his *Discourses* that 'the peoples of ancient times were greater lovers of liberty than those of our own day.'[1] That was 500 years ago. Back then, the greatest threats to 'common liberty' were kings, colonization, and conquest. In the future, there will be a fourth: code.

As well as supercharging the state, digital technology will increasingly concentrate power in the hands of the tech firms that control the technologies of power. Recall the line from Alexis de Tocqueville's *Democracy in America* (1835) that opened this Part of the book: 'Every individual lets them put the collar on, for he sees that it is not a person, or a class of persons, but society itself which holds the end of his chain.' This won't always be true in the digital lifeworld. The 'collar' of digital power, where not held by the state, will often be controlled by a very particular 'class of persons', that is, the firms that control the technology. This chapter is dedicated to understanding what that means for freedom.

My view is simple: if tech firms assume the kind of power that affects our most precious liberties, then they must also understand

and respect some of the rudimentary principles of liberty. Humans have been developing these for centuries. Under what circumstances is it permissible to restrict a person's freedom? Should we be free to harm ourselves? Should we be prevented from behaving immorally? These can't be treated as merely corporate or commercial concerns. They are fundamental questions of political theory.

Liberty and Private Power

One of the curious traits of digital technology, as we've seen, is that it can enhance and restrict our freedom at the same time. It frees us to do things that we couldn't do previously. But it restricts us according to the constraints of the code. Think for a moment about using an Apple device. It's usually a thing of beauty: smooth, seamless, and intuitive. It offers a universe of applications. But it's a universe closely curated by Apple. You can't reprogram the device to your tastes. You can only use the applications chosen by Apple, whose Guidelines for app developers say:

> We will reject apps for any content or behavior that we believe is over the line. What line, you ask? Well, as a Supreme Court Justice once said, 'I'll know it when I see it.'

Despite the somewhat arbitrary power of this clause, it feels churlish to complain. Apple devices offer plenty of choice and the system works well. The legal scholar Tim Wu, referring to this example, observes that 'consumers on the whole seem content to bear a little totalitarianism for convenience.'[2] He's right. We intuitively understand that Apple devices enhance our overall freedom even if we can't do everything we'd like with them. The question is whether the same trade-off will still make sense in the digital lifeworld, where code's empire will extend to almost every freedom we currently take for granted.

Take free speech, a freedom of the utmost sanctity. Free speech permits authentic self-expression. It protects us against powerful interests, exposing them to criticism and ridicule. It allows for the 'collision of adverse opinions', a process essential to the pursuit of truth.[3] With some exceptions, most of us would be horrified by the idea of the state censoring what we say or how we say it. We venerate the idea of the Greek *agora*, where the citizenry spoke freely, fearlessly, and on equal terms.

Now think about this. In the digital lifeworld, almost *all* of our speech will be mediated and moderated by private technology firms. That's because we will come to rely almost entirely on their platforms for communication, both with people we know and people we don't. That means tech firms will determine the *forms* of communication that are allowed (for example, images, audio, text, hologram, VR, AR, no more than 280 characters, and so forth). They will also determine the *audience* for our communication, including who can be contacted (members of the network only?) and *how* content is ranked and sorted according to relevance, popularity, or some other criterion. They'll even determine the *content* of what we say, prohibiting speech that they deem unacceptable. This involves some tricky distinctions. According to leaked documents obtained by the *Guardian*, Facebook will not remove a post saying 'To snap a bitch's neck, make sure to apply all your pressure to the middle of her throat' but it will remove one saying 'Someone shoot Trump' because the President is in a protected category as a head of state. Videos of abortions are OK, apparently, unless they involve nudity.[4]

It won't just be social media platforms that have the capacity to constrain speech. There will usually be several other technical intermediaries between speakers and their audiences, including the firms that control the hardware through which the information travels.[5] With more or less precision, each will be able to control the flow of information.

We are, in short, witnessing the emergence of a historic new political balance: we are given wholly new forms and opportunities for speech, but in exchange we must accept that that speech is subject to the rules set by those who control the platforms. It's as if the *agora* had been privatized and purchased by an Athenian oligarch, giving him the power to dictate the rules of debate, choose who could speak and for how long, and decree which subjects were out of bounds. The main difference is that algorithmic regulation of these platforms means that *thousands upon thousands* of decisions affecting our freedom of speech like this will be taken *every day*, decided and executed seamlessly with no right of appeal. Microsoft, Twitter, and YouTube, for instance, have teamed up to announce the Global Internet Forum to Counter Terrorism. Among other measures, they'll 'work together to refine and improve existing joint technical work' including using machine learning techniques for 'content detection and classification'.[6]

Now, plenty of folk will be happy to let tech firms get on with the task of regulating speech. But since this is a chapter about liberty, it's worth recalling the republican principle of freedom: *that a freedom that depends on the restraint of the powerful is no kind of freedom at all.* Each firm that controls the platforms for speech could reduce or refine our freedom of speech any time if it so desired. Like the pre-revolutionary English and Americans, we depend on their whim and fancy for the survival of our freedom of speech. Politically speaking, is this satisfactory?

It's not just speech. Consider freedom of thought generally. Already we trust tech firms to find and gather information about the world, choose what is worthy of being reported, decide how much context and detail is necessary, and feed it back to us in a digestible form. With few strings attached, we give them the power to shape our shared sense of right and wrong, fair and unfair, clean and unclean, seemly and unseemly, real and fake, true and false, known and unknown. We let them, in short, control our perception of the world. That's a pretty big deal for freedom of thought.

Now consider yet another basic freedom: freedom of movement. Self-driving cars will obviously generate valuable new affordances. Non-drivers will be able to make use of the roads. Road transportation will be safer, faster, and more energy efficient. Passengers will be able to work, eat, sleep, or socialize while in transit. In return for these affordances, however, we'll necessarily sacrifice other freedoms. The freedom (occasionally) to drive over the speed limit. The freedom (occasionally) to make an illegal manoeuvre or park on a double yellow line. The freedom to make a journey with no record of it. Perhaps even the freedom to make moral choices, like (in the case of the trolley problem described in chapter six) whether to kill the child or the trucker. Again, I don't seek to suggest that this isn't a deal worth striking. But I do suggest that we see it for what it is: a trade-off in which our precious liberties are part of the bargain.

From the perspective of freedom, there are four important differences between the power wielded by the state and that wielded by tech firms.

The first and most obvious is that the democratic state is answerable to the people and citizens have a meaningful say in the rules that govern them. Power can be held to account. The same can't usually be said of most tech firms that operate in the private sector. They make the rules; we live with them. Recall that even ownership of a device doesn't necessarily mean control over it. Most technology in the digital lifeworld will be reprogrammable from afar. Our property could be repurposed under our very noses without our consent or even our knowledge.

Second, (at least in theory) the state exists to serve the general interest. A well-functioning government generates laws and policies aimed at the common good. By contrast, tech firms, like all private companies operating in a capitalist paradigm, exist for the commercial benefit of their owners.

A third difference is that mature legal systems develop in a systematic way over time according to clear rules and canons. Private code, by contrast, develops in an ad hoc and inconsistent way. Different companies take different approaches. Facebook might censor content that Twitter deems acceptable. One app may gather your personal data; another might not. Your self-driving car might kill the child; mine might swerve and kill the trucker. Code's empire is not a unified realm, but rather a patchwork of overlapping jurisdictions. This isn't necessarily a bad thing: it could enable a kind of Digital Confederalism, where people move between systems according to whose code they prefer. More on that later.

Fourth, technology in the digital lifeworld will be mind-bogglingly complex, and therefore even more inscrutable than the workings of government. This is an important point. As Samuel Arbesman observes, a Boeing 747-400 aeroplane—already a pretty vintage piece of kit—has 6 million individual parts and 171 miles of wiring.[7] But that's child's play compared with what's to come. The future will teem with components and contraptions, devices and sensors, robots and platforms containing untold trillions of lines of code that reproduce, learn, and evolve at an ever-increasing pace. Some systems will function entirely 'outside of human knowledge and understanding'.[8] The fact that machines don't function like humans makes them inherently hard to understand. But often they don't even function according to their design either. Like parents baffled by their child's decision to get a tattoo, software engineers are often surprised by the decisions of their own AI systems. As algorithms grow more complex, they grow more mysterious. Many systems already run on thousands of lines of self-generated 'dark code', whose function is unknown.[9] In the future, even the creators of technology will increasingly be heard to say: *why* did it do that? *How* did it do that? For the rest of us, the technical workings of the digital lifeworld will be utterly opaque.

It's not just complexity that makes technology inscrutable. We're often deliberately *prevented* from knowing how it works. Code is often commercially valuable and its owners use every available means to keep it hidden from competitors. As Frank Pasquale argues in *The Black Box Society* (2015), we're increasingly surrounded by 'proprietary algorithms' that are 'guarded by a phalanx of lawyers', making them 'immune from scrutiny, except on the rare occasions when a whistleblower litigates or leaks'.[10]

In addition, there will often be times in the digital lifeworld when we aren't even *aware* that power is being exerted on us. Many technologies of surveillance operate quietly in the background. And if a news algorithm subtly promotes one narrative over another, or hides certain stories from view, how are we supposed to know? The best technology in the future won't feel obtrusive. *It won't feel like technology at all*. The risk, as Daniel Solove puts it, is that we find ourselves in a world that's as much Kafka as Orwell, one of constant 'helplessness, frustration, and vulnerability' in the face of vast, unknowable, and often unseen power.[11]

In light of the power tech firms will have to shape and limit our freedom in the future, it's worth going back to fundamental principles about what should and shouldn't be permitted in society. These principles should inform the work of those entrusted with our precious liberties.

The Harm Principle

John Stuart Mill (1806–1873) was a singular figure in the history of ideas. The son of James Mill, a well-known Scottish philosopher, the young John Stuart was deliberately isolated from other children except his siblings. His education was intense. 'I have no remembrance of the time when I began to learn Greek' he records in his *Autobiography* (1873), but 'I have been told that it was when I was three years old.'[12] He began Latin when he was eight.[13] He grew up debating with 'Mr Bentham', a friend of his father and one of the most important thinkers in western philosophy. Young John Stuart was evidently a prodigy, but Mill Senior didn't let him know it. With 'extreme vigilance' he kept his son from hearing himself praised.[14]

Mill developed into a thinker of extraordinary subtlety and range, dedicated above all to the ideal of individual liberty. As Isaiah Berlin observes, what the adult Mill feared most was 'narrowness, uniformity, the crippling effect of persecution, the crushing of individuals by the weight of authority or of custom or of public opinion'. He rejected 'the worship of order or tidiness, or even peace' and he loved 'the variety and colour of untamed human beings with unextinguished passions and untrammelled imaginations'.[15] Mill was way (way) ahead of his time. In an era defined by strict Victorian moralism[16] he fearlessly advocated individualism over 'conformity' and 'mediocrity'.[17] As befitted a man who started learning ancient Greek when he was three, he believed that the main danger of his time was that 'so few now dare to be eccentric'.[18]

Mill was a liberal, not a libertarian. He accepted that there had to be more-than-minimal restrictions on individual freedom (wise restraints, if you will) in order for society to survive. But to restrict the liberty of others, he believed, there must always be a good reason.

He came to think that only one reason could ever justify such a restriction: to prevent harm to others. This is the *harm principle*, one of the most influential ideas in western political thought. It is the centrepiece of Mill's *On Liberty* (1859):

> That the only purpose for which power can be rightfully exercised over any member of a civilised community, against his will, is *to prevent harm to others*... Over himself, over his own body and mind, the individual is sovereign.[19]

Subsequent liberal thinkers have refined the harm principle. In Joel Feinberg's formulation, for instance, only those acts that cause 'avoidable and substantial harm' can rightly be prohibited.[20]

Unfortunately, the harm principle has been wantonly violated throughout history. Since the very beginning, people have been persecuted for holding the wrong convictions, punished for making love to people of the wrong gender, and pogrommed for praying to the wrong god—none of which caused harm to anyone else. Tech firms in the digital lifeworld must do better than the powerful of the past. This is our chance to structure our systems in a way that emancipates people rather than crushing them.

Harm to Self

Imagine four scenarios.

In the first scenario, Eva writes an insulting email about Laura and then accidentally sends it to Laura herself rather than its intended recipient. (We've all been there.) Luckily for Eva, an automated alert immediately pops up: 'Our system detects that this may not be the intended recipient. Proceed/Change?' Eva gratefully corrects her error and resends the email. The email system imposed a constraint on Eva's freedom by withholding the message once she had hit send. But that constraint was temporary, minor in nature, and capable of immediate override. It rescued her from considerable embarrassment. Most of us, I reckon, would welcome this kind of interference with our freedom.

Next consider James, who is a few pounds overweight. It's late at night and, feeling peckish, he tiptoes down to the kitchen, opens the 'smart' refrigerator and removes a generous slab of pie. Salivating, and standing over the sink to prevent crumb spillage, he readies himself for the first delicious bite. Suddenly, the fridge pipes up in a loud and scornful tone: 'Do you *really need* that pie, James?' Shocked and ashamed, he drops the pie and scarpers back to bed. Like Eva's email alert, James's sassy fridge caused him to change his course of action (or, to put it in terms of power, caused him to refrain from doing something he would otherwise have done). True, the fridge didn't *force* James not to eat the pie, but the disciplinary effect of its scrutiny was just as strong. I suspect that most of us would be uncomfortable with this kind of intrusion from any person, let alone one of our kitchen appliances. It's a little too personal, a little offensive, even if it's in our own best interests. Of course, the situation would be different if James had *asked* his fridge to police his eating habits in an act of Odyssean self-restraint (see chapter ten).

Next imagine Nick, who instructs his food preparation system (call it RoboChef)[21] to make a curry according to a recipe he has prepared. Nick's recipe, however, contains a measurement error that would result in an excessively high amount of capsicum (chilli spice) in the dish—enough to give Nick an evening of acute discomfort. RoboChef detects the disproportionate figure, chooses to ignore it, and instead prepares a tasty jalfrezi with just the right amount of spice. Nick enjoys the meal. How should he feel about RoboChef's intervention? It can't be denied that RoboChef acted according to (what it correctly recognized as) Nick's interests. But there's something mildly disconcerting about the fact that it disobeyed Nick's instructions without telling him and assumed his interests without consulting him. What if the capsicum figure had not been an error? Perhaps Nick is someone who *enjoys* the sensation of overwhelming spice—and to hell with the gastric consequences! If he directly instructed RoboChef to deliver a truly, ridiculously, spicy curry—one that would be genuinely harmful to the majority of the population—would it be right for RoboChef to refuse to cook it? Put another way: should a digital system be able to deny Nick, a competent adult, the consequences of his voluntary and informed choice?

Finally, picture an elderly gentleman—call him Grahame—who suffers from a painful chronic disease. Lonely but at peace, he sincerely wishes to end his life. To that end he instructs his RoboChef to prepare a lethal cyanide broth. RoboChef refuses. Frustrated, Grahame lies down in his driveway and orders his self-driving car to reverse over his skull. The vehicle politely declines. Desperate, Grahame then tries to commit suicide by ingesting too many pills. But his 'smart' floor system detects his unconscious body on the floor and places an automated call to emergency services.[22] Against his will, Grahame's life is again saved.

In each of these scenarios a digital system intervenes to restrain the will or conduct of a human being, albeit in a way that is essentially benign. The aim is to protect the human from self-harm. We can call this Digital Paternalism. To what extent should Digital Paternalism become a feature of the digital lifeworld? Should our digital systems coddle and protect us? We know what John Stuart Mill might say. He believed that a person's 'own good, either physical or moral' is never 'a sufficient warrant' to exercise power over him. Only harm to others could justify that.[23] But Mill's principle feels a little narrow in a world where tiny restrictions on freedom could lead to real benefits. At the other extreme from Mill, Isaac Asimov's well-known First Law of Robotics provides that a robot may not injure a human being or, through inaction, allow a human being to come to harm. But, as Asimov knew well, the First Law, by itself, is not a sufficient guide to action. It doesn't tell us how far a robot must go in defence of humanity. And it doesn't say what counts as harm. Personal embarrassment? The cardiac implications of an unhealthy slice of pie? A fiendishly spicy curry? Death?

Of the four scenarios, Grahame's is the simplest but also the most vexing. In most developed countries, suicide has long ceased to be a crime. What *is* generally illegal is for one person to assist in the death or mutilation of another; hence why in most countries you can't take part in euthanasia, inject someone with heroin, or seriously wound them in a loving act of sadomasochism. Consent on the part of the harmed person is no defence. But Grahame isn't exactly asking for the assistance of any other human being. On the contrary, his death need not morally implicate any person other than himself. He has one desire—to die—and he thinks he has the technical means to make it happen. Yet the technology either refuses to assist in that goal or actively thwarts it. Isn't this a serious intrusion into Grahame's liberty? Perhaps, but it might be a mistake to think that liberty is the only value at play here. Allowing machines to assist in self-harm (or to stand by while it takes place) could threaten 'one of the great moral principles upon which society is based, that is, the sanctity of human life'.[24] Grahame's suicide, on this view, may not be a crime against himself but an offence 'against society as a whole' that cannot be tolerated, even if that means making Grahame and others in his position less free.[25]

My view is that a fruitful balance could be struck between the freedom espoused by Mill and the benefits of Digital Paternalism. As the scenarios demonstrate, at least eleven considerations could factor in to any decision about whether freedom should be restricted to prevent self-harm. How significant is the freedom being denied? By comparison, how great is the benefit being sought? Is what's being restricted a voluntary, conscious, informed choice or an accidental, reflexive, or otherwise involuntary act? Is the constraint imposed openly or covertly? Is the constraint imposed according to our express interests or just our perceived interests? Does it involve force, or influence, or manipulation (that is, does it properly leave room for freedom of choice)? Is the constraint an omission or an act? Can it be overridden? How long does it last? Is the person whose freedom is being restricted an adult or a minor? Of sound mind?

The line has to be drawn somewhere, but the question is political and not technical.

Immoral but Not Harmful

Every society has some shared sense of what counts as evil, wrong, sinful, and immoral. One longstanding question is whether we should be permitted to do things that are immoral even if they're not harmful. It's possible, for instance, for a person to have paedo-philic fantasies without ever harming a child (and indeed being repulsed by the very idea of doing so). It's possible for a man to have consensual sex with his father, in private, with no risk of harm to anyone else. Should such things be forbidden? If discovered, should they be punished? These questions were previously the sole province of law and ethics. In the digital lifeworld they will also be matters of code.

I want you to imagine a VR platform that provides an immersive experience for its users. Sight and sound are provided by a headset and earphones. Smells are synthesized using olfactory gases and oils. Physical touch is stimulated by way of a 'haptic' suit and gloves, together with customized props in the appropriate form. Sexual stimulation, when needed for the experience, is provided by 'teledildonic' apparatus designed for the task.[26] The system is even able to respond directly to brainwaves it detects using electroen-cephalography (EEG).[27] It is used in the privacy of the home. Only the user is shown their virtual adventures: not even the manufacturer receives that information.

What would you choose to do with this kind of technology? Throw the winning pass in the SuperBowl? Go toe-to-toe with Muhammad Ali? Dance with Fred Astaire?

What if, instead, you wanted to 'experience' what it was like to be a Nazi executioner at Auschwitz? Or to recreate, in first person, the final day of Mohammed Atta, one of the 9/11 hijackers? Should users be able to simulate, in virtual reality, the act of drowning a puppy or strangling a kitten? Should they be allowed to torture and mutilate a virtual avatar of their neighbour? What about the 'experience' of raping a child? Should it be possible to know, in virtual reality, what it felt like to nail Jesus to the cross?

For most, these scenarios are horrendous even to contemplate. They grossly offend our shared sense of morality. But that's why we have to think about them. The test of a society's commitment to liberty is not what it thinks about conduct within the moral mainstream, but rather what it has to say about activities considered unspeakable, obscene, or taboo.

We might well agree that choosing to experience these things in VR is itself likely to corrode people's moral character. That's a kind of harm to *self*. But no one else is actually mutilated, raped, or violated. If you believe that society has no business policing people's private morality then you can logically conclude that in VR *anything* should be permitted. So long as no harm is caused to others, people should be left to do as they please. It's not forbidden, after all, to fantasize about these things in the privacy of one's own mind. Is VR so different?

Some philosophical context is helpful in thinking this through.

It's long been seen as bad form to punish people for what goes on inside their heads. In the nineteenth century, as we've seen, John Stuart Mill argued that society has no business in prohibiting conduct that affected no one other than the individual in question. One of Mill's contemporaries, a judge called Sir James Fitzjames

Stephen, disagreed, arguing that we have every reason to be 'interested not only in the conduct, but in the thoughts, feelings, and opinions' of other people.[28] A century later, the same disagreement resurfaced in the famous Hart–Devlin debate of the 1960s. Herbert Lionel Adolphus Hart was the quintessential liberal law professor, a genial man with a colossal intellect and a gentle manner. Lord Devlin cut an altogether sterner figure. Like Sir James Fitzjames Stephen, he too was a judge. The Hart–Devlin debates were prompted by the publication in 1957 of the Report of the Committee on Homosexual Offenses and Prostitution, generally known as the Wolfenden Report. The Report's famous conclusion was that 'It is not the duty of the law to concern itself with immorality as such': 'there must remain a realm of private morality and immorality which is, in brief and crude terms, not the law's business.'[29]

Devlin disagreed. He believed that a society is a 'community of ideas', not just political ideas, but 'ideas about the way its members should behave and govern their lives'. That is to say, every society must have shared *morals*.[30] Without them 'no society can exist'.[31] To let immorality go unpunished, even immorality that causes no identifiable harm to other people, is to degrade the moral fabric that holds peoples together. 'A nation of debauchees,' he wrote in 1965, 'would not in 1940 have responded satisfactorily to Winston Churchill's call to blood and toil and sweat and tears.'[32]

H. L. A. Hart accepted that some shared morality was needed for the existence of any society, at least in order to limit violence, theft, and fraud. But he dismissed Devlin's notion that the law has any business in regulating morality. There is no evidence, Hart, suggested, that deviation from 'accepted sexual morality' by adults in private is 'something which, like treason, threatens the existence of society': 'As a proposition of fact it is entitled to no more respect than the Emperor Justinian's statement that homosexuality was the cause of earthquakes.'[33] For Hart, our personal choices, especially those made in private, have no bearing on whether we are loyal citizens. (Hart himself had no problem answering Churchill's call to service, having worked in military intelligence for most of the Second World War. Nor had the great mathematician and code-breaker Alan Turing who also worked at Bletchley Park, and who was the subject of criminal prosecution for homosexual acts.)

So how would the Hart–Devlin debate play out today?

We might firstly argue that VR is actually pretty different from pure fantasy. Its realism and sensual authenticity bring it closer to actually *doing* something than merely thinking about it. The trouble with this argument is that if you believe on principle (as Mill and Hart did) that mere immorality should never be made the subject of coercion, then to say something is *very* immoral, as opposed to merely quite immoral, doesn't take you much further.

Another argument is that if we let people do extreme things in virtual reality then they might be more likely to do them in the 'real' world. Violent sexual experiences in VR could encourage acts of sexual violence against people. This is an empirical argument that can be tested through experiment. And early research suggests that it might well be right.[34] If it is true that virtual behaviour causes real behaviour, then prohibition of certain VR experiences could be justifiable under the harm principle. That said, if my obscene VR experience involves virtual harm to me alone—say, a fantasy about being violently dominated—then it doesn't obviously follow that I would go out and inflict harm on others.

A third, more Devlinesque objection is that untrammelled liberty in VR could be simply too corrosive to shared morality or to our traditional way of life.[35] Another way of putting it is to say that our liberties should be structured so as to 'elevate or perfect human character' rather than debasing it.[36] Amusingly, one early study suggests that the experience of being able to fly in VR encourages altruistic behaviour in the real world, apparently by triggering associations with airborne superheroes like Superman.[37] Another variant on this argument is that no system of morality is inherently better than another, but that *multiple* moralities within the same community are a problem. As Devlin put it, in the absence of 'fundamental agreement about good and evil', society will 'disintegrate'. 'For society is not something that is kept together physically; it is held by the invisible bonds of common thought.'[38] In modern coinage, we might call this the problem of *fragmented morality*.

A final argument against untrammelled virtual liberty, with broader implications for all digital technology, is that those who *manufacture* the VR hardware and software ought not to be able to profit from such grotesquery. In its strong form, the argument is that tech firms should be legally prohibited from making such technologies at all, or required to code them so that certain virtual experiences are made impossible. It might, alternatively, be said that manufacturers should be given the *discretion* to choose the functionality of their VR systems—and if that means the facilitation of obscenity, then on their moral heads be it. There are two difficulties with the second approach. It doesn't solve the problem of fragmented morality. And it entirely delegates important questions of human liberty to private firms. Is it wise for matters of public morality to be reduced to questions of corporate strategy? It's questionable, to say the least, whether our freedoms should be determined by the tastes of managers, lawyers, and engineers at tech firms. How would we feel about a platform that permitted virtual violence against only one racial group; or which allowed straight sex but not gay sex? Or (thinking more broadly) a self-driving car that refused to take its passengers to a fast food drive-through because its manufacturers oppose factory farming? Or a general-purpose surgical robot that refused to perform legal abortions because its manufacturers were religious Christians?

Technology will bring new affordances, it's true, but that also means new opportunities for immoral conduct and deviancies we can't even yet imagine. Some will rejoice in the prospect. Others will shudder. Will we be free to do as we please so long as others aren't harmed? Tech firms can't be left to decide these questions on their own.

Digital Liberty

Let's pause to think through the implications of the last two chapters. What we need, I suggest, is a new collection of concepts that can explain different approaches to the future of freedom. I set out a modest selection here. See which most appeals to you.

Digital Libertarianism is the belief that freedom in the future will mean freedom *from* technology. Every line of code that exerts power is contrary to liberty. Freedom begins where technology ends. This doctrine supports the reduction of all forms of digital power, whatever their nature or origins. No one should ever be forced to use digital systems where another means would do. If I don't want 'smart' appliances in my home, I shouldn't have to install them.

Digital Liberalism is the more nuanced belief that technology should be engineered to ensure the maximum possible individual liberty for all. This is the 'wise restraints' approach. To work, it requires that code should, as far as possible, be neutral between different conceptions of the good. It should not aggressively promote one course of life over another. It should leave individuals the largest possible sphere of personal freedom to determine their paths, perhaps through individual customization of owned devices.

Digital Confederalism is the idea that the best way to preserve liberty is to ensure that people can move between systems according to whose code they prefer. If I consider one speech platform, for instance, to be unduly restrictive, I should be able to find another one. Digital Confederalism requires that for any important liberty— communication, news-gathering, search, transport—there must be a plurality of available digital systems through which to exercise that liberty. And it must be possible to move between these systems without adverse consequences.[39] In practice, a private or state monopoly over any given platform or technology could be fatal to Digital Confederalism (chapter eighteen).

By contrast, Digital Paternalism and Digital Moralism hold that technology should be designed, respectively, to protect people from the harmful consequences of their own actions and steer them away from lives of immorality. How far they do this is a matter of taste. A refrigerator that warns you about the health consequences of eating another slice of pie would constrain you less than one that physically prevented access to any more pie until you'd digested the previous helping. A VR system that placed limits on extreme obscenity would be less restrictive than one that only allowed its users to experience wholesome activities like going to virtual church or attending a virtual lecture.

Finally, Digital Republicanism is the belief that nobody should be subject to the *arbitrary* power of those who control digital technologies. At the very least, it means we must be helped to understand how the technologies that govern our lives actually work, the values they encode, who designed and created them, and what purpose they serve. More radically, authentic Digital Republicanism would require not only that we *understand* the digital systems that exert power over us, but that we actually have a hand in *shaping* them. It's not good enough to rely on the benevolence and wisdom of tech companies to make important decisions about our freedom. So long as they can arbitrarily change the rules, making technology work in their interests and not ours (the argument runs) we must consider ourselves unfree. Even in technological terms, this is actually an old idea, not a new one. As Douglas Rushkoff explains, in the early days of personal computing, 'there was no difference between operating a computer and programming one.' Computers were 'blank slates, into which we wrote our own software'.[40] We controlled our technology, not the other way round. That's not to say that Digital Republicanism means everyone should retrain as a software engineer. But it is an activist doctrine. It requires citizens to cultivate the civic virtues that will be needed to hold the state and tech firms to account: technical understanding where possible, but also vigilance, prudence, curiosity, persistence, assertiveness, and public-spiritedness. It's about ensuring that we do not become subject to rules that we can't understand, that we haven't chosen, and that could be altered at any time. Demand transparency, the slogan may run. Demand accountability. Demand participation. 'Program, or be programmed.'[41]

Liberty and Democracy

An idea has haunted the last two chapters without ever becoming explicit. It's that there is an important bond between liberty and democracy. For liberals, the nature of this bond is simple: only in a democracy can the people make sure that their liberties are not trampled on and take part in crafting the 'wise restraints that make men free'. For republicans in the Roman tradition, the bond is even closer. They believe that a restriction imposed by a democratic process is less inimical to freedom than an identical restriction imposed by a private body *by virtue of the fact that it was decided democratically*. As Rousseau puts it, 'What man loses by the Social Contract is his natural liberty ... what he gains by the social contract is civil liberty.'[42] Combining the liberal and republican approaches, what becomes clear is that democratic accountability will become more important than ever in the digital lifeworld. It will be an indispensable weapon when it comes to protecting ourselves against the growing power of the state and private firms. I said in chapter nine that the digital lifeworld would be thick with power and drenched with politics. If we care about liberty, there must also be a corresponding increase in the ability of the citizenry to hold that power to account.

And so we turn to democracy.

PART IV

Future Democracy

'As I would not be a slave, so I would not be a master. This expresses my idea of democracy.'

Abraham Lincoln

TWELVE

The Dream of Democracy

'"It's always best on these occasions to do what the
mob do."

"But suppose there are two mobs?" suggested
Mr Snodgrass.

"Shout with the largest," replied Mr Pickwick.'

Charles Dickens, _Pickwick Papers_ (1837)

So much has been written in praise of democracy that it can be
daunting even to broach the subject. We are told by leading scholars
that the term is 'sacred'.[1] It is the 'motherhood and apple pie' of
politics;[2] a 'universal value'[3] with a 'privileged aura of legitimacy'.[4]
It presses 'a claim for authority and a demand for respect.'[5] It is 'the
leading standard of political legitimacy in the current era.'[6]

In the twentieth century, democracy's prestige in the realm of
ideas was matched by its growth in practice. From just twelve at
the end of the Second World War,[7] by the late 1990s there were
120 electoral democracies in the world: more than 60 per cent of
independent states.[8]

But that's not the whole story.

For most of history, humans have been sceptical of the idea of
democracy. As a system of government it has been conspicuous
mainly in its absence. And since 2006 the number and quality of
democracies in the world has been gradually declining.[9] Many

are plagued by low turnout, political gridlock, chronic public indifference, and mistrust.[10] Public discontent is growing.[11] Politicians are widely disliked. Although Plato was being a little dramatic when he described democracy as when the poor 'win, kill or exile their opponents',[12] he was right to resist the assumption that democracy is always a good thing, that more is always better, or that democracy as he knew it (or as we currently know it) is the best possible form of government. Like any other concept, the meaning and value of democracy can be contested. Like any other system of government, it can be changed, lost, or destroyed.

In the next two chapters we look to the digital lifeworld and ask: how can the people rule? Democracy in the future will not be the same as it was in classical Athens or even in advanced twentieth-century democracies. Some aspects of the process, like campaigning and deliberation, have already been irrevocably changed by digital technologies. Looking ahead, we may see further changes to the way we prepare legislation, how we vote, even *whether* we vote. Technology could fundamentally alter what it means for humans to govern themselves. Visible on the horizon are five different systems of democracy, each made possible by digital technology: Deliberative Democracy, Direct Democracy, Wiki Democracy, Data Democracy, and AI Democracy. None of these systems is perfect. But each has aspects that are superior to what we currently have. The question for democrats in the twenty-first century will be: could some combination of these models enable a new and better way of organizing our collective life and holding power to account? Democracy may be having a hard time right now, but as the power of the state and tech firms continues to swell, it could come to be more important than ever. The digital lifeworld could yet offer democracy its finest hour. But it has to be fit for purpose.

We begin with the story of democracy, from classical Athens to the present. Next, we look at the arguments that have traditionally been made in its favour. Then we turn our focus to the digital lifeworld, looking at each of the alternative models set out above, asking first what they might entail in practice, and second, what we like and what we find problematic.

Some of the ideas in the next two chapters will be unfamiliar, even repugnant to many readers. Conscious of how much has been sacrificed in its name, we are rightly sceptical of any challenge to the idea of electoral democracy. And it can be hard to see past the way we currently do things. But remember. The way you may feel about the idea (for instance) of an AI system making political decisions on your behalf is little different from the way many of our forebears felt when they contemplated the idea that such decisions might be made by those who seemed least qualified to make them: the people.

The Story of Democracy

What is Democracy?

The term *democracy* is generally used to refer to a form of government in which ultimate political power rests with the many (the people, the masses, the multitude, the governed) rather than the few (monarchs, dictators, oligarchs). This definition embraces a range of systems, from those in which *everyone* takes part in the business of government, to more diluted systems in which rulers are merely accountable to, elected by, or rule in the interests of the ruled.[13] As is well-known, the word *demokratia* is a fudge of two Greek root words, *demos* (people) and *kratos* (rule). In addition to describing a procedure for taking collective decisions, the term *democratic* is also used to describe the social ideals that underpin that procedure. More on these ideals later. First, a brief biography of the most charismatic concept in politics.

Classical Democracy

Democracy first flourished in Athens around 500 years before the birth of Jesus Christ. The Athenian *Ecclesia* (Assembly) was both a sovereign body and a physical place. Citizens gathered there to decide all sorts of matters, from laws and taxes to questions of war and peace. The community in Athens was modest in size and tightly knit. Out of about 30,000 citizens, the Assembly was quorate with 6,000 present. That's 6,000 people *in one place*, trying to reach a unanimous decision. Where agreement was impossible, the majority prevailed. Any Athenian citizen could be selected for office, and participation was remunerated so that all had an equal chance of doing so.[14] As chronicled by Thucydides in *The Peloponnesian War* (fifth century BCE), Athens' greatest statesman, Pericles, described the system thus:[15]

> Our constitution is called a democracy because we govern in the interests of the majority, not just the few. Our laws give equal rights to all in private disputes, but public preferment depends on individual distinction and is determined largely by merit rather than rotation: and poverty is no barrier to office, if a man despite his humble condition has the ability to do some good to the city.

This was *classical democracy*, a model that is often wistfully invoked as the highest form of human self-rule. In reality it had some quite serious shortcomings. Pericles' reference to the 'whole people', for instance, was somewhat misleading. Not everyone in Athens was a citizen. In fact, only Athenian-born men older than twenty were eligible for citizenship. This excluded every immigrant, woman, and slave—who, together, outnumbered the voting citizenry by ten to one.[16]

Athenian democracy lasted for just 175 years until it was extinguished by Macedonian conquerors in 322 BCE.[17] As the classical world gave way to early Christendom, and for more than 2,000 years after the death of democracy in Athens, not only the system but the very concept of democracy was absent from human affairs. The medieval world that replaced classical antiquity was dominated by a conception of politics in which the central purpose of human life was submission to God's will. Power came from the heavens, not the people.[18] Princes and popes ruled by divine right, not popular consent. Obedience was a matter of faith, not politics.

It was not until the eleventh century, nearly a millennium and a half after the fall of Athens, that genuine systems of secular self-rule began to re-emerge in Europe. The early Italian city-republics of Florence, Siena, Pisa, and Milan were ruled not by kings but by councils drawn from wealthy families. The Athenians would not have seen these statelets as practising genuine self-government and they were certainly not democracies as we would understand them. But the idea that earthly affairs could be run by lay consuls and administrators was, at the time, revolutionary.[19]

Another reason why the early Italian city-republics did not describe themselves as democracies was that until the thirteenth century the word *demokratia* itself remained unknown to the Latin-speaking world. The Romans had never used it and it had been lost with Greek civilization. It only entered the Latin language in around 1260 when Aristotle's great work, *Politics*, was translated by the Dominican friar William of Moerbeke.[20] But after centuries in obscurity, the concept did not receive a warm welcome home. On the contrary, *democracy* came to describe a frightful political system in which the stinking masses were able to force their sordid desires on everyone else. Thomas Aquinas, the foremost scholar of the medieval world, and a contemporary of William of Moerbeke, described democracy as, 'when the common people oppress the rich by force of numbers ... like a single tyrant'.[21]

This proved to be a durable view.

If we fast-forward 500 years to seventeenth-century England, a nation riven by civil war, we find a group called the Levellers engaged in a historic struggle for popular sovereignty, an expanded franchise, and equality before the law. Their aim was quintessentially democratic: to bring the might of the government under the control of the people. Yet even the Levellers did not describe themselves as *democrats*, though the word had entered the English language in the previous century.[22] Democracy was still a term of derision used *against* the Levellers by their political enemies.[23]

It was not until the late-eighteenth century, the rumbling age of revolution in France and America, that the noun *democrat*, the adjective *democratic*, and the verb *democratize* came into mainstream usage.[24] And yet they were still uttered with mistrust. James Madison believed that, 'democracies have ever been spectacles of turbulence and contention . . . and have in general been as short in their lives as they have been violent in their deaths.'[25] In his memoirs, the eighteenth-century lothario Giacomo Casanova described seeing a horde of drunken men rampaging destructively through London (picture an English soccer match *circa* 1975) and described the members of the mob as *democratic animals*. It was not meant as a compliment.[26]

Liberal Democracy

After more than 2,000 years in the conceptual wilderness, the term *democracy* began its comeback tour at the end of the nineteenth century, not in its classical form but reborn as *liberal democracy*. The central premise of liberal democracy, which can be traced back to the seventeenth-century writings of John Locke, is that individuals should be given a wide remit to live their lives as they see fit. This means using the democratic process to keep the powerful in check (the *democratic* bit) but it also means protecting the people from *the people themselves* (the *liberal* bit). Advocates of liberal democracy believe— to repurpose the words of the Roman historian Livy—that although

'the mob' is often a 'humble slave', it can also be a 'cruel master'.[27]
What made liberal democracy unique was its rejection of the idea
that more democracy is always better. The rule of law, individual
rights, divorce of church and state, separation of powers—these are
core facets of liberal democracy but their fundamental purpose is
to *limit* untrammelled people power, not facilitate it.

Liberal democracy places strict restrictions on what the democratic sovereign can lawfully do. In classical Athens your rights
depended entirely on the good graces of the Assembly: if the
people decided you were to lose your property or die, that was it.
Socrates understood this the hard way. There was even a yearly
process, known as *ostracism*, by which the Assembly could vote to
exile any citizen for ten years. The unfortunate 'winner' of this poll
had to leave the city within ten days. The penalty for returning was
death. In a liberal democracy *ostracism* would not be possible.
Citizens' rights are carved in rock and cannot so easily be ignored.
Checks, balances, and due process will protect you even when
the majority is screaming for your blood. Unlike the Athenian
Assembly, the liberal state is forbidden by law from violating the
sacred spaces that are reserved for private life: home, family, and
some aspects of social life.[28] Though he could scarcely be called a
liberal thinker, even Niccolò Machiavelli recognized that an uncontrolled multitude and an uncontrolled prince were just as bad as
each other: 'all err in equal measure when they err without fear of
punishment.'[29]

The liberal ideal dominated democratic thinking in the twentieth
century.

Competitive Elitism

Many of the democracies that emerged in the second half of the
twentieth century were liberal democracies. A common feature was

that they were *representative* rather than *direct* systems. This meant that key decisions were taken by elected representatives rather than the assembled citizenry. A number of factors made this desirable. First, the size and scale of modern polities made institutions like the Athenian Assembly impossible. Second, the complexity of modern life was seen as ill-suited to mass-deliberation. Third, representative democracy enabled the 'best and the brightest' (ahem) to rise to the fore, creating a class of professional politicians equipped to serve the public interest. Fourth, dialogue between politicians and the public was seen as a useful generator of good ideas. Finally, representative systems were considered the best way to mediate and moderate the fickle passions of the masses, while still taking their sentiment into account. In the modern age, representative democracy has been our best answer to the question posed by Joseph Schumpeter: 'How is it technically possible for "people" to rule?'[30]

Schumpeter, a giant of twentieth-century economics, was sceptical in principle of the value of direct participation, arguing that the 'electoral mass…is incapable of action other than a stampede'.[31] And he looked at the modern democracies around him and decided that in practice, too, they were nothing like the classical archetype. For Schumpeter, what defined democracy was not popular participation or deliberation; it was the basic act of electing and ditching political leaders—and little more. On this view, the democratic process is basically the same as the market for consumer goods. In the words of the political theorist Alan Ryan:[32]

> We do not sit at home elaborating the specification of something as complex as an automobile, and then go and ask a manufacturer to build it. Entrepreneurs dream up products that they think advertisers can persuade us to want; they assemble the capital and the workforce to create these products, and then offer them to us at a price. If they have guessed right about what we can be persuaded to want, they prosper; if not, they go broke.

So it is with politicians, who think up policies, present them to the people, and succeed or fail depending on their popularity.

The name generally given to the Schumpeterian model of democracy—*competitive elitism*—sounds like a Harvard College drinking game, but is in fact the most accurate way of describing the democracies in which many of us live.[∝]

Democracy after the Internet

The internet was supposed to transform democracy. And since its widespread adoption in the 1990s it has indeed made important differences to the way that advanced democracies function. Three in particular are significant.

The first is in the field of political *campaigning*. In almost every major election, online tools are now used to raise funds, organize supporters, enforce message discipline, disseminate information, and keep tabs on activists. In recent years, political élites have also begun to exploit the remarkable potential of big data in profiling citizens, modelling their political behaviour, predicting their intentions, and targeting advertisements and organizational resources accordingly.[33] Schumpeter would have been thrilled: the process effectively mirrors the techniques used by corporations to profile and market their goods to consumers. Barack Obama's 2012 presidential re-election campaign, for instance, gathered voter information into a single database and combined it with data scraped from social media and

---

∝ Schumpeter himself, who taught at Harvard, was an interesting fellow. He claimed to have three goals in life: to be the greatest economist in the world, the best horseman in all of Austria, and the finest lover in Vienna. Wikipedia reports that he claimed to have met two of these goals but never said which two—although he did observe that there were, 'too many fine horsemen in Austria for him to succeed in all his aspirations'.[34]

elsewhere. His machine learning algorithms then tried to predict how likely each individual voter would be to support Obama, to turn up to vote, to respond to reminders, and to change his or her mind based on a conversation on a specific issue. The campaign ran 66,000 simulations of the election every night and used the results to assign campaign resources: 'whom to call, which doors to knock on, what to say'.[35] Four years later, in the 2016 US presidential election, the political consultancy Cambridge Analytica (whose services were engaged by Donald Trump) reportedly gathered a database of 220 million people—almost the entire US voting population—with psychological profiles of each voter based on 5,000 separate data points.[36] This enabled the Trump campaign to use bots (AI systems) and advertisements on social media to target individual voters with pinpoint accuracy. The result appeared to be the holy grail of political campaigning: a large-scale shift in public opinion. This new approach to data-based campaigning has been called the 'engineering of consent'[37] and more ominously, the 'weaponized AI propaganda machine'.[38]

Second, the internet has changed the relationship between government and citizens, enabling them to work together to solve public policy problems. Online consultations, open government, e-petitions, e-rulemaking,[39] crowdsourcing of ideas (as in Estonia and Finland),[40] hackathons, and participatory budgeting (as in Paris, where residents propose and vote on items of public spending)[41] are all new ways of coming up with ideas, subjecting policy to scrutiny and refinement, bringing private-sector resources to bear on big problems, and increasing the efficiency and legitimacy of government.[42] The notion of *e-government* is underpinned by one question, posed by Beth Simone Noveck: 'If we can develop the algorithms and platforms to target consumers, can we not also target citizens for the far worthier purpose of undertaking public service?'[43] The answer seems to be a tentative yes.

Third, the internet has changed the relationship between citizens and other citizens by facilitating the emergence of online associations and movements. Arab Spring activists, MoveOn, Occupy, the antiglobalization movement, 'right to die' advocates, Anonymous 'hacktivists'—all these groups have used the internet to coordinate action and protest in a way that would have been done very differently (if at all) before. The result is that old-school interest groups like trade unions, guilds, and clubs have been joined by a dazzling new array of online associations that require much smaller investments of time and money on the part of their parti-cipants. Hitting 'Like' for a cause on Facebook, or retweeting a political statement with which you agree, is much less onerous than sitting through endless dreary meetings in church basements. Online groups typically grow, mutate, and decay much faster than their offline counterparts. As a result, our political ecosystem is much more febrile than in the past. It's been described, aptly, as 'chaotic pluralism'.[44]

In my view, these developments are impressive but in the grand arc of democratic history they are not revolutionary. Online campaigning and e-government are both new ways of doing old things. They haven't changed what a campaign is, or what a government does. And there is nothing novel in the idea of democratic pluralism:[45] as Alexis de Tocqueville recognized in *Democracy in America* (1835), civil society organizations like clubs and societies have been a 'necessary guarantee against the tyranny of the majority' since at least the earliest days of the American republic.[46] If Schumpeter were still around, he would say that online campaigning has actually strengthened the *competitive elitism* model of democracy that dominated the twentieth century, and that other internet-related developments have merely chafed at its edges. If by some gruesome miracle Henry Ford were resurrected too, he might describe what we've seen so far as *faster horses*.

Why Democracy?

We need to think about the challenges and opportunities for democracy in the digital lifeworld, if possible looking past the classical, liberal, and competitive elitist models. In chapter thirteen we investigate five quite different models we might come to see. But before we get there, let's take a moment just to consider why democracy has so consistently been said to be the best form of government. That will leave us better able to assess the merits of the different forms of democracy in the future.

The first and oldest argument in favour of democracy rests on *liberty*. It is related to the republican ideal of freedom we saw in Part III. It holds that we are only truly free when we live under laws of our own making. Otherwise we are the playthings of alien powers—princes, despots, foreign occupiers—unable to set our own course or choose our idea of the good life. Uniquely, democracy enables *all* the people to be free *together*. Only when we all sacrifice a little bit of our natural liberty, by submitting to the will of everyone else, can we be masters of our own collective fate. 'Since each man gives himself to all, he gives himself to no-one,' argues Jean-Jacques Rousseau, 'and since there is no associate over whom he does not gain the same rights as others gain over him, each man recovers the equivalent of everything he loses.'[47]

The second argument in favour of democracy, also taken from the republican tradition, relies on human nature. Aristotle famously claimed that 'man is by nature a political animal'.[48] In the Athenian *polis* from which he came, the state was not merely a means for people to clump together in the same place; it existed 'for the sake of noble actions' by its citizens through participation in public affairs.[49] Taking part in politics was an integral part of being human and living a full life. Hence Pericles again: '[w]e are unique in the way we regard anyone who takes no part in public affairs: we do not call that a quiet life, we call it a useless life.'[50] A more modest version of this argument, advanced by John Stuart Mill in his *Considerations on Representative Government* (1861), is that participation in politics is an important means of self-improvement. Neither the routine drudgery of our daily work, nor the pursuit of private wealth, nor the mere 'satisfaction of daily wants' can fully cultivate our moral and intellectual faculties. But giving us 'something to do for the public' is good for us. It requires us to weigh interests other than our own and to be guided, for a time, by the common interest rather than our own selfish desires.[51] By taking part with others, we improve ourselves.

A third argument, and for many the most important, draws on the ideal of *equality*. If each human life is of equal moral worth, then political decisions should pay equal heed to everyone's interests and preferences. As such, every person within a political community ought to have an equal chance to influence the decisions that affect them, and no élite group or person should be allowed to hoard power for themselves. Democracy is a good way of ensuring this. Equal representation is also seen as a necessary gateway to other equalities, whether socio-economic or cultural (see Part V).

The fourth argument in favour of democracy is that, of all forms of government, it most often results in the creation of the best *outcomes* in terms of laws and policies. This argument has been put in various ways, but it boils down to the idea that democracy is the best way to harness the useful information and knowledge that is scattered across the individual minds of a given community. We can call this democracy's *epistemic superiority* (epistemology is the study of knowledge).[52] For starters, it is said, the people themselves are best placed to say what their own interests and preferences are, as opposed to some prince or bureaucrat deciding for them.

And even if each individual is not particularly knowledgeable, the *combined* experiences, skills, insight, expertise, intuitions, and beliefs of a community can yield a rich seam of wisdom. This idea was at the root of Aristotle's famous observation that:[53]

> the many, no one of whom taken singly is a sound man, may yet, taken altogether, be better than the few, not individually but collectively, in the same way that a feast to which all contribute is better than one supplied at one man's expense.

Aristotle's belief was rooted in empirical reality. Scholars now say that ancient Athens' success in outwitting and outlasting its non-democratic Greek rivals was the result of its unique ability to organize the knowledge of its citizenry, through the debates and votes that were part of the democratic process.[54]

Political scientists have tried to explain the epistemic superiority of democracy in two different ways. One group, called the *counters*, say that as a matter of mathematical logic a large and more diverse group of people will answer political questions better than a smaller group—even one composed of experts.[55] This line of thinking can be traced back to the seventeenth-century Dutch thinker Baruch Spinoza, who believed that 'it is almost impossible that the majority of a people, especially if it be a large one, should agree in an irrational design.'[56] Spinoza's conviction was shared by the Marquis de Condorcet, the eighteenth-century philosopher and mathematician whose famous Jury Theorem holds, in basic terms, that if a large group votes on a yes-or-no issue, the majority of voters are virtually certain to vote for the correct answer as long as: (a) the median voter is better than random at choosing correct answers, (b) voters vote independently of each other, and (c) voters vote sincerely rather than strategically. These days, this phenomenon is referred to as *the wisdom of crowds*.[57] It explains the hoary old tale in which the *average* answer of 800 participants in a competition to guess the weight of an ox was within one pound of the correct number.[58]

Another group of theorists, called the *talkers*, say that democracy is more than mere aggregation of individual opinions. It is democracy's *deliberative* element that leads to better legislative outcomes. Public airing of political issues allows ideas and information to be shared, biases and vested interests to be revealed, and reason and rationality to triumph over ignorance and prejudice. 'The unforced force of the better argument' ultimately leads to the best outcomes.[59]

Whether you are a counter or a talker, the argument that we are smarter *together* than we are as individuals is an important one. It contradicts the belief, held for much of human history, that the filthy hordes have no business interfering with complex matters of state.

The fifth argument in favour of democracy concerns *stability*: since a democratic system is more likely than other forms of government to be *perceived* as legitimate, it is least likely to collapse under the strains of governing. De Tocqueville, describing the fledgling American democracy, spoke of the 'sort of paternal pride' that protects a democratic government even when it messes up.[60]

The final argument for democracy is that—for all its faults—it is the best way of preventing tyranny and corruption. So long as the people retain control of the levers of power, they are likely to spare themselves the worst excesses of mad kings and bad dictators. Government 'of the people' and 'by the people' is most likely to be 'for the people' too. This was the point stressed at the end of the last chapter.

Every one of these arguments is open to challenge. For example, if *liberty* is what matters, can it really be said that a minority group in a democracy is 'free' if it lives under harsh rules dictated by the majority? Rousseau's cheery answer that those persons are 'forced to be free' is unlikely to be to everyone's taste.[61] If *equality* is so important, then isn't it more important for a system to be *liberal*—enshrining equal rights under the law—than *democratic*, where a majority could pass laws treating certain groups unequally? Do we accept that the crowd—which sentenced Socrates to death, elected Hitler, and often seems to be irrational, capricious, or xenophobic—is as wise as claimed?[62] Doesn't the recent trend toward authoritarianism suggest that democracy isn't as stable or strong a bulwark against tyranny as we might have hoped? To these objections, the democrat can always reply in the manner of Churchill: that democracy is indeed flawed—but less flawed than all other systems. But it seems unsatisfactory. Can we do better?

THIRTEEN

Democracy in the Future

'Is a democracy, such as we know it, the last improvement possible in government? Is it not possible to take a step further towards recognizing and organizing the rights of man?'

Henry David Thoreau, *Resistance to Civil Government* (1849)

We turn now to the future of democracy. The digital lifeworld will offer some interesting opportunities for those who want their system of government to combine the values of liberty, equality, human flourishing, epistemic superiority, stability, and protection from tyranny. But it will also present some challenges to democracy as we have traditionally understood it. This chapter is structured around five distinct conceptions of self-rule, some old and some new: Deliberative Democracy, Direct Democracy, Wiki Democracy, Data Democracy, and AI Democracy.

We begin with Deliberative Democracy, an ancient but increasingly fragile form of self-government.

Deliberative Democracy

Deliberation is the process by which members of a community rationally discuss political issues in order to find solutions that can be accepted by all (or most) reasonable people. In an ideal process of deliberation, everyone has the same opportunity to participate on equal terms, and anyone can question the topic or the way it is being discussed.[1] Political debate has always been messy, but deliberation is seen as an important part of the process because it pools knowledge and information, encourages mutual respect, allows people to change their views, exposes who is pursuing their own interest, and increases the hope of consensus rather than simply totting up the ayes and noes. Supporters of deliberation say it is the only grown-up way of accommodating reasonable moral disagreements.[2] 'Deliberative Democrats' go further, arguing that deliberation is not just one part of democracy but an *essential* part of it: only decisions taken with the benefit of genuine public deliberation can claim the mantle of democratic legitimacy. The classical Athenians would have agreed.

The arrival of the internet prompted a great deal of optimism about the future of Deliberative Democracy. Cyberspace would become a vibrant forum for political debate. Great multitudes of dispersed individuals, not just a few mass media outlets, would create and exchange reliable political information. Rather than passively absorbing information, citizens would participate in discussion, debate, and deliberation.[3] Unfortunately, it hasn't really turned out that way. Though there are more opportunities than ever for ordinary citizens to have their say, the result hasn't been an increase in the quality of deliberation or political discourse more generally. On the contrary, politics feels as divisive and ill-informed as it did in the past, possibly even more so.[4] Without a change in course, there is a risk that the quality of our deliberation could wither still further in the digital lifeworld. This is the result of four threats: perception-control, fragmented reality, online anonymity, and the growing threat posed by bots.

Perception-Control

We've already seen that in the future, how we perceive the world will be increasingly determined by what is revealed or concealed by digital systems. These systems—news and search services, communication channels, affective computing, and AR platforms—will determine what we know, what we feel, what we want, and what we do. In turn, those who own and operate these systems will have the power to shape our political preferences. The first threat to Deliberative Democracy, therefore, is that our very perceptions are increasingly susceptible to control, sometimes by the very institutions we would seek to hold to account. It's hard to contribute rationally when your political thoughts and feelings are structured and shaped for you by someone else.

Fragmented Reality

The second threat comes from the disintegration and polarization of public discourse.[5] People tend to talk to those they like and read news that confirms their beliefs, while filtering out information and people they find disagreeable.[6] Technology increasingly allows them to do so. If you are a liberal who uses Twitter to follow races for the US House of Representatives, 90 per cent of the tweets you see (on average) will come from Democrats; if you are a conservative, 90 per cent of the tweets you see will typically come from Republicans.[7] In the early days of the internet it was predicted that we would personally customize our own information environment, choosing what we would read on the basis of its political content. Increasingly, however, the work of filtering is *done for us* by automated systems that choose what is worthy of being reported or documented, and decide how much context and detail is necessary.

Problematically, this means that the world I see every day may be profoundly different from the one you see.

'You are entitled to your own opinion,' said the US Senator and ambassador to the United Nations Daniel Moynihan, 'but you are not entitled to your own facts.'[8] In the digital lifeworld, the risk is that rival factions will claim not only their own opinions but their own facts too. This is already becoming a problem. When deliberation takes place over digital networks, truth and lie can be hard to distinguish. As Barack Obama put it, 'An explanation of climate change from a Nobel Prize-winning physicist looks exactly the same on your Facebook page as the denial of climate change by somebody on the Koch brothers' payroll...everything is true and nothing is true':[9]

> Ideally, in a democracy, everybody would agree that climate change is the consequence of man-made behavior, because that's what ninety-nine per cent of scientists tell us...And then we would have a debate about how to fix it...you'd argue about means, but there was a baseline of facts that we could all work off of. And now we just don't have that.

The term *fake news* was initially used to describe falsehoods that were propounded and given wide circulation on the internet. Now even the term *fake news* itself has been drained of meaning, used as a way to describe anything the speaker disagrees with. Although some social media platforms have taken steps to counter it, the nature of online communication (as currently engineered) is conducive to the rapid spread of misinformation. The result is so-called *post-truth politics*. Think about for this a moment: in the final three months of the 2016 US presidential campaign, the top twenty *fake news* stories on Facebook generated more shares, reactions, and comments than the top twenty stories from the major news outlets combined (including the *New York Times*, *Washington Post*, and *Huffington Post*).[10] A poll in December 2016 found that 75 per cent of people who saw *fake news* headlines believed them to be true.[11]

Two further factors exacerbate the problem of *post-truth politics*. The first is that, in addition to *filtering*, individualized political *messaging* from political élites will mean that the information you receive from a given candidate or party will not be the same as the information I receive. Each will be tailored to what we most want to hear.[12] Second, our innate tendency toward group polarization means that members of a group who share the same views tend, over time, to become more extreme in those views. As Cass Sunstein puts it, 'it is precisely the people most likely to filter out opposing views who most need to hear them.'[13] I refer to the dual phenomenon of polarization and *post-truth politics* as *fragmented reality*. (It's related to the idea of *fragmented morality* discussed in chapter eleven.)

If the digital lifeworld falls victim to fragmented reality we'll have fewer and fewer common terms of reference and shared experiences. If that happens, rational deliberation will become increasingly difficult. How can we agree on anything when the information environment encourages us to disagree on everything? 'I am a great believer in the people,' Abraham Lincoln is supposed to have said. 'If given the truth, they can be depended upon to meet any national crisis. The great point is to bring them the real facts.'

Who will bring us the real facts?

Who Goes There?

The cause of deliberation is not helped by a third threat, that many online platforms allow us to participate anonymously or pseudonymously. This encourages us to behave in a way that we wouldn't dream of in face-to-face interactions. The sense that our actions can't be attributed to us, that nobody knows what we look like, that these are not real people, and that this is not the real world, combine to cause many of us to behave atrociously.[14] It's like slipping on J. R. R. Tolkien's One Ring (or if you're being posh, Plato's Ring of Gyges)[15] that makes us invisible and free to do as we please. The fact that it's often technically possible for authorities to discover who we are doesn't help the quality of deliberation.

The Athenians would have scoffed at the idea that deliberation could take place without revealing who you were. In the Assembly you couldn't possibly conceal your identity, allegiance, or interests. That would defy the whole point. In the digital lifeworld, though, more and more discourse will take place in a digital medium— with no meaningful distinction between online and offline. This raises an important question: should deliberation be seen as a *private* act, done anonymously by individuals in pursuit of their own self-interest; or should it be treated as something *public*, done in the open by members of a community in pursuit of the common good? If the latter, then we need to code our digital platforms in a way that reflects that ideal. Some work is already being done on this, discussed in a few pages' time.

To Be, or Bot to Be?

If you want people to hate your enemies, one strategy is to masquerade as those enemies and say repulsive things. On Twitter, so-called 'minstrel accounts' do this by impersonating minority groups and spewing out stereotypes and invective. To counter them, accounts like @ImposterBuster track down minstrel accounts and reveal them as the frauds they are. Impersonating the enemy is a venerable political technique. Forgeries like the *Protocols of the Elders of Zion* have circulated for centuries as a means of rousing hatred against Jews. But there's one critical difference today: *neither the minstrel accounts on Twitter nor @ImposterBuster are humans.*[16] Both are AI bots that have 'learned' to mimic human speech.

Bots have only just begun to colonize online discourse, but they are swiftly rising in significance. One 2017 study estimates that 48 million (9 to 15 per cent of accounts) on Twitter are bots.[17] In the 2016 US presidential election, pro-Trump bots using hashtags like #LockHerUp flooded social media, outgunning the Clinton campaign's own bots by 5:1 and spreading a whopping dose of *fake news*. It's estimated that around one-third of all traffic on Twitter in the buildup to the EU Brexit referendum came from bots. Almost all were for the Leave side.[18] Not all bots are bad for deliberation: so-called HoneyPot bots distract human 'trolls' by using provocative messages to lure them into endless futile online debate.[19] But by and large, bots' impact so far has not been benign.

Can Deliberative Democracy survive in a system where delibera-tion itself is no longer the preserve of human beings? It's possible that human voices could be crowded out of the public sphere altogether by bots that care little for our conversational norms. In the future, these won't just be disembodied lines of code: they could look and sound like humans, endowed with faces and voices and extraordinary rhetorical gifts. How can we, with our feeble brains and limited knowledge, participate meaningfully in deliberations if our views are instantaneously ripped to shreds by armies of bots armed with a mil-lion smart-ass retorts? Advocates of bots might put it differently: why spend time deliberating when increasingly sophisticated bots can debate the issues faster and more effectively on our behalf?

The bots that are used to fix errors on Wikipedia may offer a small indicator of what the world might look like if bots were con-stantly arguing among themselves. Behind the scenes, it turns out that many of these simple software systems have been locked in ferocious battle for years, undoing each other's edits and editing each other's hyperlinks. Between 2009 and 2010, for instance, a bot called Xqbot undid more than 2,000 edits made by another called

Darknessbot. Darknessbot retaliated by undoing more than 1,700 of Xqbot's own changes.[20]

In due course we could see the automation of deliberation itself. It's not an especially appetizing prospect.

Epistemic Epidemic

The prospects look a little bleak for Deliberative Democracy, and not just for its claim to epistemic superiority. Think of the argument from *liberty*: can we really claim to govern ourselves as free citizens if the laws we choose are based on lies and fabrications? Or the argument from *equality*: how can we have an equal chance of influencing the decisions that affect our lives if the deliberative process is at the mercy of whoever has the most sophisticated army of bots? Would Aristotle and John Stuart Mill really think we were ennobling or improving ourselves by fighting like animals over the truth of the most basic facts? Would a regime elected on the basis of *fake news* really be *stable* if its survival depended on the masses never discovering the truth?

Importantly, the problems described in this section don't *need* to be part of the digital lifeworld. We can find technical solutions. Social network proprietors are slowly taking steps to regulate their discussion spaces. Software engineers like those at *loomio.org* are trying to create ideal deliberation platforms using code. The Taiwanese *vTaiwan* platform has enabled consensus to be reached on several matters of public policy, including online alcohol sales policy, ride-sharing regulations, and laws concerning the sharing economy and Airbnb.[21] Digital fact-checking and troll-spotting are rising in prominence[22] and the process of automating this work has begun, albeit imperfectly.[23] These endeavours are important. The survival of deliberation in the digital lifeworld will depend in large part on whether they succeed. What's clear is that a *marketplace* of ideas, attractive though the idea sounds, may not be what's best. If content is framed and prioritized according to how many clicks it receives (and how much advertising revenue flows as a result) then truth will often be the casualty. If the debate chamber is dominated by whoever has the power to filter, or unleashes the most ferocious army of bots, then the conversation will be skewed in favour of those with the better technology, not necessarily the better ideas. Deliberative Democracy needs a forum for civil discussion, not a marketplace of screaming merchants.

That said, we shouldn't think that the challenges facing Deliberative Democracy are purely technical in nature. They raise philosophical problems too. One is the problem of extreme speech. In most democratic societies it's accepted that some limits on free speech are necessary when such speech would pose such an unacceptable threat to other freedoms or values. Words of violence that have nothing to do with politics, incitement to criminality, and threats fall into this category. But there's no consensus on where the line should be drawn. The First Amendment to the US Constitution, for instance, grants an unusual amount of protection to speech that would be illegal elsewhere. To deny the Holocaust is a crime in Austria, France, and Germany, but lawful in the US. In Europe there are strict laws prohibiting hate speech against racial, religious, and ethnic groups, while in the US neo-Nazis can happily wave swastikas in a town of Holocaust survivors. In Britain you can be arrested for *recklessly* publishing a statement that's *likely* to be understood by *some* of the public as an *indirect encouragement* to commit an act of terror. In the US, the same speech would have to be *directed* to *inciting or producing imminent lawless action* and *likely to incite or produce* such action.[24]

In the digital lifeworld, as we've seen, those who control digital platforms will increasingly police the speech of others. At present, tech firms are growing more bold about restricting obviously hateful speech. Few among us will have shed a tear, for instance, when Apple removed from its platform several apps that claimed to help 'cure' gay men of their sexuality.[25] Nor when several content intermediaries stopped trafficking content from right-wing hate groups after white supremacist demonstrations in Charlottesville in mid-2017. (The delivery network *Cloudfare* terminated the account of the neo-Nazi *Daily Stormer*.[26] The music streaming service Spotify stopped providing music from 'hate bands'.[27] The gaming chat app Discord shut down accounts associated with the Charlottesville fracas. Facebook banned a number of far-right groups with names like 'Red Winged Knight', 'White Nationalists United', 'Right Wing Death Squad', and 'Vanguard America'.)[28]

But what about when Facebook removed the page belonging to the mayor of a large Kurdish city, despite it having been 'liked' by more than four hundred thousand people? According to Zeynep Tufekci, Facebook took this action because it was unable to distinguish 'ordinary content that was merely about Kurds and their culture' from propaganda issued by the PKK, a group designated as a terrorist organization by the US State Department.[29] In Tufekci's words, it 'was like banning any Irish page featuring a shamrock or a leprechaun as an Irish Republican Army page'.[30]

My purpose is not to critique these individual decisions, of which literally millions are made every year, many by automated systems. The bigger point is that the power to decide what is considered so annoying, disgusting, scary, hurtful, or offensive that it should not be uttered at all has a significant bearing on the overall quality of our deliberation. It's not clear why so-called 'community guidelines' would be the best way to manage this at a systemic level: the ultimate 'community' affected is the political community as a whole. To pretend that these platforms are like private debating clubs is naïve: they're the new *agorae* and their consequences affect us all.

So as a political community—whether we personally use a particular platform or not—we need to be vigilant about the way that speech is policed by digital intermediaries. That does mean accepting that some restrictions on speech (wise restraints, if you will) will be necessary for deliberation to survive in the digital lifeworld. The idea of unfettered freedom of speech on digital platforms is surely a non-starter. Some forms of extreme speech should not be tolerated. Even in the nineteenth century John Stuart Mill accepted that certain restrictions were necessary. In his example, it's acceptable to tell a newspaper that 'corn-dealers are starvers of the poor' but not acceptable to bellow the same words 'to an excited mob assembled before the house of a corn-dealer'.[31] Mill understood that we certainly shouldn't be squeamish about rules that focus on the *form* of speech as opposed to its *content*. Just as it's not too burdensome to refrain from screaming in a residential area at midnight, we can also surely accept that online discourse should be conducted according to rules that clearly and fairly define who can speak, when, for how long, and so forth. In the digital lifeworld this will be more important than ever: Mill's 'excited mob' is much easier to convene, whether physically or digitally, using the technologies we have at our disposal.

Another difficult issue of principle relates to *fragmented reality*. It would be easy to blame *post-truth politics* on digital technology alone. But the truth (!) is that humans have a long and rich history of using deceit for political purposes.[32] Richard Hofstadter's 1963 description of the 'paranoid style' in public life—'heated exaggeration, suspiciousness, and conspiratorial fantasy'—could have been meant to describe today.[33] So could Hannah Arendt's observation in 'Truth and Politics' (1967): 'No one has ever doubted that truth and politics are on rather bad terms with each other.'[34] So too could George Orwell's complaint, in his diary of 1942, that:[35]

> We are all drowning in filth. When I talk to anyone or read the writings of anyone who has any axe to grind, I feel that intellectual honesty and balanced judgment have simply disappeared from the face of the

earth . . . everyone is simply putting a 'case' with deliberate suppression of his opponent's point of view, and, what is more, with complete insensitiveness to any sufferings except those of himself and his friends.

Today's problems have developed because of technology, no doubt, but also in part because of a political and intellectual climate that is itself hostile to the idea of objective *truth*. Influential postmodernist and constructivist thinkers have long argued that the notion of *truth* is nonsense. Beliefs are just plain beliefs; no more, no less. What we think is true is what we agree with, or what 'society' tells us to believe, or the product of language games without any basis in objective reality.[36] Or if there is an objective reality out there, it's so vaporous and elusive that there's no point in trying to catch it. Foucault even argued that *truth* itself was an instrument of repression:[37]

> we *must* speak the truth; we are constrained or condemned to confess or to discover the truth. Power never ceases its interrogation, its inquisition, its registration of truth: it institutionalises, professionalises and rewards its pursuit.

This isn't the only school of thought in the academy, but it has a vocal following.

I suggest that if you introduce technologies capable of rapidly disseminating falsehoods into a political ecosystem in which *truth* is not seen as a paramount political virtue—or is even seen in some quarters as a vice—then there are going to be grave consequences for the quality of public debate. We need to confront some hard questions about the relationship between democracy and truth.

First, is democracy's purpose to pool our collective wisdom so that we might find our way toward a discoverable 'truth' (an instrumentalist perspective)? Or for practical purposes, is what is 'true' and 'false', 'right' or 'wrong', simply what the multitude decides at any given time? This question has troubled many people since the election of Donald Trump and the Brexit referendum in the United Kingdom. The people have spoken—but could they be wrong? A related question is whether democracy should be seen as the process by which the community determines—on the basis of agreed underlying facts—what to do, or whether it should be seen as the process by which the community determines what the *underlying facts themselves are*? The writer Don Tapscott suggests that Holocaust denial could be countered by means of 'algorithms that show consensus regarding the truth'.[38] But what if an ignorant majority at a certain time didn't believe that the Holocaust happened? Would that mean that, for political purposes, *it didn't happen*? Surely not.

These are (sigh) questions on which people can disagree, but I don't accept that they are questions of pure theory, of interest only to philosophers—and they certainly can't be left to tech firms to answer. I know which side I favour. As Matthew D'Ancona puts it, truth is a 'social necessity . . . a gradual and hard-won achievement' that acts as a 'binding force' not just in politics, but in science, law, and commerce.[39] It's one thing to say that the politician whose 'truth' is accepted by the public is likely to be the winner of a democratic election. It's another to say that such a victory was legitimate or desirable—if it turns out that that so-called 'truth' was not true at all.

Direct Democracy

In a Direct Democracy, the people vote directly on the issues rather than electing politicians to decide for them. A show of hands, a heap of ballots, a roar of acclamation: it's democracy at its purest.

And yet for most of history it has been a fiction.

As we have seen, the vast majority of democracies have been *indirect* or representative, operating within a competitive elitist framework. There have been many reasons for this, but foremost among them is the problem of practicality: 'One can hardly imagine,' says Rousseau, 'that all the people would sit permanently in an assembly to deal with public affairs.'[40] In the digital lifeworld, however, the people will not need to 'sit permanently' in order to sustain Direct Democracy. It will be technically feasible for citizens to do as much real-time voting as they like on a wide range of issues.

It's not hard to conceive of a daily notification on your smartphone (or whatever replaces it) listing the issues up for decision each week—whether a new building development should go ahead, whether a new school curriculum should be adopted, whether we should commit more troops to a conflict—accompanied by a brief AI-generated introduction to the issues with punchy summaries of the arguments for and against. You could cast your vote from your bed or on the train. Voting apps for private use already exist[41] and although internet voting is not yet secure or transparent enough for general use, it's reasonable to expect that it might be in the future. Some think that the eventual solution might come from Blockchain technology, with its promise of unhackable encryption.[42]

So Direct Democracy may become practicable, but is it desirable?

The arguments in favour are well-known. Direct democracy would allow genuine *self*-government on an *equal* basis. Everyone could perform meaningful public service and thereby enhance their moral faculties. It would bring the unmediated wisdom of the crowd to a vastly greater array of public policy decisions. And because the people would buy into political decisions, knowing them to be truly their own, the system would be stable and secure. Perhaps best of all, Direct Democracy would mean no more need for politicians. Marx would have been overjoyed. He once wrote that electoral democracy was no more than deciding once every three or six years 'which member of the ruling class was to misrepresent the people in Parliament'.[43] Good riddance to them!

And yet.

In our heart of hearts, do we really trust ourselves to make unfiltered decisions on complex matters of public policy? Is that a burden we wish to bear? We all have our areas of interest, to be sure, but is my deep and rich knowledge of *Monty Python* really going to be useful in deciding between different schemes of environmental regulation?

It's not just that most of us have a limited knowledge of public policy, or even that we sometimes vote irrationally. On one level, it's actually irrational to vote *at all*: since polls are usually decided by many thousands or millions of votes, each of us only has a tiny say in the eventual outcome. Why bother? This has been called the 'problem of rational ignorance' and it's a big challenge for Direct Democracy.[44] In practical terms, it means that people may not actually end up participating. Is an exhausted working parent really going to set aside precious minutes in the day to consider the merits of a new regulation on financial derivatives? Should he or she have to?

We should also reflect carefully before consigning politicians to the scrapheap. Perhaps there is something to be said for having a class of professional politicians to do the day-to-day work of governing for us, sparing us the hassle and worry. Sunstein argues that democracy in America was *never* based on the idea that Direct Democracy was desirable but merely infeasible. On the contrary, for the founding fathers, 'good democratic order' involved 'informed and reflective decisions, not simply snapshots of individual opinions'.[45] That's why James Madison favoured '*the total exclusion of the people in their collective capacity* from any share' in the government.[46]

A possible middle way lies in a system of *partial* Direct Democracy. We need not ask *all* the people to vote on *all* the issues *all* the time.

Citizens could whittle down the issues on which they wish to vote, by *geography* (I want to vote on issues affecting London, where I live), *expertise* (I want to vote on issues relating to the energy industry, which I know a lot about), or *interests* (I want to vote on agricultural issues which affect my livelihood as a farmer). (I know that few farmers live in London but you get the point.) Such a system would radically fragment the work of national government, but in pursuit of the broader ideal of a more authentic democracy. On the other hand, the risk would remain that particular areas of public policy could be hijacked by local or specialized interest groups. And as ever, the apathetic or disconnected might be left behind.

A more unorthodox system would be one in which we could *delegate* our votes on certain issues, not just to politicians but to anyone we liked. Instead of abstaining on issues we don't know or care about, we could give our proxy vote to people who *do* know or care about them. On matters of national security, for instance, I might want a serving army officer to vote on my behalf; on questions of urban planning, I might want a celebrity architect to cast my vote; on healthcare I might delegate my say to a consortium of nurses, doctors, and patient groups. A digital platform for this model of democracy has already been pioneered by the creators of DemocracyOS[47] and used by various political parties in Europe. It's called *liquid democracy*.[48] The idea has a long vintage. Back in the nineteenth century, John Stuart Mill observed in 'Thoughts on Parliamentary Reform' (1859) that there is no one who, 'in any matter which concerns himself, would not rather have his affairs managed by a person of greater knowledge and intelligence, than by one of less'.[49] A suitably constituted system of liquid democracy in the digital lifeworld could balance the need for legitimacy, stability, and expertise.

Direct Democracy would mark a radical break from Schumpeterian competitive elitism—by eliminating or radically reducing the need for elected politicians. Another model would achieve the same result by involving the populace in the work of drafting legislation: Wiki Democracy.

Wiki Democracy

Imagine that instead of sending delegates to constitutional conventions, the entire eighteenth-century population of the American states had tried to write the Constitution together. How would they have done it? Perhaps they would have gathered in a massive stretch of countryside somewhere. The noise would have been cacophonic. Even the finest orators would have been drowned out by the din. Few attendees would have known what was going on at any given time. There would probably have been a festival atmosphere, with much taking of drink, revelry, occasional stampedes, and perhaps some copulation and outbursts of brawling. Rival drafts of the document would have circulated at the same time. No doubt some would have been defaced, torn up, and lost as every citizen tried to have his or her say. Chaos.

Until fairly recently it was not feasible for a large group of strangers to collaborate efficiently or meaningfully in the production of content, let alone to draft a precise and sensitive set of rules to govern their collective life. This has now changed. The internet has given rise to a new way of producing content in which individuals who have never met can cooperate to produce material of great sophistication. Although there are fewer successful examples than some predicted, the most famous is Wikipedia, the online encyclopaedia whose content is written and reviewed by anyone who wishes to contribute. The other often-cited exemplar is open-source (or 'free') software, including the operating system Linux that runs on tablets, televisions, smartphones, servers, and supercomputers around the world. The code is curated by nearly 12,000 contributors, each working on the premise that any technical problem—no matter how difficult—can be solved if enough people are working on it. Where it is undertaken without top-down control, this kind of activity has been called *commons-based peer production* or *open-source production*.[50] Where there's more central direction and control, it tends to be called *crowdsourcing*.

In the digital lifeworld it will be possible, using commons-based peer production or crowdsourcing, to invite the citizenry directly to help set the political agenda, devise policies, and draft and refine legislation. Advocates of this sort of democracy, or variants of it, have called it *wiki-government*, *collaborative democracy*, and *crowdocracy*.[51] I refer to it as *Wiki Democracy*.

Small experiments in Wiki Democracy have already been tried with some success. As long ago as 2007, New Zealand gave citizens the chance to participate in writing the new Policing Act using a wiki.[52] In Brazil, about a third of the final text of the Youth Statute Bill was crowdsourced from young Brazilians and the Internet Civil Rights Bill received hundreds of contributions on the *e-Democracia Wikilegis* platform.[53] These were carefully planned exercises within closely confined parameters. There is potential for more as digital platforms grow increasingly sophisticated.

Like Direct Democracy, Wiki Democracy would reduce the role of elected representatives. And in a Wiki Democracy we would not merely be asked to say yes or no to a set of pre-ordained questions decided by someone else; instead we would have the chance to shape the agenda ourselves in a richer and more meaningful way. Wiki Democracy also enjoys some of the same epistemic advantages of Direct Democracy, in that it would draw on the wisdom of crowds generally and (where appropriate) experts in particular.

In a full-blown Wiki Democracy, as in a Direct Democracy, there would have to be flexibility in how and to what extent individuals contributed. The policymaking process could be broken down into various parts (diagnosis, framing, data-collection, drafting and refining legislation, and so forth) and each part could be guided by the groups and individuals most willing or best placed to contribute.[54] In a world of code-ified law (see chapter six) the code/law could theoretically be reprogrammable by the public at large, or by persons or AI systems delegated to undertake the task for them.

However, the idea of full-blown Wiki Democracy is beset with difficulties. More than any other model of democracy, it places serious demands of time and attention on its participants. Not everyone would feel comfortable editing a law. Even fewer would feel comfortable tinkering with code. The result could be a rise in apathy and a decline in legitimacy, as Wiki Democracy slid into a Wiki Aristocracy of the learned and leisured classes.

It could also give rise to delay and gridlock without an obvious mechanism for taking and sticking to decisions. Unlike Direct Democracy, which is inherently decisive, there is no clear end-point to the growth and evolution of a collaborative process. Linux and Wikipedia are constantly changing. 'Discourses,' as Jürgen Habermas put it in the twentieth century, 'do not govern.'[55] The same might be said of wikis—at least if they are left open-ended.

It's also unclear how well a wiki could function in circumstances where the basic aims of collaboration were themselves contested. At least on Wikipedia the overall goal is clear: to produce verifiable encyclopaedic content. It's apparent when contributors are trying to advance that goal and when troublemakers are seeking to undermine it. But when it comes to laws, there will always be reasonable disagreement on the goal. How do I contribute to a new wiki law legalizing drugs if I don't believe that drugs should be legalized at all? By deleting the whole statute?

That a wiki can be refined and adapted over time, much like the common law, is desirable. But the common law moves at a stately pace while a wiki may change thousands of times each second. Jaron Lanier rightly invites us to imagine 'the jittery shifts' of wiki law: 'It's a terrifying thing to consider. Superenergized people would be struggling to shift the wording of the tax code on a frantic, never-ending basis.'[56]

The practical problems with Wiki Democracy seem overwhelming. But they are only fatal if we try to defend a model of *pure* Wiki Democracy without any checks or balances. To do that would be nonsensical. With a proper constitution (perhaps not one that can be altered by anyone at the click of a button) a Wiki Democracy could be built on the basis of clear rules about which laws may be edited, when and by whom, what they may or may not contain, and so forth. This would not be the first time that humans have had to consider how the chaos of unbridled democracy can be harnessed into something stable and useful. Fluctuating and unstable majorities, the tyranny of a dominant class, uncertainty of the law: these were exactly the sorts of problems that animated John Locke in the seventeenth century and those in the liberal democratic tradition. The fact that Wiki Democracy would need brakes, controls, checks, and balances, does not make it illegitimate or impossible.

Data Democracy

We have seen that one of the main purposes of democracy is to unleash the information and knowledge contained in people's minds and put it to political use. But if you think about it, elections and referendums do not yield a particularly rich trove of information. A vote on a small number of questions—usually which party or candidate to support—produces only a small number of data points. Put in the context of an increasingly quantified society, the amount of information generated by the democratic process—even when private polling is taken into account—is laughably small. Recall that in 2020 there will be 40 zettabytes of data in the world—the equivalent of about 3 million books for every living person. Soon we'll generate the same amount of information every couple of hours as we did from the dawn of civilization until 2003.[57] This data will provide a log of human life that would have been unimaginable to our predecessors. This prompts the question: if the world contains reams of rich data for every citizen, why would we govern on the basis of a tick in a box every few years? A new and better system of synthesizing information in society must be possible. Drawing on the work of Hiroki Azuma and Yuval Noah Harari,[58] we can call such a system Data Democracy.

In a Data Democracy, ultimate political power would rest with the people but some political decisions would be taken on the basis of data rather than votes. By gathering together and synthesizing large amounts of the available data—giving equal consideration to everyone's interests, preferences, and values—we could create the sharpest and fullest possible portrait of the common good. Under this model, policy would be based on an incomparably rich and accurate picture of our lives: what we do, what we need, what we think, what we say, how we feel. The data would be fresh and updated in real time rather than in a four- or five-year cycle. It would, in theory, ensure a greater measure of political *equality*—as it would be drawn from everyone equally, not just those who tend to get involved in the political process. And data, the argument runs, doesn't lie: it shows us as we are, not as we *think* we are. It circumvents our cognitive biases. We are biased, for example, toward arguments that favour our own special interests. We tend to dismiss things that are inconsistent with our worldview. We view the world through frames drawn by élites. We dislike being inconsistent, even when changing our minds would be the rational thing to do. We are overly influenced by others, especially those in authority. We like to conform and be liked by others. We prefer our intuitions to reason. We favour the status quo.[59]

Machine learning systems are increasingly able to infer our views from what we do and say, and the technology already exists to analyse public opinion by processing mass sentiment on social media.[60] Digital systems can also predict our individual views with increasing accuracy. Facebook's algorithm, for instance, needs only ten 'Likes' before it can predict your opinions better than your colleagues, 150 before it can beat your family members, and 300 before it can predict your opinion better than your spouse.[61] And that's on the basis of a tiny amount of data compared to the amount that will be available in the digital lifeworld.

In short, the argument in favour of Data Democracy is that it would be a *really* representative system—more representative than any other model of democracy in human history.

It's true that governments already use data in order to make policy decisions.[62] The rise of 'civic data' is a welcome development. A Data Democrat, however, would say that there is a difference between using data sporadically as a matter of *discretion* and using it all the time as a matter of *moral necessity*. We wouldn't be happy with an electoral democracy where elections were held on an ad hoc basis when ruling élites felt like it. By the same token, if democracy is about taking into account people's preferences, then using data is something that *must* happen, and not merely a sign of good governance. A government that ignores data, on this argument, is as bad as one that ignores how the people vote. The more data in government, the more 'democratic' the system.

Pausing for a moment, there are some clear problems with a pure model of Data Democracy. On a practical level, the system would depend on data of a decent quality, uncorrupted by malfeasance or bot-interference. This isn't something that can be guaranteed.

On a more philosophical plane, we know that democracy is not *just* about epistemic superiority. Those who see democracy as being based on *liberty* would argue that Data Democracy reduces the important role of human will in the democratic process. A vote is not just a data point: it is also an important act of consent on the part of the voter. By *consciously* participating in the democratic process, we agree to abide by the rules of the regime that emerges from it, even if we occasionally disagree with those rules. A Data Democrat might respond that human will could be incorporated into a system of Data Democracy—perhaps through the conscious act of agreeing (or refusing) to submit certain data to the process. A more strident retort might be that if Data Democracy produced dramatically better outcomes than electoral democracy then it would have its own legitimacy by virtue of those outcomes.

Another argument against Data Democracy is that by making our *entire lives* an act of subconscious political participation, the system deprives us of the benefits of conscious political participation. Which 'noble actions', to return to Aristotle, does it allow us to perform? In what way does it help us to flourish as humans? Democracy is about more than the competent administration of collective affairs.

The most powerful argument against Data Democracy is that data is useless in making the kind of political decisions that are often at stake in elections. A democratic system needs to be able to resolve issues of reasonable moral disagreement. Some of these are about scarce resources: should more be spent on education or on health-care? Others concern ethics: should the infirm be allowed the right to die? It's hard to see how even the most advanced systems—even those that can predict our future behaviour—could help us answer these questions. Data shows us what *is*, but it doesn't show what *ought to be*. In a country where the consumption of alcohol is strictly prohibited, data revealing a low rate of alcohol consumption reflects only the fact that people obey the law, not that the law itself is right.

This difficulty would not have troubled Auguste Comte, who believed that all human behaviour was pre-determined by 'a law which is as necessary as that of gravity'.[63] But most of us would not see *prediction* as a substitute for moral reasoning. A system of Data Democracy would therefore need to be overlaid with some kind of overarching moral framework, perhaps itself the subject of democratic choice or deliberation. Or to put it more simply, Data Democracy might be more useful at the level of policy rather than principle.

Data Democracy is a flawed and challenging idea but democratic theorists cannot sensibly ignore it. The minimal argument in favour is that by incorporating elements of it we could dramatically improve the democratic processes we already have. The stronger version of the argument holds that Data Democracy could ultimately provide a more desirable political system than electoral democracy. The question is: which aspects are worthwhile and which are not?

AI Democracy

What role will artificial intelligence come to play in governing human affairs? What role *should* it play? These questions have been floating around since the earliest computing machines. In the twentieth century, reactions to the first question tended to involve dark premonitions of humankind languishing under the boot of its robotic overlords. Reflection on the second question has been somewhat limited and deserves more careful thought.

We know that there are already hundreds, if not thousands, of tasks and activities formerly done only by humans that can now be done by AI systems, often better and on a much greater scale. These systems can now beat the most expert humans in almost every game. We have good reason to expect not only that these systems will grow more powerful, but that their rate of development will accelerate over time.

Increasingly, we entrust AI systems with tasks of the utmost significance and sensitivity. On our behalf they trade stocks and shares worth billions of dollars, report the news, and diagnose our fatal diseases. In the near future they will drive our cars for us, and we will trust them to get us there safely. We are already comfortable with AI systems taking our lives and livelihoods in their (metaphorical) hands. As they become explosively more capable, our comfort will be increasingly justified.

Aside from tech, in recent decades we have also become more interested in the idea that some political matters might be best handled by experts, rather than dragged through the ideological maelstrom of party politics. 'Experts' are sometimes derided and frequently ignored, but the increased prominence of central bankers, independent commissions, and (in some places) 'technocratic' politicians is testament to the fact that we don't always mind them taking difficult, sober, long-term decisions on our behalf. Since Plato, in fact, countless political theorists have argued that rule by benevolent guardians would be preferable to rule by the masses.

In the circumstances, it's not unreasonable, let alone crazy, to ask under what circumstances we might allow AI systems to partake in some of the work of government. If Deep Knowledge Ventures, a Hong-Kong based investor, can appoint an algorithm to its board of directors, is it so fanciful to consider that in the digital lifeworld we might appoint an AI system to the local water board or energy authority? Now is the time for political theorists to take seriously the idea that politics—just like commerce and the professions—may have a place for artificial intelligence.

Vox Populi, Vox AI

What form might AI Democracy take, and how could it be squared with democratic norms?

In the first place, we might use simple AI systems to help us make the choices democracy requires of us. Apps already exist to advise us who we ought to vote for, based on our answers to questions.[64] One such app brands itself as 'matchmaking for politics',[65] which sounds a bit like turning up to a blind date to find a creepy politician waiting at the bar. In the future such apps will be considerably more sophisticated, drawing not on questionnaires but on the data that reveals our actual lives and priorities.

As time goes on, we might even let such systems *vote on our behalf* in the democratic process. This would involve delegating authority (in matters big or small, as we wish) to specialist systems that we believe are better placed to determine our interests than we are. Taxation, consumer welfare, environmental policy, financial regulation—these are all areas where complexity or ignorance may encourage us to let an AI system make a decision for us, based on what it knows of our lived experience and our moral preferences. In a frenetic Direct Democracy of the kind described earlier in this chapter, delegating your vote to a trusted AI system could save a lot of hours in the day.

A still more advanced model might involve the central government making inquiries of the population *thousands of times each day*, rather than once every few years—without having to disturb us at all.[66] AI systems could respond to government nano-ballots on our behalf, at lightening speed, and their answers would not need not be confined to a binary *yes* or *no*. They could contain caveats (*my citizen supports this aspect of this proposal but not that aspect*) or expressions of intensity (*my citizen mildly opposes this but strongly supports that*). Such a model would have a far greater claim to taking into account the interests of the population than the competitive elitist model with which we live today.

In due course, AIs might also take part in the legislative process, helping to draft and amend legislation (particularly in the far future when such legislation might take the form of code itself: see chapter six). And in the long run, we might even allow AIs, incorporated as legal persons, to 'stand' for election to administrative and technical positions in government.

AI systems could play a part in democracy while remaining subordinate to traditional democratic processes like human deliberation and human votes. And they could be made subject to the ethics of their human masters. It should not be necessary for citizens to surrender their moral judgment if they don't wish to.

There are nevertheless serious objections to the idea of AI Democracy. Foremost among them is the transparency objection: can we really call a system *democratic* if we don't really understand the basis of the decisions made on our behalf? Although AI Democracy could make us freer or more prosperous in our day-to-day lives, it would also rather enslave us to the systems that decide on our behalf. One can see Pericles shaking his head in disgust.

In the past humans were prepared, in the right circumstances, to surrender their political affairs to powerful unseen intelligences. Before they had kings, the Hebrews of the Old Testament lived without earthly politics. They were subject only to the rule of God Himself, bound by the covenant that their forebears had sworn with Him.[67] The ancient Greeks consulted omens and oracles. The Romans looked to the stars. These practices now seem quaint and faraway, inconsistent with what we know of rationality and the scientific method. But they prompt introspection. How far are we prepared to go—what are we prepared to sacrifice—to find a system of government that *actually* represents the people?

Democracy's Day

Back to the tentative thesis from the Introduction: when a society develops new technologies of information and communication, we might expect political changes as well. This applies even to a concept as venerable as democracy. The classical model in ancient Greece, the liberal and competitive elitist models in the modern era: these were all models of 'democracy' but tailored to the conditions of their time. The digital lifeworld will challenge us to decide which aspects of democracy are most important to us. Do we care enough about deliberation to save it from the threats that face it? If we value liberty and human flourishing, then why not a system of Direct Democracy or Wiki Democracy? If we want the best outcomes and equal consideration of interests, then is there a place for Data Democracy and AI Democracy?

I started this chapter by observing how much has been written about democracy. As the digital lifeworld comes into view, it turns out that there is still a great deal more thinking and debating to do. In a future thick with new and strange forms of power, we'll need a form of democracy that can truly hold our masters to heel.

PART V

Future Justice

'Each time a man stands up for an ideal, or acts to improve the lot of others, or strikes out against injustice, he sends forth a tiny ripple of hope, and crossing each other from a million different centers of energy and daring those ripples build a current that can sweep down the mightiest wall of oppression and resistance.'

Robert F. Kennedy

FOURTEEN

Algorithms of Distribution

'A well-ordered society is...one that is effectively regulated
by a public conception of justice'
John Rawls, A Theory of Justice (1971)

The pursuit of justice can inspire us to great acts of sacrifice and
devotion. The sight of injustice can move us to unspeakable sadness
and rage. The political theorist John Rawls called justice the 'first
virtue' of social institutions.[1] Martin Luther King taught that it lay
at the end of the arc of the moral universe. The ideal of justice has
never been the exclusive preserve of politics or law. Alone among
political concepts, it has a sacred standing, forever inscribed in the
commandment of Deuteronomy: 'Justice, and justice alone, you
shall pursue.'

Justice has been the subject of a great deal of political theory
but it does not lend itself easily to dispassionate analysis. In our
day-to-day lives it's most often present as a sense or a feeling, even
an instinct. Its home is the gut, not the brain. For those struggling
at the sharp end of the world's injustices, even the finest treatises
on the subject must seem curiously cold and detached. War and
famine, squalor and hunger, ignorance and disease, hatred and
bigotry—these are matters of the heart as well as the head. The
richest eight people on the planet possess the same wealth as the

poorest half of humanity.[2] One per cent of the world's population controls half its assets.[3] No political theory, however eloquent, can make full sense of the towering injustice that exists in the world. No scholarly argument can match the primal urgency of Marx's old battle-cry: '*I am nothing and I should be everything.*'[4]

Yet political theorists are not entitled to the luxury of getting carried away. There's already enough bluster and confusion out there. Sometimes the cause of justice is better served by an open mind than a clenched fist. It can be studied calmly, even clinically. It can even be studied *algorithmically*. The aim of a theory of justice, as Rawls put it, is to 'clarify and to organize our considered judgments about the justice and injustice of social forms'.[5] My aim in this part of the book is similar, if more modest: to clarify and organize our ideas about social justice from the viewpoint of the digital lifeworld.

The next three chapters are all about what I call *algorithmic injustice*. The argument is that questions of distribution and recognition—the very soul of social justice—will increasingly be settled by algorithms contained in code. Chapter seventeen is about technological unemployment: what does political theory have to say about a world without enough work? Finally, chapter eighteen looks at the risk that wealth in the digital lifeworld could increasingly funnel into the hands of a tiny group of people and corporations, and asks how a change in our understanding of *property* could stop that from happening.

What is Social Justice?

The precise meaning of the term *social justice* is contested, but broadly speaking it's used to describe the idea that everyone should get what they're entitled to—no more and no less. It can be distinguished from its sister concepts in law—criminal justice (the punishment of those who break the law) and civil justice (the resolution of legal disputes between parties) although they all spring from the same conceptual pasture.

Justice and equality are not necessarily the same thing. Many ancient and medieval thinkers believed that humans were inherently *unequal*, not only in talent and strength but in their fundamental worth and moral status. On this view, serfs, slaves, and women should not be treated equally: rather, they should live quietly under the yoke of their masters and betters.[6] That's what justice meant in those times. The principles were said to be derived from tradition, or from the 'law of nature', or even from the word of God. 'It is thought that justice is equality,' trolled Aristotle in the *Politics*, 'and so it is, but... only for those that are equal.'[7]

These days, by contrast, social justice is closely associated with the idea of equality. *Egalitarianism* is the belief that people should be treated as equals in important respects, although egalitarians themselves don't always agree on what these are. Some will say that justice requires an equalization of wealth, for instance, while others will say it only requires equal opportunities to acquire wealth. Most theories of social justice share at least one egalitarian belief: that all humans are equal in fundamental worth. Few nowadays would agree with Aristotle that some lives are worth less than others. This consensus, rare among political theorists, is called *the egalitarian plateau*.[8] In the next few chapters I do not depart from the egalitarian plateau. I take it as self-evident that all humans are essentially of equal cosmic value, and ask what the consequences might be for social justice in the digital lifeworld. As we'll see, justice doesn't always translate to material equality. There may be good reasons, *just* reasons, why inequalities in wealth or status should still be permitted. And justice certainly doesn't mean treating everyone the same.

Justice in Distribution

There are two main ways to think about social justice. *Justice in distribution*, as the name suggests, is about how assets should be distributed in society. *Justice in recognition* is about how humans see and treat each other, including our systems of social status and esteem. We begin in this chapter with distribution.

What gap, if any, should there be between rich and poor? How much tax should the ultra-wealthy pay? Who should benefit most from government spending? These are classic questions of distributive justice. They prompt a range of possible responses.

Let's say that you don't mind about the gap between the rich and the poor but you do believe that no one should go hungry or live on the streets. Everyone need not have the *same*, but everyone must have *enough*.[9] If this is your view, you're a *sufficientarian*.[10]

Many would argue that the neediest in society have a stronger moral claim to resources than the rich. So if the government has half a million dollars to play with, better to build a youth centre in a deprived area than a leisure complex in a prosperous suburb. Some go a step further, arguing as a general rule that the less well-off you are (even if you are not genuinely needy), the stronger your moral claim to society's resources.[11] This kind of belief is usually called *prioritarian*. Like sufficientarians, prioritarians are not concerned with the overall gap between rich and poor, as long as the less well-off get priority.

Others say that distributive justice demands yet more, namely *equality of opportunity*, meaning that irrelevant factors like gender, sexuality, or ethnicity should not count against people in their pursuit of valuable things like a job or education.[12] Equality of opportunity is closely linked with another conception of justice, *justice as desert*. It means that people should get what they deserve.

If you work hard and play by the rules then you should be rewarded. But if you laze around all day then you should get a smaller share of the pie. (Justice as desert differs from a less well-known principle, *justice as dessert*, in which everyone receives a pudding of their choice.) Combining the *equality of opportunity* and *justice as desert* approaches, you might say that a just distribution of resources is one in which everyone has *an equal opportunity to get what they deserve*.

Many egalitarians question whether qualifications or other 'object-ive' indicia of merit are actually just guides to what people deserve. Consider a brilliant and hardworking kid from the wrong side of the tracks who leaves high school with patchy grades, despite her best efforts. Her less-gifted cousin might graduate from a private school with better grades and a range of extra-curricular baubles. If both apply for the same college place, the rich student is plainly better *qualified*, but does she really deserve the place more? In this instance, true equality of opportunity requires more than merely abstaining from racism or bigotry. It calls for affirma-tive action, taking into account the socio-economic conditions in which the cousins' qualifications were earned (or not earned, as the case may be).

Another perspective on desert is to say that instead of looking at *merit* to decide what people deserve, we should try to reward *morally* deserving or socially useful conduct, as opposed to work whose value is only valuable in commercial terms. On this view, nurses and teachers should earn just as much as financiers and corporate lawyers.

From a still more radical position, we might question whether any of us actually deserve the qualities that determine our financial success in life. The fact that I am ugly and talentless is no more my fault than the fact that you are beautiful and gifted. It's also bad luck for me that the market, left to itself, will tend to distribute more of the good stuff to beautiful and gifted people than to ugly and talentless ones. The *luck egalitarian* tradition sees this as a matter of injustice: we shouldn't be punished or rewarded for matters beyond our control. That includes genetics and upbringing. If the luck egalitarians are right, then the genetically well-endowed should be stripped of their unearned riches and the less fortunate should be compensated for their bad luck in the lottery of life.[13]

One objection to luck egalitarianism is that it fails to take into account different work ethics. If you and I are equal in talent, but I work harder to make the most of myself, then surely I deserve more than you! Perhaps. But only if you believe that the desire and capacity for hard work are themselves things we choose or deserve, rather than attributes we are born with or the product of upbringing. If they too are a matter of luck, then (on the luck egalitarian argument) neither of us deserves more than the other.

So does this mean that the talentless and lazy should receive the same share of society's resources as the talented and industrious? Surely benefits and burdens should still be allotted according to choices for which people can be held morally responsible. If I take a gamble and it pays off, I should receive the rewards of that risk—but likewise I must be held responsible for my bad bets and pay the price when they go wrong.

Alternatively, it might be said that if talent and hard work are to be harnessed for the benefit of society, then they need to be *incentivized* with higher rewards even if that generates some inequality. In the absence of economic incentives, it's often said, people would be less inclined to work hard, take risks, or innovate. The economy would stagnate and we would all be condemned to receive an equal but slender slice of a very stale economic pie (that's a metaphor for the wealth of a society, not a return to *justice as dessert*). John Rawls made a claim of this type in *A Theory of Justice* (1971), arguing that social and economic inequalities are permissible so long as they are to the greatest benefit of the least advantaged.

It's okay if the gap between rich and poor grows, but only if the result is that the poor get richer faster than any other group. Rawls' principle brings us back to the first issue touched on in this section—whether distributive justice is more concerned with *absolute* deprivation or *relative* inequality. There are certainly reasons why relative inequality matters. It can erode the common ties of humanity that bind us together. It can cause the less well-off to feel misery and shame. And it can prevent the poor from participating fully in social and political life.[14] That's why many radical egalitarians say that to ensure equal opportunity for children, the state must equalize *outcomes* for adults, whose wealth determines the privilege of their offspring.[15] They seek a society in which everyone has the same.

On the other side of the political spectrum, many are hostile to the very idea of egalitarianism, at least insofar as it requires the redistribution of social goods. One view, advanced by Edmund Burke, David Hume, and Friedrich Hayek, is that the pursuit of equality would require a disastrous amount of state intervention in our lives.[16] In Hume's opinion:[17]

> Render possessions ever so equal, men's different degrees of art, care, and industry will immediately break that equality. Or if you check these virtues, you reduce society to the most extreme indigence; and instead of preventing want and beggary in a few, render it unavoidable to the whole community.

Other non-egalitarians, like Robert Nozick, say that any artificial redistribution of wealth unjustly interferes with individuals' over-riding right to their own *property*. For Nozick—possessed of a remarkable mind and equally remarkable eyebrows—the justice of any distribution of assets depends only on *how* it came about, and has nothing to do with *how much* anyone has. The only question is whether a particular resource was properly acquired and freely transferred (that is, given and received in exchange for something, or as a gift). If so, then the state has no right to take it from me and give it to you, no matter how deserving you may be. 'We are not in the position of children who have been given portions of pie by someone who now makes last minute adjustments to rectify careless cutting.'[18] (Like all good theorists, Nozick enjoyed a dessert metaphor—but would have felt queasy at the thought of an egalitarian gateau.) Blockchain technology, the most advanced system yet known for securely recording the complete ownership trail of a piece of property, would probably have fascinated Nozick. The lawful chain of ownership, recorded immutably on the Blockchain, would itself demonstrate that justice had been done—irrespective of how much everyone ended up with.

The Market and the State

How do questions of distributive justice play out in the real world? By what mechanisms do we decide how social goods are actually distributed? The answer is through the interplay of the *market system* and the *state*.

In modern times the default way of distributing resources is using the *market system*. In a pure market system there is no central decision-maker directing where resources should be allocated. Instead, the work of distribution is done by every participating individual. Resources are privately owned, swapped for other goods, traded for money, and exchanged for labour. As Adam Smith famously observed in *Wealth of Nations* (1776), the market system works on the basis that people will generally pursue their own self-interest: 'It is not from the benevolence of the butcher, the brewer, or the baker, that we expect our dinner, but from their regard to their own interest. We address ourselves, not to their humanity but to their self-love.'[19]

When it comes to distributing economic resources, the market is reasonably efficient. It allows for resources to find their way to the places where they are wanted most, at least as indicated by how much buyers are prepared to pay. (This is, of course, a imperfect measurement of demand: the billions who lack electricity in their homes, and would like to have it, would say that the market system does not really reflect their demand.) The market gives people a degree of freedom to do as they please with the resources at their disposal. The tragic flaw, however, is that markets generally distribute resources *efficiently* but not necessarily *justly*. There is nothing in the logic of the market to ensure that the poorest get enough, or that they get priority, or that there is equality of opportunity, or that people get what they deserve (whichever principle of distributive justice you prefer). Allocation according to how much people are willing to pay will not mean that resources get to where justice requires that they go. Some kind of external intervention in the market system is always needed to ensure a semblance of egalitarian justice.

That's where the state comes in.

One of the state's main economic functions is to regulate the market, specifying what can be owned, bought, sold, and under what conditions (including, for instance, the rule that *people* should not be bought and sold). This prevents some of the most egregious injustices. The state often does a lot of its own spending too, enabling it to direct resources to the places required by justice (for example: welfare payments for the unemployed or sick).

One of the great political debates of the twentieth century concerned the extent and nature of state intervention in the market economy. At one extreme, some favoured a 'command economy' in which the state did most of the spending and determined what people earned, what was produced, and at what price. At the other end of the spectrum were those who supported a 'night watchman state' that did as little regulating and spending as possible. There are a range of options between these poles.

Algorithmic Distribution

What is the future of distributive justice? I suggest that, in a substantial departure from the past, *algorithms* will play four vital roles. But we should recap briefly before I introduce them. *Code*, you may recall, refers to the instructions given to hardware—the physical stuff of technology—telling it what to do. Digital systems can only operate according to their code. An *algorithm* is any set of instructions for performing a task or solving a problem. An algorithm can be expressed in code, that is, in programming language that digital systems can understand. *Machine learning algorithms* are algorithms capable of 'learning' knowledge and skills, including how to find patterns, create models, and perform tasks. Some have the potential to act in ways that their human creators did not foresee- and many are high-performing in ways that do not necessarily mimic human intelligence.

Algorithms and Distribution

First of all, in the digital lifeworld algorithms will increasingly be used, along with the market and the state, to determine our access to important social goods. This makes algorithms a new and important mechanism of distributive justice.

Let's begin with access to *work*. Work is (for now, at least) the main way that most of us get the money we need to live and flourish. Even now, applications for jobs are routinely decided by algorithms. It is claimed that seventy-two per cent of resumés 'are never seen by human eyes'.[20] Instead, algorithms are used to scan applications and determine which candidates have the requisite skills and experience. Other algorithms are used to make determinations about candidates' character and ability outside the formal application process.

They do this by chomping through whatever personal data is available and drawing their own conclusions. The data used may have no direct connection with employment. A candidate's online browsing activity, or the 'quality' of their personal connections on Facebook, may determine whether their application succeeds or fails.[21] Machine learning algorithms can also be used to discover personal traits—habits, addictions, weaknesses—that human recruiters assessing paper applications could never possibly have known about. Algorithms, in short, will decide whether millions of people access the most precious thing the market has to offer: a livelihood.

Once a person has a job, it will become more common for their work itself to be measured, monitored, and assessed using algorithms. They're already used to predict when employees are likely to resign; it's expected that they will be used, more and more, to scrutinize general performance and determine pay and promotions.[22] The world's largest hedge fund, Bridgewater Associates, is building an AI system to run the day-to-day management of the firm, including the scoring and dismissal of staff.[23]

Algorithms will also play an increasing role in determining our access to other resources of critical importance. They're already used to decide our credit scores and whether we would make good tenants.[24] Insurance companies use algorithms to estimate when we are going to die. Automated predictions of health risk (perhaps based on data relating to our food consumption) could come to determine how much we pay in premiums.[25] Data on our driving skills might be used to determine our access to motor insurance.[26]

Work, loans, housing, insurance: these are not luxuries. They're essential social goods. All rational citizens want them, regardless of whatever else they want.[27] Those who have them live comfortable and prosperous lives. Without them, life can be very difficult. How they are distributed, and according to what criteria, are of central importance to the future of distributive justice.

Using algorithms and data to make these decisions is not inherently a bad thing. On the contrary, it's possible that carefully crafted algorithms could eliminate the biases and prejudices of human decision-makers. With regard to work, for instance, affirmative action algorithms could be used to broaden the pool of successful applicants from beyond the usual colleges and institutions. When it comes to loans, housing, and insurance, algorithms could be used to widen access for those who need or deserve it most. My point, at this stage, is simpler: *code* (embodying algorithms) is an increasingly important mechanism of distributive justice. It demands close political attention.

As time goes on, we'll learn more about how algorithms interact with the market and the state, formerly the key mechanisms of distributive justice. Algorithms plainly won't substitute for them but they will affect their functioning in interesting and important ways. In the examples already laid out, algorithms determine the terms of access to social goods that were previously distributed according to market forces or state intervention. This is a significant change.

Algorithms and Participation

Next, algorithms intervene as *participants* in the market, buying and selling (among other things) billions of dollars' worth of financial products. This has distributive consequences: the share of overall wealth enjoyed by financial institutions, for instance, has ballooned since the introduction of automated trading.[28]

Algorithms and Information

Third, algorithms are increasingly used to determine the *information* available to purchasers. In the United States, more than 80 per cent of consumers search online before buying a product. The results they are shown directly delimit the set of options from which they choose in making their purchases.[29] Often algorithms will apply class distinctions: online shopping platforms routinely display advertisements for payday loans to less well-off groups.[30] Every algorithm of this kind will benefit some groups more than others. The question is, whose interests should they prioritize? The seller or the buyer? Rich or poor? These are quintessential questions of distributive justice.

Algorithms and Price

Finally, algorithms increasingly intervene in the most fundamental machinery of the market economy: the price mechanism. Customers may be charged more or less depending on where they live (outlets like Staples charge differentially for the same product depending on the postcode of the buyer),[31] the time of day (gas stations charge more in hours of peak demand),[32] and the weather outside (vending machines can use algorithms to vary prices according to the heat).[33] These are simple examples but the possibilities are much more radical. Research suggests that if Netflix took into account its customers' online behaviour (5,000 variables, including how frequently they visit IMDB and Rotten Tomatoes) it could increase its profits by more than 12 per cent.[34] An extreme end result would be 'person-specific pricing' whereby algorithms are used to charge customers the precise maximum they are prepared to pay at that moment.[35] This kind of practice was previously impossible. Vendors never had enough information about buyers and it wasn't feasible to change prices so rapidly. Not so in the digital lifeworld. When I buy my groceries using a digital application it might be hard to know if the price displayed has been tailor-made for me, based on my past spending habits and whatever else the vendor has learned about me.

Algorithmic interference with the price mechanism raises profound questions of distributive justice. Is it just for people to pay different prices for the same thing? The market already fails to guarantee that the least well-off will get enough or receive priority in the allocation of scarce social goods. Nor does it ensure equality of opportunity, or that people get what they deserve. Person-specific pricing could make this worse. A sick man might be prepared to pay his entire life's savings for a simple operation to save his life, but does that mean that's what he *should* pay? Perhaps algorithms could be used for redistributive purposes, with the rich paying a little more, and the poor paying a little less, for the same goods.

This could also been seen as an issue of *liberty*. The price of goods determines, to a meaningful extent, what we can and can't do in life.[36] According to the old quip, the Ritz is open to rich and poor alike—but we know that, in reality, only the rich are free to enter because they have the resources to pay. When we mess with prices, we also mess with liberty.

For centuries, we have debated the economic rules that ought to apply to the market and the state. We must now do the same for code. What should be seen as acceptable, what should be regulated, and what should be forbidden? More on that in chapter sixteen.

Now, the second type of justice: recognition.

FIFTEEN

Algorithms of Recognition

'In the central highlands in Africa where I work, when
people meet each other walking...on the trails, and
one person says, "Hello, how are you, good morning,"
the answer is not, "I'm fine, how are you?" The answer
translated into English is this: "I see you." Think of that.
"I see you." How many people do all of us pass every day
that we never see?'

Bill Clinton

There is more to social justice than the distribution of *stuff*. The
slave forced to kiss the dirt at his master's feet; the worker who
cowers before his screaming boss; the wife made to submit to her
husband's cruel demands—these instances offend our conscience, but
not necessarily because one party is rich and the other is poor. The
injustice comes from a different set of inequalities—of *status* and
esteem. This is a second way to think about social justice: *justice in
recognition*.[1]

From this point of view, regardless of how much anyone *has*, it is
unjust for some to be *seen* or *treated* as superior or inferior on the
basis of their birth, class, gender, race, occupation, age, disability,
culture, or other such characteristics. Humans have a deep-seated
desire to be treated with dignity and for their way of life to be
regarded with respect.[2] Axel Honneth calls this 'the struggle for
recognition'.[3] For many, this struggle is more important than the

fight for distributive justice. It's one reason why, to the bafflement of political élites, less well-off voters frequently reject social welfare policies that appear to be in their economic interests. Revulsion at being pitied or patronized—lowered in the eyes of others—can outweigh purely economic concerns.

The idea of recognition has its roots in a German tradition of political thought that emphasizes the role of the community in helping individuals to reach their full potential. It takes from G. F. W. Hegel the paradoxical idea that, in order to flourish as individuals, we first need the recognition and respect of others. And it takes from Immanuel Kant the idea that every individual deserves to be recognized as an autonomous and morally responsible agent, capable of authoring the laws and rules that govern them. The German language has a special word for describing the failure to give someone due recognition, *Mißachtung*, which sounds like a sneeze and is usually translated as 'disrespect' but which neatly embraces a wide range of interpersonal wrongs: humiliation, insults, degradation, abuse.[4]

Injustices in recognition take two forms: objective and intersubjective. *Objective* injustices are caused by *hierarchies*. In Aristotle's society, for instance, slaves and women were held to be inferior by birth. This was duly reflected in their social status. *Intersubjective* failures of recognition occur when individuals fail to see each other as beings of equal moral worth. Are you one of those who turns their nose up at the dress, habits, and accents of the *lower orders*? When you look at people who are different from you, do you recognize your shared humanity or do fear, disgust, even hatred, swell in your heart? When you walk past a gang of youths lurking on the street corner, do they appear to you as fellow human beings—or as *animals* or *scum*?

Justice in recognition is about seeing and treating each other as peers. Generations of LGBTQ+ people have struggled against (objective) rules that deprived them of the chance to serve in the military and the right to marry and adopt children. They have also fought to have their love and desires recognized (intersubjectively) as of equal worth to those of others—and not as sinful or shameful. Movements like these have never been about the mere redistribution of wealth. Their point, in Elizabeth Anderson's words, has been 'to create a community in which people stand in relations of equality to others.'[5] The opposite of this ideal is the belief that some people's lives are intrinsically less valuable. This belief can be found in the doctrines of racism, sexism, and xenophobic nationalism.[6]

Another consequence of failures of recognition is *oppression*. As Iris Marion Young explains, this concept embraces a variety of ills. Sometimes it means *exploitation*, when one group makes another work in unjust conditions. Sometimes it means *marginalization*, when certain groups (such as the dependent elderly) are condemned to lives of loneliness and exclusion. Marginalized groups must prove their respectability anew every day, while those with the right dress, accent, skin colour, and qualifications have their probity taken for granted, their needs more attentively serviced by police, waiters, public officials, and so forth. *Cultural* oppression allows dominant groups to have their art, music, and aesthetics seen as 'normal' while minority cultures are stereotyped, ignored, or suppressed. The fact that Muslim characters are disproportionately cast as terrorists on television is an instance of cultural oppression. The problem couldn't be solved with a redistributive tax.

In extreme cases, failures of recognition spill over into physical violence: women sexually assaulted in public and in private, French Jewish schoolchildren forced to use armed escorts for fear of attack, African-American men made the victims of police brutality.[7] In the United States, people with disabilities are around two-and-a-half times more likely to be victims of violent crimes than those without disabilities.[8]

Treating people with equal concern and respect doesn't necessarily mean treating them the *same*. In fact, it's partly about recognizing that people aren't all the same; that we all have unique identities, needs, and attributes that make us who we are. The just way to treat a vulnerable elderly person will be different from the just way to treat a wayward young man. And if a particular culture is facing difficulties, then we may justly divert extra resources to support it.[9]

There is some debate in the academic literature about whether distribution or recognition is the more important dimension of social justice. We don't need to decide: both are plainly important and they're often intimately connected. If you're at the bottom of the social hierarchy, in the 'wrong' caste or class, then your lowly status will probably spill over into a lack of job opportunities and a smaller share of economic resources. Likewise, if you are destitute and living on the street then your social status is unlikely to be high. Failures of distribution and recognition are often two sides of the same injustice.

Law and Norms

How does the struggle for recognition actually play out? What mechanisms do we use to determine people's rank, status, and esteem? The two most important are *laws* and *norms*.

Laws that prohibit women from voting, or people of colour from using public facilities, or gay and lesbian couples from marrying, give rise to unjust hierarchies. The Nuremberg Laws introduced in Germany in 1935 enshrined Nazi racialist doctrines into law, forbidding sex and marriage between Jews and non-Jews and declaring that only those of 'German' blood could be citizens. Laws like these are unjust because they endow some groups with valuable rights while depriving others of the same, on the basis of distinctions between groups that cannot be justified according to recognized principles of justice.

Laws can also enshrine rules of caste, class, honour, and rank that have historically played an important part in the politics of recognition. But such rules can also exist outside the law. Like the systems of nobility that dominated European life for centuries, they tend to come with an intricate code of titles, forms of address, routines, and rituals. Members of royal and noble households are greeted with magnificent unction as 'Your Majesty' or 'Your Excellency', with accompanying bonanzas of fawning, bowing, curtseying, cap-doffing, and ring-kissing. In less self-aware times, the upper classes would routinely address their inferiors with generic names that barely acknowledged their personhood: 'boy', 'girl', 'woman', 'slave'.[10]

These days, overtly unjust legal hierarchies are increasingly rare and the old rules of social rank and role are generally viewed with embarrassment. But many of the old hierarchies have ossified into norms of economic, workplace, family, and social life. Norms are tricky because they lurk, often unspoken, in our traditions and customs. They lead us, sometimes unconsciously, to treat each other in ways that are unjust. The fact that men are routinely paid more than women for the same work is sometimes the result of deliberate discrimination, but it's more often caused by a confluence of unspoken assumptions and behaviours in the workplace. It's also a good example of where a failure of recognition can lead to an injustice of distribution.

Algorithms and Recognition

In the future, as well as distributing resources, algorithms will increasingly be used to identify, rank, sort, and order human beings.

This means that the struggle for recognition will play out in code as well as in laws and norms.

Algorithms are significant in the struggle for recognition in three ways.

Digital Disrespect

At some point we've all been treated rudely by an official or patronized by a customer service agent. Nowadays, though, only humans can treat us disrespectfully or inhumanely. In the digital lifeworld this will no longer be the case. In New Zealand in 2016 a man of Asian descent had his online passport application rejected because the automated system said that his eyes were closed in the photograph he uploaded.[11] This is a hint of what might be to come: a world in which the struggle for recognition, for so long confined to relationships among humans, comes to include our relationships with the digital systems that surround and interact with us every day. More than a third of Americans admit to abusing their computers, verbally or even physically.[12] We already get angry when our tech doesn't work properly. Now imagine the rage and indignity of being dismissed, ignored, or insulted by an 'intelligent' digital system—particularly if that maltreatment arises because of race or gender or some other arbitrary characteristic.

Digital Ranking

Second, in the future there will be new methods of bestowing praise, honour, prestige, and fame. Followers, friends, favourites, likes, and retweets have already become a new currency by which people's thoughts and activities are scored, measured, and compared.[13] Fame, celebrity, publicity, and recognition itself are increasingly sought and received on social media platforms. There is nothing revolutionary about holding and sharing such opinions about each other, but the technology means we can express them more often, more efficiently, and more precisely. The big difference is that algorithms increasingly determine how these systems of rank and order function, choosing who is seen and who remains hidden; who is in and who is out; which content goes viral and which is doomed to obscurity. In the digital lifeworld there will be many 'inequalities of visibility'—and some among us will be quite invisible.[14] It used to be the role of political, legal, cultural, and social élites to determine matters of visibility, status, and esteem. Increasingly, in the future, it'll be done by algorithms. Again, this isn't inherently a bad thing. The question (for chapter sixteen) is whether the new way will lead to more justice than the old.

In the same vein, our lives in the digital lifeworld will be increasingly *rateable*. We'll be able to give holistic personal scores based on trustworthiness, reliability, attractiveness, charm, intelligence, strength, fitness, and whatever else is considered desirable. From the perspective of the struggle for recognition, this may make you feel a little uneasy. Is it wise to have our popularity, our social *worth*, so starkly and publicly quantifiable? Is it prudent to develop systems that rank and rate people on the basis of their perceived merits— not just because those rankings may be wrong or unfair, but because the very act of assigning *scores* to *people* distorts our ability to see their lives as of equal value? The risk with personal rating systems is that they encourage us not to seek *equal* social status but instead *favourable* status compared with our peers. This is a regrettable but common human foible: the self-esteem and social status we derive from income, for instance, depends much more on how much *others* in our peer group are earning rather than the objective amount we earn.[15] Constant social comparison would only exacerbate this regrettable trait. In the good old days we could go home and bitterly reflect on our inadequacies in peace; now there's always some smug schoolfriend on Facebook humblebragging about his latest promotion. The digital lifeworld could hold much worse horrors.

Digital Filtering

Finally, as we saw in chapter eight, in the digital lifeworld technologies of perception-control will serve to filter *how* we perceive other people. Digital systems—including but not limited to AR—will increasingly insert themselves between people, determining what they know of each other. What could be more influential on the struggle for recognition?

Thinking Algorithmically about Justice

Some of the ideas in the last two chapters may seem heretical to those who have spent a lifetime thinking about justice exclusively in terms of the market, the state, laws, and norms. But is it surprising that code should touch on issues of social justice? Rival models of distribution (who gets what, under what terms, in what proportions) and of recognition (who ranks higher or lower, who matters more, who less) are themselves *algorithmic* in nature. Recall the Sumerian clay tablets from the Introduction. They contained the earliest recorded algorithms in human history, and they were directly concerned with a problem of distribution: how best to divide a grain harvest between different numbers of men.[16] At their core, competing visions of justice are essentially alternative algorithms justified by rival principles. Now these algorithms are making their way into code, and they'll soon touch every aspect of our lives. Our task is to make sure that the algorithms that come to define the digital lifeworld can be justified according to principles of justice. To see how matters might go wrong, we turn now to the idea of algorithmic injustice.

SIXTEEN
Algorithmic Injustice

'We cannot know why the world suffers. But we can know
how the world decides that suffering shall come to some
persons and not to others.'

Guido Calabresi and Philip Bobbit, *Tragic Choices* **(1978)**

In the digital lifeworld, social engineering and software engineering
will become increasingly hard to distinguish. This is for two reasons.
First, as time goes on, algorithms will increasingly be used (along
with the market and the state) to determine the distribution of
important social goods, including work, loans, housing, and insurance. Second, algorithms will increasingly be used (along with
laws and norms) to identify, rank, sort, and order human beings.
Distribution and recognition are the substance of social justice, and
in the future they'll increasingly be entrusted to code. That's been
the subject of the last two chapters. To put it mildly, it's an important development in the political life of humankind. It means that
code can be used to reduce injustice or it can reproduce old iniquities
and generate new ones. Justice in the digital lifeworld will depend,
to a large degree, on the algorithms that are used and how they are
applied. I speak of the *application* of algorithms because often it's
the alchemy of algorithms and data *together* that yields unjust results,
not algorithms alone. This is explained further below.

In the next few pages I offer a broad framework for thinking about *algorithmic injustice*: where an application of an algorithm yields consequences that are unjust. We start with two main types of algorithmic injustice: data-based injustice and rule-based injustice. Then we examine what I call the *neutrality fallacy*, the flawed idea that what we need are algorithms that are *neutral* or *impartial*. At the end of the chapter we see that, lurking behind all the technology, most algorithmic injustice can actually be traced back to the actions and decisions of *people*, from software engineers to users of Google search.

The Rough and Ready Test

Before getting down to specifics, there's an easy way of telling whether a particular application of an algorithm is just or unjust: *Does it deliver results that are consistent or inconsistent with a relevant principle of justice?* Take the example of an algorithm used to calculate health insurance premiums. To be consistent with *sufficientarian* principles, for instance, such an algorithm would have to focus resources on the most deprived parts of the community, ensuring that they had an acceptable minimum standard of coverage. If instead it demanded higher premiums from those with conditions prevalent among the poor, then obviously its effect would be to make it *harder* for the poor to secure cover. From a sufficientarian perspective, therefore, that application would be unjust. Simple! This rough and ready test takes a *consequentialist* approach, meaning it doesn't try to assess whether an application of code is *inherently* virtuous or right. Nor does it require close technical analysis of the algorithm itself. It simply asks whether an application of an algorithm generates results that can be reconciled with a given principle of justice. This is just one way of assessing algorithmic injustice. One of the tasks for political theorists will be to find more.

'Algorithmic Discrimination'

Different types of algorithmic injustice are sometimes lumped together under the name 'algorithmic discrimination'. I avoid this term, along with the term *algorithmic bias*, because it can lead to confusion. *Discrimination* is a subtle concept with at least three acceptable meanings. The first is *neutral*, referring to the process of drawing distinctions between one thing and another. (If I say you are a highly discriminating art critic, I am praising your acuity and not calling you a bigot.) The second relates to the drawing of apparently *unjust* distinctions between groups—like a father who refuses to allow his child to associate with children of other ethnic backgrounds. The third sense is *legalistic*, describing rules or acts contrary to a particular law that prohibits the less favourable treatment of specified groups. In my work as a barrister, I often act in cases involving allegations of discrimination.

It's easy to see from these distinctions that not all discrimination is unjust, and not all injustice is unlawful. In English law, for instance, employers are prohibited from discriminating on the basis of age, disability, gender reassignment, pregnancy and maternity, race, religion and belief, sex, and sexual orientation—but they *can* discriminate on the basis of *class* (so long as that doesn't fall foul of some other law). Class-based discrimination is not illegal but it's arguably still unjust. The distinction is important because journalists and lawyers too often reduce issues of algorithmic injustice to a question of legalistic discrimination—is this application of code permissible under US or European law? Important though this sort of question is, the approach of political theory is broader. We need to ask not only what is legal, but what is *just*; not only what the law *is* but also what it *should be*. So let's take a step back and look at the bigger picture.

There are two main categories of algorithmic injustice: data-based injustice and rule-based injustice.

Data-Based Injustice

Injustice can occur when an algorithm is applied to *data* that is poorly selected, incomplete, outdated, or subject to selection bias.[1] Bad data is a particular problem for machine learning algorithms, which can only 'learn' from the data to which they are applied. Algorithms trained to identify human faces, for instance, will struggle or fail to recognize the faces of non-white people if they are trained using majority-white faces.[2] Voice-recognition algorithms will not 'hear' women's voices if they are trained from datasets with too many male voices.[3] Even an algorithm trained to judge human beauty on the basis of allegedly 'neutral' characteristics like facial symmetry, wrinkles, and youthfulness will develop a taste for Caucasian features if it is trained using mostly white faces. In a recent competition, 600,000 entrants from around the world sent selfies to be judged by machine learning algorithms. Of the forty-four faces judged to be the most attractive, all but six were white. Just one had visibly dark skin.[4] The image-hosting website Flickr autotagged photographs of black people as 'animal' and 'ape' and pictures of concentration camps as 'sport' and 'jungle gym'.[5] Google's Photos algorithm tagged two black people as 'Gorillas'.[6] No matter how smart an algorithm is, if it is fed a partial or misleading view of the world it will treat unjustly those who have been hidden from its view or presented in an unfair light. This is *data-based injustice*.

Rule-Based Injustice

Even when the data is not poorly selected, incomplete, outdated, or subject to selection bias, injustice can result when an algorithm applies unjust rules. There are two types: those that are overtly unjust, and those that are implicitly unjust.

Overtly Unjust Rules

An overtly unjust rule is one that is used to decide questions of distribution and recognition according to criteria that (in that context) are unjust on their very face. A robot waiter programmed to refuse to serve Muslim people *because* they are Muslim; a security system programmed to target black people *because* they are black; a resumé-processing system programmed to reject female candidates *because* they are female—these are all applications of overtly unjust criteria. What makes them overtly unjust is that there is no principled connection between the personal characteristic being singled out (religion; race; sex) and the resulting denial of distribution or recognition (a plate of food; access to a building; a job).

Overtly unjust rules are most obvious when they relate to characteristics like race and sex that have typically been the basis of oppression in the past, and which are irrelevant to the context in which the rule is being applied. But there are many other criteria that might also result in injustice. Take ugliness: if I own a nightclub, would it be unjust to install an automated entry system (call it *robobouncer*) that scans people's faces and only grants entry to those of sufficient beauty? Real-life bouncers do this all the time. Would it be unjust for a recruitment algorithm to reject candidates on the basis of their credit scores, notwithstanding their qualification for the job? In turn, would it be unjust for a credit-scoring algorithm to give certain individuals a higher score if they are friends with affluent people on Facebook?[7] These examples might not amount to discrimination under the current law, but they are arguably still unjust, in that they determine access to an important social good by reference to criteria other than the directly relevant attributes of the individual.

Rules can be overtly unjust in various ways. There's the problem of *arbitrariness*, where there's no relationship between the criterion being applied and the thing being sought. Or they fall foul of the *group membership* fallacy: the fact that I am a member of a group that *tends* to have a particular characteristic does not necessarily mean that I share that characteristic (a point sometimes lost on probabilistic machine learning approaches). There's the *entrenchment* problem: it may well be true that students from higher-income families are more likely to get better grades at university, but using family income as an admission criterion would obviously *entrench* the educational inequality that already exists.[8] There's the *correlation/causation* fallacy: the data may tell you that people who play golf tend to do better in business, but that does not mean that business success is *caused* by playing golf (and to hire on that basis might contradict a principle of justice which says that hiring should be done on merit). These are just a few examples—but given what we know of human ignorance and prejudice, we can be sure they aren't the only ones.

Implicitly Unjust Rules

An implicitly unjust rule is one that does not single out any specific person or group for maltreatment but which has the indirect effect of treating certain groups less favourably than others. A rule of recruitment that required candidates to be six feet in height with a prominent Adam's Apple would obviously treat women less favourably than men, despite the fact that it makes no mention of sex.

Implicitly unjust rules are sometimes used as fig-leaves for overt sexism or racism, but often injustice is an unwanted side-effect. Imagine, for instance, a recruitment algorithm for software engineers that gives preference to candidates who began coding before the age of eighteen. This rule appears justifiable if you believe, as many do, that early experience is a good indicator of proficiency later in life. It does not directly single out any social group for less favourable treatment and is therefore not overtly unjust. But in practice it would likely damage the prospects for female candidates who, because of cultural and generational factors, may not have been exposed to computer science at a young age. And it would limit the chances of older candidates who did not grow up with personal computers at home. Thus a rule that seems reasonable can indirectly place certain groups at a disadvantage.[9]

Unjust rules can be subtle in form. Let's return for a moment to face-identification algorithms, discussed earlier in the context of data-based discrimination. These will be common in the digital lifeworld as powerful systems identify and interact with us on a daily basis. Think of the horror of trying to interact with a digital system that failed to recognize your face because of its colour, or because of features like scarring or disfigurement. Yet even the simple business of *locating* a human face, let alone *identifying* it as belonging to a particular person, can be fraught with peril. One method is to use algorithms that locate 'a pattern of dark and light "blobs", indicating the eyes, cheekbones, and nose'. But since this approach depends on the contrast between the colour of a face and the whites of its eyes, it could present difficulties for certain racial groups depending on the lighting. An 'edge detection' approach instead tries to draw distinctions between a face and its surrounding background. Again, this might still pose problems, depending on the colour of the person's face and the colour of the backdrop. Yet another approach is to detect faces depending on a pre-programmed palette of possible skin colours—a technique that might be less conducive to offensive outcomes but which would require considerable sensitivity on the part of the programmer in defining what colours count as 'skin' colours.[10] You might ask what all the fuss is about, but if a *human* consistently failed to recognize certain groups because of their race, it would be seen as an obvious failure of recognition and an affront to the dignity of the person on the receiving end. Whether it is worse to be treated offensively by a machine or by a human is not immediately obvious. Both are matters of justice.

Take another tricky example. The Princeton Review offers online SAT tutoring courses and its software charges different prices to students depending on their zip code. The aim seems to be that that the rich should pay more for private tuition. In more deprived areas the course might cost $6,000, while in wealthier ones it could be up to $8,400. On its face, this rule might be justifiable according to recognized principles of justice: it gives priority to the less well-off and appears to encourage equality of opportunity. But it also has the side-effect of charging more to Asian-American students, who (in statistical terms) tend to be concentrated in wealthier areas and are therefore almost twice as likely as other groups to be charged a higher price.[11] What makes this a tricky example is that it demands a trade-off between two conflicting principles of justice: is it more important that we give educational advantages to the poor, or that we avoid treating certain ethnic groups less favourably than others? Your answer may reasonably differ from mine.

The example in the previous paragraph serves as a helpful reminder that not all rules drawing distinctions between groups are necessarily unjust, even if they treat some less favourably than others. Sometimes 'discrimination' can be justified according to principles of justice. A few years ago, trying to decide what to do with my life, I considered the idea of joining the army. Like many delusional young men I was drawn to the idea of joining an élite British regiment, the Special Air Service (SAS). Pen and notepad in hand, I eagerly sat down to study the SAS recruitment website. There I learned that 'Many try to get into the Special Air Service regiment. Most of them fail. Out of an average intake of 125 candidates, the gruelling selection process will weed out all but 10.' 'Excellent,' I told myself, 'Sounds like a challenge.' Reading on, I learned that the selection process involves three stages. The first is 'endurance', a three-week fitness and survival test ending with a 40-mile schlep, carrying a 55lb (25 kg) backpack, through the notoriously hostile terrain of the Brecon Beacons and Black Hills of South Wales. Those who survive can then look forward to the 'jungle training' stage in Belize, which 'weeds out those who can't handle the discipline' in order to find 'men who can work under relentless pressure, in horrendous environments for weeks on end'. Sounds great. After their sojourn in Belize, cadets can relax in the knowledge that the third stage of the process, 'Escape, Evasion, and Tactical Questioning', only involves being brutally interrogated and made to stand in 'stress positions' for hours at a time with white noise 'blasted' at them. 'Female interrogators', the website patiently explains, may even 'laugh at the size of their subject's manhood'.

I became a lawyer instead.

The SAS recruitment process is plainly discriminatory in the neutral sense of the word. It distinguishes, first and foremost, between genuine warriors and corpulent couch-dwellers who enjoy the thought of wearing a cool uniform without any of the effort. More directly, in seeking 'men' with the requisite qualities it entirely excludes women from the opportunity to serve in the unit. The first distinction is plainly a legitimate form of 'discrimination'. The second is debatable: some might say if women can pass the stages, why shouldn't they join the SAS? Women now play combat roles in various armies around the world. My point, besides the fact that most lawyers make lousy soldiers, is that identifying 'discrimination' is only the start of the conversation. That's why, to labour the point, the term isn't always helpful. The real question for an implicitly or overtly unjust rule is always whether its results can be justified according to principles of justice.

The Neutrality Fallacy

One of the most frustrating things about algorithms is that even when they apply rules that are studiously neutral between groups they can still result in injustice. How? Because neutral rules reproduce and entrench injustices that already exist in the world.

If you search for an African-American sounding name on Google you are more likely to see an advertisement for instantcheckmate. com, a website offering criminal background checks, than if you enter a non-African-American sounding name.[12] Why is this? It could be because Google or instantcheckmate.com have applied an overtly unjust rule that says African-American sounding names should trigger advertisements for criminal background checks. Unsurprisingly, both Google and instantcheckmate.com strongly deny this. What instead seems to be happening—although we don't know for sure—is that Google decides which advertisements should be displayed by applying a neutral rule: if people who enter search term X tend to click on advertisement Y, then advertisement Y should be displayed more prominently to those who enter search term X. The resulting injustice is not caused by an *overtly* unjust rule or poor quality data: instead we get a racist result because people's previous searches and clicks have themselves exhibited racist patterns.

Something similar happens if you use Google's autocomplete system, which offers full questions in response to the first few words typed in. If you type 'Why do gay guys...', Google offers the completed question, 'Why do gay guys have weird voices?' One study shows that 'relatively high proportions' of the autocompleted questions about black people, gay people, and males are 'negative' in nature:[13]

> For black people, these questions involved constructions of them as lazy, criminal, cheating, under-achieving and suffering from various conditions such as dry skin or fibroids. Gay people were negatively constructed as contracting AIDS, going to hell, not deserving equal rights, having high voices or talking like girls.

These are clear cases of algorithmic injustice. A system that propagates negative stereotypes about certain groups cannot be said to be treating them with equal status and esteem. And it can have distributive consequences too. For instance, more advertisements for high-income jobs are shown to men than to women. This necessarily means an expansion in economic opportunity for men and a contraction in such opportunity for women.[14]

What appears to be happening in these cases is that 'neutral' algorithms, applied to statistically representative data, reproduce injustices that already exist in the world. Google's algorithm auto-completes the question 'Why do women...' to 'Why do women talk so much?' because so many users have asked it in the past. It raises a mirror to our own prejudices.

Take a different set of examples that are likely to grow in importance as time goes on: the 'reputation systems' that help to determine people's access to social goods like housing or jobs on the basis of how other people have rated them. Airbnb and Uber, leading lights of the 'sharing economy', rely on reputation systems of this kind. There are also ways of rating professors, hotels, tenants, restaurants, books, TV shows, songs, and just about anything else capable of quantification. The point of reputation systems is that they allow us to assess strangers on the basis of what other people have said about them. As Tom Slee puts it, 'reputation is the social distillation of other people's opinion.'[15] You would tend to trust an Airbnb host with a five-star rating more than an alternative with only two stars.

Reputation systems are relatively young and are likely to become more common in the digital lifeworld. Services like reputation. com already exist to help you get better scores.[16] I've suggested the possibility that our access to goods and services in the digital life-world might eventually be determined by what others think of us. Recall the Chinese example: more than three dozen local govern-ments are compiling digital records of social and financial behaviour in order to rate their citizens, so that 'the trustworthy' can 'roam everywhere under heaven' while making it hard for the 'discredited' to 'take a single step'.[17]

The algorithms that aggregate and summarize people's ratings are generally said to be *neutral*. People do the rating; the algorithm just tots up the ratings into an overall score. The problem is that even if the *algorithm*s are neutral, there's ample evidence to suggest that the humans doing the actual scoring are not.[18] One study shows that on Airbnb, applications from guests with distinctively African-American names are 16 per cent less likely to be accepted as compared with identical guests with distinctively white names. This is true among landlords big and small, from individual property owners to larger enterprises with property portfolios.[19] As Tom Slee observes:[20]

> reputation . . . makes it difficult for John to gain a good reputation—no matter how trustworthy he is—if he is a black man trying to find work in a white community with a history of racism, or difficult for Jane the Plumber's skills to be taken seriously if the community has traditional norms about women's roles.

So it is that neutral algorithms can reproduce and institutionalize injustices that already exist in the world.

As time goes on, digital systems learning from humans will pick up on even the most subtle of injustices. Recently a neural network unleashed on a database of 3 million English words learned to answer simple analogy problems. Asked *Paris is to France as Tokyo is to [?]*, the system correctly responded *Japan*. But asked *man is to computer programmer as woman is to [?]* the system's response was *homemaker*. Asked *Father is to doctor as mother is to [?]*, the system replied *nurse*. *He is to architect* was met with *she is to interior designer*. This study revealed something shocking but, on reflection, not all that surprising: that the way humans use language reflects unjust gender stereotypes. So long as digital systems 'learn' from flawed, messy, imperfect humans, we can expect neutral algorithms to result in more injustice.[21]

These examples are troubling because they challenge an instinct shared by many—particularly, I have noticed, in tech circles— which is that *a rule is just if it treats everyone the same*. I call this *the neutrality fallacy*. To be fair, it has a long history. It originates with the Enlightenment ideal of universality—the idea that differences between people should be treated as irrelevant in the public sphere of politics.[22] This ideal evolved into the contemporary belief that rules should be *impartial* as between people and groups. Iris Marion Young, writing at the end of the twentieth century, described impartiality as 'the hallmark of moral reason' at that time.[23] Those who unthinkingly adopt the neutrality fallacy tend to assume that code offers an exciting prospect for justice precisely because it can be used to apply rules that are impersonal, objective, and dispassionate. Code, they say, is free from the passions, prejudices, and ideological commitments that lurk inside every flawed human heart. Digital systems might finally provide the 'view from nowhere' that philosophers have sought for so long.[24]

The fallacy is that neutrality is not always the same as justice. Yes, in some contexts it's important to be neutral as between groups, as when a judge decides between two conflicting accounts of the same event. But the examples above show that treating disadvantaged groups the same as everyone else can in fact reproduce, entrench, and even generate injustice. The Nobel Peace Prize winner Desmond Tutu once remarked, 'If an elephant has its foot on the tail of a mouse and you say that you are neutral, the mouse will not appreciate your neutrality.' His point was that a neutral rule can easily be an implicitly unjust rule. To add insult to injury, the neutrality fallacy gives these instances of injustice the veneer of *objectivity*, making them seem natural and inevitable when they are not.

The lesson for technologists is that justice sometimes demands that different groups be treated differently. This idea underpins affirmative action and the subsidizing of minority arts. And it should underpin all our efforts to avoid algorithmic injustice. An application of code should be judged by whether the results it generates are consistent with a relevant principle of justice, not by whether the algorithm in question is neutral as between persons. 'Neutrality,' taught the Nobel laureate Elie Wiesel, 'helps the oppressor, never the victim.'

A Well-Coded Society

Algorithmic injustice already seems to be everywhere. But just think about the sheer amount of code there will be in the digital lifeworld, the awesome responsibilities with which it will be entrusted, and the extensive role it will play in social and economic life. We're used to online application forms, supermarket self-checkout systems, biometric passport gates in airports, smartphone fingerprint scanners, and early AI personal assistants like Siri and Alexa. But the future will bring an unending number of daily interactions with systems that are radically more advanced. Many will have a physical, virtual, or holographic physical presence. Some will have human or animal-like qualities designed to build empathy and rapport.[25] This will make it all the more hurtful when they disrespect, ignore, or insult us.

As the reach and remit of code grows, so too does the risk of algorithmic injustice.

If we're going to grant algorithms more control over matters of distribution and recognition, we need to be vigilant. But it can sometimes be hard to know *why* a particular application of code has resulted in injustice. In the past, discriminatory intent was hidden in people's hearts. In the future it may be hidden deep within machine learning algorithms of mind-boggling size and complexity. Or it might be locked up in a 'black box' of code protected by confidentiality laws.[26]

Another difficulty is that potential injustice seems to lurk everywhere—in bad data, in implicitly unjust rules, even in neutral rules—waiting to catch us out. Regrettably so. But the responsibility falls to *people* to create a world in which code is an engine of opportunity and not injustice. It's too easy to treat algorithms, particularly machine learning algorithms, as disembodied forces with their own moral agency. They're not. The fact that machines can increasingly 'learn' does not absolve us of the responsibility to 'teach' them the difference between justice and injustice. Until AI systems become independent of human control, and perhaps even then, it'll be for humans to monitor and prevent algorithmic injustice. This work cannot be left to lawyers and political theorists. Responsibility will increasingly lie with those who gather the data, build the systems, and apply the rules. Like it or not, software engineers will increasingly be the social engineers of the digital lifeworld. It's an immense responsibility. Unjust applications of code will sometimes creep into digital systems because engineers are unaware of their own personal biases. (This isn't necessarily their fault. The arc of a computer science degree is long, but it doesn't necessarily bend toward justice.) Sometimes it'll be the result of wider problems in the culture and values of firms. At the very least, when it's learning *machines* coming up with the rules and models, their outputs must be carefully examined to see if they are overtly or implicitly unjust in that context. Failure to do so will result in algorithmic injustice. In chapter nineteen we look at some other possible measures to avoid such injustice, including the regulation of tech firms and auditing of algorithms. But why not consciously engineer systems with justice in mind—whether equal treatment, equality of opportunity, or whatever other principle might be applicable to that particular application? Code could offer exciting new prospects for justice, not merely another threat to worry about.

We need a generation of 'philosophical engineers' of the kind imagined by Tim Berners-Lee, and this cohort must be more diverse than today's. It can't be right that algorithms of justice are entrusted to an engineering community that is overwhelmingly made up of men.[27] African-Americans receive about 10 per cent of computer science degrees and make up about 14 per cent of the overall workforce, but make up less than 3 per cent of computing occupations in Silicon Valley.[28] At the very least, a workforce more representative of the public might mean greater awareness of the social implications of any given application of code.

For now, it's time to leave algorithmic injustice and turn to another potential challenge for social justice in the digital lifeworld: technological unemployment.

SEVENTEEN

Technological Unemployment

'it is the machine which possesses skill and strength in
place of the worker, is itself the virtuoso, with a soul of its
own in the mechanical laws acting through it'
Karl Marx, *Fragment on Machines* (1861)

Do you work to live or live to work? In the future you may do
neither.

AI systems can already match or outperform humans in a range of
capabilities: translating languages, recognizing faces, mimicking
speech, driving vehicles, writing articles, trading financial products,
diagnosing cancers. Once perceived as cold and lifeless, they can
tell if you are happy, confused, surprised, or disgusted—sometimes
by reading signals not detectable by the human eye. The economic
consequences will be profound. Between 1990 and 2007, before the
introduction of smartphones—industrial robots alone eliminated up
to 670,000 American jobs.[1] Between 2000 and 2010, the US
lost about 5.6 million manufacturing jobs, 85 per cent of which
losses were attributable to technological change.[2] In 2016, analysts
at McKinsey estimated that 'currently demonstrated technologies'
could be used to automate 45 per cent of the tasks that we pay
people to do.[3] That's just using today's technologies. In the digital
lifeworld we can expect systems of radically greater capability and
sophistication.

Digital systems can increasingly perform many tasks that we thought only humans could do. Soon they may be able to perform almost all of them. Eventually they may be able to do them better than us. And of course, they've always been able to do many tasks that we can't.

I said in the Introduction that predicting technological change is inherently risky and controversial. Trying to predict the economic consequences of technological change is even trickier. For personal reasons, if nothing else, however, I have a go in the next two chapters. In 2015 Richard Susskind (the world's leading authority on legal technology, also my dad) and Daniel Susskind (one of the world's most acclaimed young economists, also my brother) together published *The Future of the Professions*, in which they predicted the gradual replacement of human professionals by digital systems.[4] Daniel Susskind publishes in 2020 a new book on the topic which, with fraternal pride, I expect to become the definitive work. It would be an act of family betrayal if I didn't at least consider the idea of technological unemployment in this book.

(I know, I know, we're a strange family.)

Clan loyalty aside, no responsible citizen can now ignore the prospect, accepted by growing ranks of economists, that in the future there may not be enough human work to go around. I call this the *technological unemployment thesis*. I don't seek to assess in detail the economic case for and against it. I acknowledge that there are respectable thinkers who think it's flawed. My view, however, is that if it's even *partly* correct then the consequences will be so profound that we can't afford to wait and see. What's the harm in sharpening our intellectual tools while we still have time?

The analysis in this chapter is organized in four stages. We begin with the technological unemployment thesis itself. Next, we look at the *work paradigm*, the idea that we need work for income, status, and wellbeing. We then consider three responses to technological unemployment from within the work paradigm: treating work as a scarce resource, giving people a right to work, and trying to resist automation altogether. Finally, we dare to challenge the work paradigm itself, asking whether we could build a world where income, status, and wellbeing can be enjoyed in the absence of general employment. It may be that economic upheaval requires intellectual upheaval too.

Technological Unemployment

The Thesis

The technological unemployment thesis predicts that developments in technology will eventually cause large-scale human unemployment. In basic outline, it runs as follows.

What we think of as 'jobs' are really just bundles of economically useful tasks. As time goes on, machines will increasingly rival and surpass humans in their ability to perform those tasks. Instead of hiring humans it will be more economical for firms to use machines. The people currently paid to perform those tasks will find, eventually, that their services are no longer required.[5]

In the first stage of technological unemployment, there will be less overall work to do but still enough to go around. Laid-off workers may be able to retrain and find new jobs. Mid-career education and training will be essential to their prospects of finding work. But as time goes on, the jobless will find it increasingly hard to get the training or resources needed to perform the shrinking number of tasks that remain. If you're a steelworker from the north of England and you lose your job to a robot, it's not much help to know that Google is hiring software engineers in Palo Alto.[6]

As unemployment grows, competition for work will be fierce. With a few exceptions for superstars in high demand, wages for most of those still employed will fall, because there will usually be a group of desperate souls at the factory gates willing to work for less.

In due course, when machines achieve the requisite levels of capability and reliability, it will make no economic sense for firms to employ humans for most economic tasks. As a result, there will no longer be enough human work to go around. Eventually only a small number will be able to find paid work. By this stage, to borrow the young Karl Marx's phrase from 1844, the 'worker has become a commodity, and it is a bit of luck for him if he can find a buyer'.[7]

Widespread unemployment will not cause the economy to stagnate. On the contrary, automation will enable firms to make great savings and efficiencies. Their profits will be reinvested as capital or passed on to consumers in lower prices.[8] The economy will grow. In the past, we might have expected such growth to create more jobs for humans, because more demand for goods and services normally means more demand for the people who provide them. But not in the future, because eventually any extra tasks created would themselves be done more efficiently by machines than humans.

Overall, the economic pie will grow (that's right, folks, the dessert metaphors are back) but human workers will receive an ever-smaller slice. On an extreme outcome, the vast majority of people of working age could be unemployed. But the effects would be radical if even *half* or a *third* of the working-age population was unable to find work.

Who Will Be the First to Go?

It's intuitive to assume that lower-educated workers will be hardest-hit by technological unemployment. It currently costs about $25 an hour to pay a human welder and about $8 an hour to use a robot.[9]

Supermarket check-out staff face the prospect of 'smart' stores that run without human check-out staff and shelf-stackers.[10] Truckers, of whom there are 3.5 million in the United States alone, could be superseded by self-driving vehicles that can trundle for weeks without rest. Millions of low-paid workers in the food and beverage industry could be replaced by systems that can do every task from patty to plate:[11]

> Whereas a fast food worker might toss a frozen patty onto the grill, Momentum Machines' device shapes burgers from freshly ground meat and then grills them to order—including even the ability to add just the right amount of char while retaining all the juices. The machine, which is capable of producing about 360 hamburgers per hour, also toasts the bun and then slices and adds fresh ingredients like tomatoes, onions, and pickles only after the order is placed. Burgers arrive assembled and ready to serve on a conveyer belt.

The evidence, however, suggests that technological advance in the twentieth century actually boosted the number of low-education tasks in the economy as well as the number of high-education tasks. It was middle-education tasks that declined in number.[12] What explains this? The answer lies in one of Daniel Susskind's most important observations: that the level of education required by a human to perform a task is not a reliable guide to whether a machine will find that task easy or difficult. Machines outperform humans not by 'thinking' or working like them, but by using computational and robotic methods that are quite unhuman. This explains why we're closer to automating the work of lawyers than the work of make-up artists.

Because we don't know exactly *how* machines will perform certain tasks, it's therefore hard to say definitively who will be the first to lose out. According to McKinsey, it's easier to automate 'physically predictable activities' than unpredictable ones. The hardest tasks to automate are those that involve managing and developing people, and applying expertise to decision making, planning, and creative work.[13] But, as the Susskinds argue in *The Future of the Professions*, it looks like even the most complex professional work can be broken down and automated in the long run, including many aspects of those jobs said to require a *human touch*.[14]

As time goes on, however, the question won't just be one of automation of tasks currently done by humans. The desired outcomes might be attained in entirely different ways. For instance, the fact that we no longer need human farriers (horseshoe-makers) because cars have replaced horses and carriages makes it wholly irrelevant whether the work of a farrier could be done better by a machine. People don't want surgeons; or even machines that do the work of surgeons better than humans. They want health. If advance in another field of medicine (such as nanotechnology) makes the work of certain surgeons redundant, then it is irrelevant whether the work of human surgeons could be done better by machines.[15] Human redundancy will come in various forms; automation of tasks that we currently do is only one of them.

So what does political theory have to say about a world of widespread unemployment?

The Work Paradigm

People in the ancient world disliked the idea of working for a living. Work appears in the Old Testament as a kind of divine retribution: 'By the sweat of your brow' God berates the sinful Adam and Eve, 'you will eat your food until you return to the ground' (Genesis 3:19). As Kory Schaff observes, the ancient Greeks understood the practical benefits of work but believed that the key to human flourishing was avoiding it. A 'mechanical or commercial life', writes Aristotle, 'is not noble, and it militates against virtue.' Better, as Schaff puts it, to dedicate one's days to contemplation, statecraft, and warfare.[16]

Philosophers in the Christian tradition came to see hard work as a path to salvation, encouraging thrift, honesty, and self-discipline. Work also played an important role in the liberalism of John Locke and Adam Smith. Only through labour, they believed, could humans forge and hew the natural world into instruments of value and progress. Marx, by contrast, contended that capitalists sucked value from their workers without giving them a proper share of the rewards.[17]

So why do we work today? For three main reasons: income, status, and well-being.

Income

First and foremost, we work to live. For most of us, selling our productive powers is the only way to earn the money that keeps us fed and sheltered. We rarely grow our own food or build our own shelter. Instead we work so we can pay for those things (themselves the product of other people's labour). Most of us also hope that if we work hard there'll be a little left over to pay for pleasurable treats, luxuries, and experiences.

Status

We also work to satisfy a deep-seated need for status and the esteem of others. Gainful employment brings recognition and prestige. We feel pride when we receive positive feedback on our work from bosses, customers, and clients. Unemployment, by contrast, can lead to stigma and shame.

Wellbeing

Finally, some are fortunate to do work that contributes to their wellbeing. Such work might offer the chance of intrinsic satisfaction from a job well done. Or it could be rewarding because it brings value to others: I think of my mum, first a nurse and now a psychotherapist, who'll do anything to look after her patients. (Not every Susskind writes about technology.) For others, work offers the chance for self-improvement and a platform to cultivate skills and faculties. When people are out of work, they often speak of feeling 'useless' or 'on the scrap-heap'.[18]

Work also helps us to manage our fragile relationship with time. From our earliest years we experience the passage of time in a structured way: morning, afternoon, evening, night, each with its own routines and activities. In adulthood, our experience of time is closely connected with the routine of work. The unstructured languor of unemployment can lead to boredom and dislocation, which over long periods can become a 'major psychological burden'.[19]

Still thinking about psychological wellbeing, work can allow us to associate in a purposeful way with people outside our immediate family and gives us space to relieve our competitive urges. Even inane conversations round the office water-cooler permit us to observe other people and share stories, experiences, and knowledge of the world. This is part of keeping a grip on reality itself.[20] In *Civilization and its Discontents* (1930) Sigmund Freud claims that no other technique for the conduct of life binds the individual 'more closely to reality':[21]

> in his work he is at least securely attached to a part of reality, the human community. Work is no less valuable for the opportunity it and the human relations connected with it provide for a very considerable discharge of libidinal component impulses, narcissistic, aggressive, and

even erotic, than because it is indispensable for subsistence and justifies existence in a society.

The Work Paradigm

There is, I believe, an assumption in modern society that work is *needed* for income, status, and wellbeing. I call this the *work paradigm*. In a world where most adults work in some capacity, the nature of work is closely related to social justice. It's connected with distribution (of income and wellbeing) as well as recognition (in the form of status and esteem). Under the work paradigm, therefore, the prospect of a world without work is naturally disconcerting. How should we respond?

Three Responses Within the Work Paradigm

Accepting for a moment the idea that paid work is something we need, there are three possible responses to the problem of technological unemployment: to treat work as a *scarce resource*, to give people a *right to work*, or to *resist* automation altogether.

Scarce Resource

The first possible response would be to continue with business as usual, treating work as a resource (albeit an increasingly scarce one) to be distributed among the population according to a mechanism of our choice, the most obvious being a *labour market* like the one we have now. Through the interplay of market, state, and algorithms, tasks could be allotted to the most deserving or meritorious candidates. An alternative mechanism for distributing scarce work might be a *rota*, with work rotated around the citizens so everyone gets their turn.

The obvious difficulty with this approach is that there would still be a majority without work at any given time and those lucky enough to have a job would probably face severely depressed wages. This response would not properly address the challenge of technological unemployment.

Right to Work

A second approach would be to introduce some kind of *right to work* together with a scheme of artificial 'make-work' that provided everyone with tasks to keep them busy. Many unemployed workers in the past have claimed such a right. The *'droit au travail'*, as Jon Elster notes, was the 'battle cry' of the 1848 revolution in France.[22] Ultimate responsibility for any right to work would have to rest with the state, perhaps entitling citizens to redress if insufficient work was forthcoming. Public authorities could offer work directly or issue vouchers redeemable with firms for that purpose.

The notion of a right to work is normally seen as economically nonsensical because it would be a colossal drain of resources. In the digital lifeworld, however, it might be possible to skim some profits from firms and spend them on schemes that put people to work. As well as taxing their profits (or potentially instead of doing so) the state could place a legal duty on firms to provide a certain amount of make-work in proportion to their economic output.

All things considered, however, a right to work of this kind would be deeply problematic. Producing enough make-work and paying people for it would be massively inefficient and perhaps even impossible, particularly if the scheme tried to cater to each person's conception of worthwhile work. Moreover, the status and esteem associated with winning and keeping a job would be rather undermined by the knowledge that it was guaranteed as a right, and by the knowledge that the work itself was fundamentally useless.[23]

Resistance

A third possible response would be to try to resist automation altogether through some kind of global technological moratorium. I don't see this as a practical possibility.

After the Work Paradigm

The three responses just described share a common trait: they accept the work paradigm and seek to find ways of maintaining it. This may be the wrong approach, intellectually and practically. Technological unemployment could in fact present an opportunity to dismantle the work paradigm and replace it with a different set of ideas. In short, it would mean weakening or even sundering the connection between work on the one hand, and income, status, and wellbeing on the other.

1. Weaken the Connection between Work and Income

The first step would be to recognize that there is no *necessary* connection between work and income. Even under the current system, work does not guarantee a living. One in four employed adults in the United States earns wages below the official poverty line. Almost half are eligible for food stamps.[24] The converse is also true: there need not be a connection between non-employment and destitution. In past eras, losing your job may have meant homelessness and starvation. Today, most advanced political systems have some kind of collective safety net.

The digital lifeworld offers the chance to take the next step, weakening or even breaking the connection between work and income altogether. At its simplest, this could mean taxing some of the profits of firms and redistributing them among the general public. Hence Bill Gates' proposal for a 'Robot Tax' by which firms would be taxed for their use of machines with the proceeds going to fund employment opportunities elsewhere.[25]

Another increasingly popular notion is a universal basic income (UBI) paid in cash to everyone 'with no strings attached'.[26] On the radical model advocated by Philippe van Parijs, a UBI of about a thousand dollars a month would be available to every citizen with no means test or qualifying obligations.[27] Such a system would differ from the 'make-work' model above because it would not require people to work. On the contrary, how people chose to satisfy their needs would be a matter for them.

The idea of a UBI is not new. It has long been discussed in the context of a functioning labour market like we have today. Theorists have argued about whether a UBI would discourage work, or whether people would seek to 'top up' their UBI with additional income, or whether a UBI would prepare people well for re-entry into the job market. These questions (like human workers themselves) would be redundant in a world of technological unemployment. The function of a UBI in the digital lifeworld would be to *replace* the labour market and not to augment it. And if it's going to be their main source of income, people in advanced economies might need more than a thousand bucks a month.

There is room for debate about what a UBI might look like. It makes sense, for instance, that payments should not be contingent on looking for work if there's not enough work out there for everyone. It probably makes less sense that a UBI should be paid to each citizen regardless of their other means of income. Those who earn massive rents from land or capital arguably don't have the same moral claim to public funds as others. They would also have less use for the cash, and be less likely to spend it, making it economically inefficient to hand it to them. There could also be ways of topping up a UBI through economically productive work (if there is any for humans to do). If you have a rare medical condition that requires extra funds, perhaps you could be entitled to a higher basic income. Or if you do good deeds like caring for the elderly, perhaps you ought to receive a bonus (even if a machine could have performed that care better).[28] Rewards could incentivize self-improvement, with extra income to those who learn new skills (the value of such skills not being measured in economic terms).

To sever the connection between work and income would be radical. But luck egalitarians would see it as a step forward for social justice. For too long, they would say, people's wealth has been determined by their unearned talents and capacity for work. Marx himself described the right to work as an 'unequal right for unequal labour' because some are better equipped, 'physically or mentally', to participate in the labour market.[29] Isn't there something degrading, after all, in having to offer ourselves up as commodities on the labour market? Owners of capital have been earning without labouring for centuries. Might it not be progress if the rest of us could do the same?

2. Sever the Connection between Work and Status

The second step would be to challenge the assumption, perhaps the deepest and most entrenched in the work paradigm, that only gainful work is deserving of status and esteem while unemployment is a source of stigma and shame. 'Most people in modern competitive market societies', says the political theorist Richard Arneson, 'believe that the failure of an able-bodied person to earn his keep is degrading.'[30] We are taught to believe in the dignity of work, no matter how disgusting or dangerous that work is. Long-term unemployment is often thought to suggest personal incompetence or moral decay. Idleness is seen as disgraceful. 'What is a man,' asks Hamlet, 'If his chief good and market of his time | Be but to sleep and feed? A beast, no more.'[31]

When listening to public debate about unemployment, it sometimes feels like we haven't shaken off the Victorian moralism of Herbert Spencer, the patron saint of grumpy old men in pubs:[32]

> 'They have no work,' you say. Say rather that they either refuse work or quickly turn themselves out of it. They are simply good-for-nothings, who in one way or other live on the good-for-somethings—vagrants and sots, criminals and those on the way to crime, youths who are burdens on hard-worked parents, men who appropriate the wages of their wives, fellows who share the gains of prostitutes; and then, less visible and less numerous, there is a corresponding class of women.

The duty to work, and work *hard*, is usually referred to as the *work ethic*, the name itself cunningly suggestive of the idea that work and ethical conduct are intertwined. The work ethic is one of the most pervasive and accepted doctrines of our time—so much so that it often goes without explaining or justifying. But now is as good a time as any to ask whether people's status should depend, as much as it does, on their economic contribution, rather than (say) their goodness, kindness, or public-spiritedness.

The work ethic is hard to reconcile with a world of technological unemployment. It's one thing to insist that people have a duty to work even when the work is hateful; but it's downright sadistic to say they have such a duty when there isn't even enough work to go around. We might also doubt whether the work ethic *could* survive in a world where the majority do not work. Would there still be a stigma to losing your job if everyone else was unemployed too?[33] The term 'unemployed' would lose some of its explanatory force, and the 'un' part would lose much of its stigma.

Technological unemployment prompts us to think about building an economy where status and esteem are associated with traits other than economic productivity. This might not be a bad thing.

3. Sever the Connection between Work and Wellbeing

The third and final intellectual step would be to challenge the connection between work and wellbeing. It's true that some jobs are fun, safe, creative, educative, uplifting, and meaningful, but many others are drudgerous, dangerous, repetitive, stultifying, depressing, and pointless. How effectively a job satisfies a particular person's needs will depend partly on their subjective preferences—one teacher may love her vocation; another may despise it—but some jobs are plainly worse than others.

It's not just that not all work enhances the human condition. On the contrary, work often *harms* the worker. Consider this account:

> The pace of work is unremitting. Workers are reprimanded for 'time theft' when they pause to catch their breath after an especially difficult job. They are subjected to ever-increasing quotas, threatened daily with discharge, and eventually fired when the required pace gets too high for them to meet—a fate of the vast majority... But not before they suffer injury on the job: workers have to get on hands and knees hundreds of times per day, a practice that leaves few unscathed. [The employer] forces them to sign papers affirming that their injuries are not work-related, or they are given demerits that can lead to discharge. [On one occasion the employer] allowed the indoor heat index to rise to 102 degrees. When employees asked to open the loading doors to

> let air circulate...[the employer] refused, claiming this would lead to employee theft. Instead, it parked ambulances outside, waiting for employees to collapse from heat stroke. When they did, they would be given demerits for missing work, and fired if they accumulated too many. [The employer] didn't care, because regional unemployment was high, and they had hundreds of applicants to replace the fallen workers...

Do these workers benefit physically or psychologically from their work? The extract reads like it's from the pages of Friedrich Engels' *Condition of the Working-Class in England* (1845), which exposed the squalorous horrors of working life under Victorian capitalism.[34] But it's actually from 2017, authored by the American scholar Elizabeth Anderson. The employer it describes is Amazon.[35]

A UBI, or something like it, would free humans from having to do awful work. It could fulfil the dream sketched by Oscar Wilde in 1891, of 'All unintellectual labor, all monotonous, dull labor, all labor that deals with dreadful things and involves unpleasant conditions' being 'done by machinery':[36]

> Machinery must work for us in the coal mines, and do all sanitary services, and be a stoker of steamers, and clean the streets, and run messages on wet days and do anything that is tedious or distressing.

So what would we do all day? Perhaps nothing. As Michael Walzer observes, for many people the opposite of work is leisure in the form of idleness. The Greek word for leisure, *scholé,* has the same etymological origins as the Hebrew word *Shabbat,* which means 'to cease' or 'to stop'.[37] But it's far from clear that a life of idleness would be good for humans, even if such a lifestyle were freed from the stigma currently attached to unemployment. 'If all the year were playing holidays,' cries Shakespeare's Prince Hal, 'To sport would be as tedious as to work.'[38]

The psychological needs said to be met by work—to feel a sense of achievement, to structure our day, and to interact meaningfully with other humans—probably won't disappear. On a libertarian approach, we might be left to decide for ourselves what to do with our days. On a paternalistic approach, the state could decide and specify mandatory non-economic activities, either as a prerequisite to receiving state funds or at pain of some other punishment.

It might turn out that what some of us need is in fact *work* or something like it! But when we surmise this, we aren't talking about horrendous work down a mine. We're talking about *good work* of the fun, safe, creative, educative, uplifting, and meaningful kind. Under the work paradigm, the joyous experience of good work is something a few are lucky enough to be *paid for*. In the digital lifeworld it may become something that people choose to *buy* with their allotted share of society's resources. As for any leftover *bad work* that still needs to be done by humans, citizens might have a duty to do it when called upon, like military conscription. Bad work could even be allotted on the basis of moral desert as a way of punishing criminals or wrongdoers.

Is work the only way to meet our need for a sense of routine or to connect meaningfully with people outside the family? In simple societies where natural conditions meant that few had to work for a living, these needs were met by 'rituals, religious and community practices'.[39] Today, when we're not working, we have hobbies, leisure pursuits, public and voluntary service, clubs and associations, and time with friends and family. Plenty of people happily abstain from paid work. As explained in Part III, in the digital lifeworld we might have much freer access to new and exciting experiences, whether in VR or otherwise. It's for political theorists, economists, social psychologists, and the like to identify our needs with precision and suggest what kind of activity, other than work, might satisfy them. That, rather than trying to salvage the work paradigm, could be a more productive use of their intellectual energy.[40]

The Next Chapter

The work paradigm insists on the necessity of work for income, status, and wellbeing. Yet who among us, waking at dawn and facing a long day of toil, has not quietly yearned for a life without labour? It might be time to challenge the idea that work is something we need. The analysis can't stop there, however. If it's right that the wealth that currently flows to labourers will be increasingly redirected toward those who own the technologies that replace them, then we need to examine that radical redistribution of wealth and consider whether it can be justified. That's the work of chapter eighteen.

EIGHTEEN

The Wealth Cyclone

'Ill fares the land, to hastening ills a prey,
Where wealth accumulates and men decay'
Oliver Goldsmith, *The Deserted Village* (1770)

It's fair to say that *property* lacks the pizazz of cooler political concepts like *liberty*, *equality*, and *fraternity*. Property is what tenants rent from landlords, not what revolutionaries cry as they leap over the barricades. Yet property has been one of the most bitterly contested concepts in political history. In the last century, leftists demanded that large swathes of the private sector be brought into public ownership. On the libertarian right, theorists like Robert Nozick argued that taxation was 'on a par with forced labor' because it forced individuals to work for the benefit of someone else.[1] The cleft between left and right concealed a deeper disagreement about the nature of property: what it truly meant to *own* something, and what could rightfully be done with things that were owned. This chapter describes a possible future in which vast amounts of wealth become concentrated in the hands of a tiny economic élite. It also proposes some ways that we might prevent this from happening.

From the earliest days of the digital revolution, people had a vague but palpable sense that the distribution of computing power would have significant consequences for social justice. As Tim Wu explains, the invention of the personal computer was an 'unimaginable' moment: 'a device that made ordinary individuals sovereign over information by means of computational powers they could tailor to their individual needs'.[2] Until then, computing power had been the preserve of large companies, governments, and university laboratories.[3] Steve Wozniak, who co-founded Apple with Steve Jobs, saw computers as 'a tool that would lead to social justice.'[4]

The idea of property will assume pre-eminent importance in the digital lifeworld, largely because (in economic terms) it will be more worthwhile to *own* things than to *do* things. Those on the wrong side of the ownership/labour divide could face real hardship. Property and social justice will be closely intertwined.

The analysis in this chapter is based on some fairly mainstream economic theory. I do not presume to develop or depart from that theory. On the contrary, I bring in only as much as is necessary to explain the problems that we're trying to untangle. Let's get started.

Capital in the Digital Lifeworld

In the digital lifeworld, those who *own* things will grow richer faster than those who *do* things.

There are two ways to earn a living. The first is through *labour*: wages, salaries, and bonuses received in exchange for productive work. The second is through *capital*: assets that produce wealth. Examples of capital include land that yields rents, shares that yield dividends, industrial machinery that yields profits, and intellectual property that yields royalties.[5] Since the early 1980s, the share of overall income yielded by capital has steadily grown in relation to the share earned by labour. In *Capital in the Twenty-First Century* (2013), Thomas Piketty predicts that the rate of return on capital will continue to outpace overall economic growth.[6] If that's right, then it means, on average, that those who own capital will enjoy higher returns than those who labour for a living. Inequality between labourers and capital-owners will grow over time.[7]

Looking to the digital lifeworld, it's useful to distinguish between a few types of capital: good old-fashioned capital, productive technologies, and data. Each of these will help to generate wealth in a different way.

Good Old-Fashioned Capital

Good old-fashioned capital—land, shares, industrial machinery, and so forth—will be an important source of income in the digital lifeworld. The value of a particular item of capital will always depend on how *productive* and how *scarce* it is. The more productive and scarce it is, the greater the wealth it is likely to generate.[8]

In *The Second Machine Age* (2014) Andrew McAfee and Erik Brynjolfsson suggest that in the future, production will depend less on physical assets and more on intangible ones like intellectual property, organizational capital (business processes, production techniques, and the like), and 'user-generated content' (YouTube videos, Facebook photos, and online ratings). They also emphasize the importance of so-called 'human capital'.[9] Elsewhere they suggest that 'ideas' will grow in economic importance, and the 'creators, innovators, and entrepreneurs', capable of generating 'new ideas and innovations' will reap 'huge rewards'.[10] I agree with McAfee and Brynjolfsson on the importance of intellectual property: patenting something lends it an artificial scarcity which, with a bit of luck, can send its value through the roof. That's why Microsoft took out more than 2,500 patents in 2010 compared with a few hundred eight years earlier.[11] It's also prudent to expect that organizational capital and brilliant ideas could make the difference between competing capital-owners. As for 'user-generated content' I prefer the broader category of *data* discussed below.

Depending on what you think of the technological unemployment thesis discussed in chapter seventeen, however, you may be sceptical of McAfee and Brynjolfsson's insistence on the importance of human capital (people's skills, knowledge, and experience). Yes, an educated and agile workforce will fare better in the initial stages of technological unemployment, when displaced workers are scrambling to secure new jobs. But if there aren't enough jobs for humans to do, *no matter how skilled or well-trained they are*, then the overall economic importance of human capital will decline even if a few superstar innovators still make a killing.

Productive Technologies

If the technological unemployment thesis is right, or even partially right, then the wealth that currently flows to labourers will increasingly be redirected toward the owners of the *labour-saving* technologies that replace them. For the same reason, it will pay to own *capital-augmenting* technologies that make good old-fashioned capital itself more productive. These *productive technologies* (both labour-saving and capital-augmenting) are likely to include intangible things like machine learning algorithms and software platforms, as well as hardware like robots, drones, sensors, 'smart' appliances, industrial machines, nanobots, 3D printers, and servers. Of course, no productive technology will guarantee a permanent economic advantage for its owner. The edge gained by automating a factory evaporates as soon as other factories do the same.[12] Patents can protect inventions for a time, but the pace of technological development means that even the most powerful systems are eventually superseded—particularly if it takes years to get a patent.

There will be a ferocious fight among capital-owners to develop and purchase the most profitable productive technologies.

Data

Data may become one of the most important forms of capital in the digital lifeworld.[13] As we know, it's been called a 'raw material of business', a 'factor of production', and 'the new coal'.[14] Why is it so precious? It's partly that it can be used for targeted advertising. It's also valuable in the work of various other industries, from agricultural technology to management consultancy. The supreme economic importance of data in the digital lifeworld, however, will derive from its role in building AI systems. Machine learning algorithms can't learn without access to plentiful data. You can't train an AI system to identify melanomas without hundreds of thousands of images of melanomas. You can't train an AI system to predict the outcome of legal cases without feeding it thousands of precedents (a learning process I also underwent as a junior lawyer). Data will be the economic lifeblood of the digital lifeworld. Whoever controls its flow will wield considerable economic clout.

The Key Distinction

Stepping back, the key economic distinction in the digital lifeworld will be between those who own capital and those who do not. Capital-owners will have the opportunity to accrue more and more wealth; those with only their labour to sell will find it increasingly hard to make ends meet. 'For whosoever hath,' says Matthew 13:12, 'to him shall be given, and he shall have more abundance: but whosoever hath not, from him shall be taken away even that he hath.'

Made With Concentrate

It's not just that wealth will increasingly flow to those who own capital and away from those who labour. The class of capital–owners itself could shrivel into a tiny élite. Even in our time, wealth is already gathering in the hands of ever-fewer firms, themselves employing ever-fewer people.

The general trend in American economic life is toward concentration in the hands of a few big players. In the last two decades, nearly three–quarters of industries have seen an increase in concentration.[15] There are four dominant airlines, four cable and internet providers, four main commercial banks. Perhaps most troubling of all, virtually all of America's toothpaste is made by just two manufacturers.[16]

Concentration in the tech industry is particularly striking.[17] The vast majority of mobile-based social media traffic runs through platforms owned by Facebook (including Instagram, WhatsApp, and Facebook Messenger).[18] Likewise a preponderance of search advertising revenue goes to Google's parent company, Alphabet.[19] Google's mobile software, Android, has more than three–quarters of the smartphone market.[20] Amazon accounts for nearly half of all online retail.[21]

A common strategy among these firms is to acquire huge cash reserves which allow them to extend their commercial advantage by buying and absorbing rival start-ups. In the decade to July 2017, Alphabet, Amazon, Apple, Facebook, and Microsoft together made 436 acquisitions worth $131 billion.[22] 'Not since the turn of the twentieth century,' writes Jonathan Taplin, 'when Theodore Roosevelt took on the monopolies of John D. Rockefeller and J. P. Morgan, has the country faced such a concentration of wealth and power.'[23]

Growing in wealth, big tech firms increasingly raid entirely new markets. Remember when Amazon was an online bookstore? In 2017 it acquired the organic food-retailer Wholefoods and its 400 physical grocery stores. Remember when Google was just a search engine? In the year to mid-2017, its parent company Alphabet acquired, among many other interests, Owlchemy Labs (a VR studio), Eyefluence (eye tracking and VR technology), Cronologics (a smartwatch startup), and Urban Engines (location-based analytics).

The general trend is toward digital services that try to serve as many of our needs as possible. In China, WeChat has evolved into a sort of Renaissance App, enabling its billion users to 'hail a taxi, order food delivery, buy movie tickets, play casual games, check in for a flight, send money to friends, access fitness tracker data, book a doctor appointment, get banking statements, pay the water bill, find geo-targeted coupons, recognize music, search for a book at the local library, meet strangers... follow celebrity news, read magazine articles, and even donate to charity', all on one platform.[24]

We see, even now, an unprecedented concentration of wealth in the hands of a few tech firms. And these firms employ surprisingly few people. As Klaus Schwab notes, in 1990 the three biggest companies in Detroit had a combined market capitalization of $36 billion and employed more than a million people. In 2014, the three biggest companies in Silicon Valley had a market capitalization of about $1.09 trillion, about thirty times higher, but employed only 137,000 people, about ten times fewer.[25] When Google acquired YouTube in Google in 2005, its valuation meant that each employee was worth more than $25 million. This was considered remarkable at the time. Then in 2012 Facebook acquired Instagram for $1 billion when it had just thirteen employees, giving it a valuation of around $77 million per employee. When Facebook acquired WhatsApp in 2014, for a whopping $19 billion, WhatsApp had just fifty-five employees, meaning a price of $345 million for every employee.[26]

This book is not about Amazon, Alphabet, Facebook, Microsoft, or Apple. I do not know if these firms will dominate the economy of the digital lifeworld as they do today's. But there are structural reasons why digital technology is likely to facilitate the concentration of more and more wealth in the hands of fewer and fewer people and corporations. Probably the most important, besides automation, is the *network effect*. The economy is increasingly interconnected in a web of overlapping networks, which have a number of important characteristics. Firstly, they are generally united by *standards*: shared rules or practices that set the rules of cooperation among members. Secondly, the more people adopt a given standard (by joining the network and abiding by its rules) the more valuable that standard becomes.[27] Metcalfe's Law holds that the value of a network increases exponentially with the number of nodes attached: doubling the number of nodes means quadrupling the value, and so on. This means ever-increasing pressure on non-members to join. Failure to be part of a popular social network can be seen as weird and eccentric. In business, failure to secure a sales platform on Amazon can mean devastation for a retailer. Finally, a networked economy rewards first movers. If you can get out ahead of the competition, every additional user/member will contribute to the acceleration of growth—and before long it'll be too late for others to catch up. An upstart rival to Facebook may offer superior functionality but it'll be worthless as a social network if it doesn't reach a critical mass of members.

The big tech firms mentioned above have all benefitted from the network effect. As soon as Microsoft Windows became the standard operating system for personal computers, it was always going to take decades for other firms to establish a rival standard.

Facebook sits at the apex of the social network it provides (often called a 'platform'), setting the standards in the code that constitutes the platform. It can determine, coordinate, and mediate the activity of the network members while hoovering up more and more precious data, which in turn increases in value the more that is gathered. The combined effect? It becomes almost impossible to challenge the behemoths. Competition is probably not, as is sometimes claimed, a click away—and even if it were, the likely outcome would be the replacement of one dominant platform with another.

Although networks benefit platform-owners and those who set network standards, that's not the only group they benefit. They also make it possible for others to *dominate* networks, at least temporarily, with the help of powerful digital technology. If people and technology are connected in a seamless web, then those with the best technology will always have an advantage. Take the example of financial trading, which now largely takes place online. The rise of automated and high-frequency trading has caused an explosion in financial activity—mostly to the disadvantage of human traders.[28] As Jaron Lanier explains:[29] 'if you have a more effective computer than anyone else in an open network [then] Your superior calculation ability allows you to choose the least risky option for yourself, leaving riskier options for everyone else.' Lanier's point may remind you of the discussion of bots and democracy in chapter thirteen. If deliberation takes place over an open network and one group brings a horde of powerful AI bots to argue their case, then they'll end up dominating the discussion. It's like bringing a gun to a knife-fight.

The network effect, and the ability to dominate a network using powerful digital technology, partly explain why tech and finance have grown more than any other sector since the 1980s, rising from about 10 per cent to 40 per cent of market capitalization.[30]

Whether you own the platform or dominate the network, the point is the same: in an increasingly networked economy, those with the most productive digital technologies will do better and better. And the possible gains are astronomical. Meanwhile, those without capital will see their economic position declining over time.

The Wealth Cyclone

What does the increasing importance of capital, together with a structural tendency toward concentration, mean for the distribution of wealth in the digital lifeworld? The worst-case scenario is a perfect storm of ever-growing inequality between a tiny rich élite and a poor majority:

> Over time, more wealth flows to capital-owners and less to workers. Using productive technologies, capital-owners control an increasing amount of economic activity while employing a shrinking number of workers. Capital comes to be concentrated in the hands of a small number of firms, themselves controlled by a small number of people. Relying on the network effect, these firms accrue huge cash reserves that enable them to acquire more capital and expand into new markets. With the data they gather, they develop AI systems of astonishing capacity and range. Large swathes of the population—former workers, failed capitalists—find themselves with no capital and no way to earn a living. Those with a stake in the successful firms enjoy a growing slice of an expanding economic pie. The wealth gap between the tiny élite of owners and the rest expands radically.

It's a Wealth Cyclone: a swirling vortex centred on an increasingly narrow point, growing stronger over time as it sucks in everything around it and destroys the stragglers in its path. The injustice in such an economic system is obvious, whether you care about equality of outcomes or merely equality of opportunity. While it's not unusual in historic terms for the gains from capital to exceed the gains from labour, the difference in the digital lifeworld will be that the traditional path from non-ownership to ownership, labour, may itself disappear over time. Abraham Lincoln described the capitalist system as he envisaged it in 1859: 'The prudent, penniless beginner in the world *labors for wages awhile, saves a surplus with which to buy tools or land for himself*... and at length hires another beginner to help him.'[31] What hope for the 'prudent, penniless beginner' in the digital lifeworld, who can't even sell his labour to begin with?

The Private Property Paradigm

To see how the Wealth Cyclone might be averted, we need first to look deeper at the logic that underpins our system of property.

Imagine a time before anyone could hold a piece of the world in their hands and say, 'this is mine'. The Stoics, in ancient Greece, believed that the world was originally owned by everyone.[32] John Locke, probably the most important philosopher of property, also believed that God gave the world 'to mankind in common' but that 'every man has a property in his own person'.[33]

At some point long ago, humans began to divide the earth and its contents among themselves. For Locke the idea of ownership was born when humans first applied their 'labour' to things in the world, 'removing them out of that common state'.[34] Once the idea of property took hold of the human imagination, it could not be forgotten. More and more of the world came to be owned by humans, not always by savoury means. For the young Jean-Jacques Rousseau, this marked humanity's downfall:[35]

> The first man who, having enclosed a plot of land, took it into his head to say *this is mine* and found people simple enough to believe him, was the real founder of civil society. What crimes, wars, murders, what miseries and horrors would the human race have been spared, had

someone pulled up the stakes or filled in the ditch and cried out to his
fellow men: 'Do not listen to this imposter! You are lost if you forget
that the fruits of the earth belong to all and the earth to no one!'

Some say that most of today's privately owned property can be traced
back to diabolical acts of plunder in the past. In the first volume
of *Capital* (1867) Karl Marx writes that, 'conquest, enslavement,
robbery, murder... play the great part'.[36]

The system of property we have today has four basic characteristics. First, to state the obvious, property tends to refer to *things*.
Real property is land and buildings. *Personal* property is moveable
things like cars, books, and jewellery. *Intangible* property covers things
that can be owned but which don't have physical form—shares,
interests, debts. *Intellectual* property is a type of intangible property.
It mainly refers to human creations protected by patents, copyrights,
and trademarks. The second characteristic of private property is
that it's easily *alienable*: it can be transferred through sale or gift
according to rules that are well known. Third, there are clear
rules about what you can and can't do with the things you own.
(By and large, full ownership of personal property means you can
do what you like with it, subject to the rights of others.) Finally,
people's property rights cannot lightly be violated. The Roman
statesman Cicero wrote that 'the first care' of the 'man in administrative office' was to ensure 'that everyone shall have what belongs
to him and that private citizens suffer no invasion of their property
rights by the state'.[37] Together, these four characteristics make up
what I call the Private Property Paradigm.

The Private Property Paradigm is just one way of structuring
a system of ownership. As I noted way back in chapter four,
property in ancient Babylon included people as well as things.
And in early Greek and Roman law, property was virtually inalienable as it 'belonged' not to individuals but to families, including
dead ancestors and unborn descendants.[38] J. K. Rowling's *Harry*

Potter novels contain the delightful idea of *goblin property*. As Aaron Perzanowski and Jason Schultz explain it, 'goblins are skilled metalsmiths. And they are deeply attached to the items they craft, regarding themselves as the true owners of those items, even after their sale.' They quote Rowling:[39]

> Goblin notions of ownership, payment, and repayment are not the same as human ones... To a goblin, the right full and true master of any object is the maker, not the purchaser. All goblin-made objects are, in goblin eyes, rightfully theirs... They consider our habit of keeping goblin-made objects, passing them from wizard to wizard without further payment, little more than theft.

Why, then, do we have the Private Property Paradigm rather than some other system? Several points can be made in its favour. First of all, it is said to encourage and foster prosperity. No one would work hard or take risks if they knew that their earnings could be snatched away at any moment. 'A person who can acquire no property', says Adam Smith in *Wealth of Nations* (1776), 'can have no other interest but to eat as much and to labour as little as possible.'[40] It's also said that the right to private property is needed to prevent interference from the state in one's private affairs. 'We are rarely in a position', writes Hayek in *The Constitution of Liberty* (1960), 'to carry out a coherent plan of action unless we are certain of our exclusive control of some material objects... The recognition of property is clearly the first step in the delimitation of the private sphere which protects us against coercion.'[41] Put more positively, private property is necessary for human flourishing. Owning property allows us to exercise our will in the real world of things. This argument is usually attributed to G. W. F. Hegel, who, in the German tradition, saw fit to explain it as impenetrably as possible: 'The person must give himself an external sphere of freedom in order to have being as Idea. The person is the infinite will, the will which has being in and for itself, in this first and as yet wholly abstract determination. Consequently, this sphere distinct from the will, which may constitute the sphere of its freedom, is likewise determined as immediately different and separable from it.'[42] (No, me neither.)

There is a high degree of consensus over the value of the Private Property Paradigm, and it seems to be hardening over time: the eye-watering top income rates of tax in some developed countries in the 1980s would be politically radioactive today. It's also reflected in law throughout the developed world. The first protocol to the European Convention on Human Rights, for instance, provides that no one 'shall be deprived of his possessions except in the public interest and subject to the conditions provided for by law and by the general principles of international law'.

Most of us have made our peace with the idea that the state will take some of our property in the form of tax, although we hope that the money is spent efficiently. In times of war or emergency, we may accept that the state has the right to requisition private land or vehicles. But otherwise, we know that what's ours is ours.

Our commitment to the Private Property Paradigm comes with certain consequences for distributive justice, however. Although it allows for economic growth and prosperity, it also makes the redistribution of wealth more difficult. This is because it reduces the acceptable scope and extent of state interference in the market. 'Wherever there is a great property,' wrote Adam Smith, 'there is great inequality. For one very rich man, there must be at least five hundred poor.'[43] Karl Marx and Friedrich Engels, who were open to the idea of abolishing private property itself, wrote in the *Communist Manifesto* (1848):

> You are horrified at our intending to do away with private property. But in your existing society, private property is already done away with

for nine-tenths of the population; its existence for the few is solely due to its non-existence in the hands of those nine-tenths.[44]

What would they have made of the Wealth Cyclone?

The Future of Property

Thomas Jefferson once observed that 'Stable ownership is the gift of social law, and it is given late in the progress of society.'[45] He believed that in a self-governing society, the people must decide which system of property rights is most favourable to their purposes. That's possibly why he left property off the list of inherent and inalienable rights that otherwise included life, liberty, and the pursuit of happiness.[46]

I submit that a Wealth Cyclone generating massive and systemic inequality would not be consistent with the demands of social justice. We may need to devise a new way of organizing property rights. The challenge is to find a system that preserves the economic wonders unleashed by technology while allowing more people to share in them. In the next section, I sketch out six possible alterations to the Private Property Paradigm. These are not the only options but they may provide a helpful starting point.

The New Property Paradigm

Tax on Capital

One way to counteract the Wealth Cyclone would be to levy a tax on capital or the profits earned by it. In *Capital in the Twenty-First Century*, Piketty argues for a 'progressive global tax on capital' as the 'ideal' way to avoid 'an endless inegalitarian spiral' and regain control over 'the dynamics of accumulation'.[47] A Robot Tax of the kind proposed by Bill Gates could be targeted at productive technologies.[48] There may even be ways of taxing the usage or flow of data. Whichever the chosen model, the principle is that some of the wealth generated by the ownership of capital should be skimmed off and spent for the benefit of those who have no capital to their name. This public spending could take the form of a universal basic income (UBI) of the kind described in chapter seventeen.

Conceptually, the idea of taxing capital is not a radical departure from the Private Property Paradigm. Taxes take a share of the wealth generated by capital rather than the capital itself. Many forms of capital are already taxed in this way. In the digital lifeworld, however, there would have to be a difference in scale, mainly because the takings would have to pay for a lot more than they do currently. Today, prosperous tech firms frequently pay *less* tax than other companies.[49] In 2016, for instance, the European Union estimated that Apple Ireland paid an effective tax rate of less than 0.01 per cent on profits of more than $100 billion.[50] This kind of practice will become increasingly unsustainable.

It's important to recognize, however, that even a UBI paid for by a tax on capital could still leave colossal inequalities between those who own capital and those who do not. For the sufficientarian, who cares only that everyone has *enough*, this might be satisfactory. But for those who care about reducing inequality, it would have to be a very high tax indeed. Some say that the 'B' in UBI should be removed and taxes on capital and productive digital technologies should be used pay for a *high* quality of life for everyone. This is the idea of 'fully automated luxury communism'.[51] Somewhere between these positions, taking into account the importance of financial incentives to capital-owners, it might be said that any tax on capital should be set at such a rate that it guarantees the

largest return that can be spent for the benefit of the least well-off.
A 95 per cent tax that stifled production would not fit the bill.

State Ownership of Capital

A second approach, departing sharply from the Private Property
Paradigm, would be to bring certain capital assets under the direct
ownership of the state, perhaps through some form of compulsory
purchase. What could be said in favour of such a drastic step? To
start with, state ownership need not encompass all forms of capital,
as in a true 'command economy', but only the most important
ones. Imagine, for instance, a postage system that delivers goods
through a network of automated drones. Or a fleet of land, water,
and air-based automated vehicles that serve as a system of public
transportation. You wouldn't necessarily want these 'utilities' to be
owned and operated privately. What if they refused to serve rural
areas? What if they excluded the poor by setting their prices too
high? Some forms of capital, like these, will be essential for the
common wealth of the digital lifeworld. Rich and poor alike will
rely on them for the economy to function. That makes them
decent candidates for nationalization, so that their operations can
be directed for the benefit of all. One can also imagine discrete
items of productive digital technology, like industrial-scale 3D
printers, being held in public ownership.

The potential problems with state ownership, however, are legion.
We know that public administration can be incompetent, inefficient,
corrupt, and unaccountable (although so too can private ownership).
Entrepreneurs may not bother to innovate in the private sector
if they think that the fruits of their hard work will be confiscated
by the state. The twentieth century taught us that sweeping state
ownership can lapse into authoritarian dictatorship. Too much concentration of wealth and power in the hands of any body—private

or public—is dangerous. That's why some have instead argued for a cooperative model, whereby consumers or workers collectively own or govern the digital systems that generate wealth.[52]

Rights of Use and Profit

A less intrusive approach than outright nationalization might be to give individuals rights of use and profit through a system of *usufructuary* rights. A *usufruct*, which sounds like the kind of candy your grandfather gives you, is a lesser kind of property right than full ownership. Instead of giving you the full title to a thing (enabling you to sell it or exclude others from using it), it gives you the right to *use* and *profit* from it along with others who share the same right. Early property rights often took this form. They gave people the legal right to use land but not to 'own' it for themselves.[53] Usufructuary rights are powerful because they attach to the capital itself. The right survives even if the capital changes hands.

Imagine the owner of a powerful facility that provides, via the cloud, important computing resources like processing power, data storage, and AI software. That firm will probably do very well for itself in the digital lifeworld, possessing as it does a powerful resource of productive technology. A system of usufructuary rights would permit non-owners—small businesses, communities—to use that technology, some of the time, for their own entrepreneurial purposes. The owner would retain legal title and most of the profits, while the community would gain valuable rights of access. There could, of course, be rules about the extent and purposes of third-party use. Perhaps use would only be permitted outside normal working hours and not in competition with the capital-owner. Such rules could be enforced using blockchain technology.

A system of usufructuary rights would work well with wealth-generating platforms like Improbable, founded in 2012, which allow users to simulate complex scenarios in vast virtual worlds. Improbable has already been used to simulate the outbreak of an infectious disease, the effects of a hurricane on a stretch of inhabited coastline, and—most volatile of all—the fluctuations of the UK housing market.[54] It's an invaluable economic asset. Should access to it be limited to big firms and national governments? A system of usufructuary rights would be less burdensome to capital-owners than full nationalization while preserving many of the advantages associated with private ownership. At the same time, it would help to neutralize the Wealth Cyclone by allowing the asset-poor to enjoy some of the most important benefits of ownership without actually having to buy anything (or indeed having 'to clean, to repair, to store, to sort, to insure, to upgrade, to maintain').[55]

Commons

Another way to hold capital is in a *commons*—as a shared pool of resources owned by no one and subject to no property rights (or very restricted ones). As James Boyle explains in his masterful book *The Public Domain* (2008), some assets are already in the commons because they are not capable of being owned. The English language is an example. Other assets, like the works of Shakespeare, are in the commons because the property rights attached to them have expired.[56] In the digital lifeworld we might create a much larger and richer commons of shared assets.

The most obvious candidates for commons-based ownership are intangibles (information goods) like ideas, inventions, designs, blueprints, books, articles, music, designs, films, images, and software. They're well-suited because they don't get *used up* like other types of capital: one person's use of an information good doesn't stop another person from using it again (or even simultaneously). As the other Susskinds argue, it may eventually be possible to digitize and automate the work of professionals like doctors and lawyers, meaning that even 'practical expertise' could be held in a commons.[57] In the field of AI and machine learning, Google's TensorFlow already offers an open source library of computational resources. IBM, Microsoft, and Amazon have also made some of their machine learning systems available via the cloud. These are positive developments.

Commons-based ownership of capital is not a new idea, however, and it's controversial. Rival groups disagree over the role of intellectual property rights—mainly patents and copyrights—that place artificial legal restrictions on the copy or use of human creations. One group, call them the *propertarians*, advocate for strong property rights. They say that if people are allowed to reproduce expensive goods at low cost there will be no incentive to produce them in the first place. Why write a great textbook, they ask, if everyone will download it for free while you receive nothing? Why dissipate precious capital if there's no prospect of earning a return? The effect of the internet, they argue, is to destroy the market for goods that are cheap to copy. Propertarians disdain the idea that resources could be left to the commons. With sombre faces they tell of the 'tragedy of the commons', when land was held collectively and nobody bothered to put in place the flood protections, drainage systems, or systems of crop rotation needed to preserve its value. Everyone hoped that someone else would do it and no one did.[58] The only solution, they say, is for the state to step in with the 'market-making' device of intellectual property law, i.e. rights excluding others from using the goods unless they pay.[59] Intellectual property law encloses information goods into discrete items of property, such that each individual patent- or copyright-owner is incentivized to do the best they can with what they have.

On the other side of the debate are a group I call the *commonsists*, a vanguard of scholar-activists armed with handworn copies of the

US constitution and a healthy dose of sass. They argue forcefully against the strengthening of intellectual property rights and what Boyle calls the 'second enclosure movement':[60]

> What if we had locked up most of twentieth-century culture without getting a net benefit in return? What if the basic building blocks of new scientific fields were being patented long before anything concrete or useful could be built from them? What if we were littering our electronic communication space with digital barbed wire and regulating the tiniest fragments of music as if they were stock certificates?

The truth, as Boyle explains, is that we are doing all of these things. Digital Rights Management ('DRM') technology is now commonplace and trying to circumvent it is a crime.[61] Despite the early promise of the internet, completely free access to things like books, music, and films is still quite rare.

The commonsist position is that intellectual property rights may be necessary but the 'goal of the system' should be 'to give the monopoly only for as long as necessary to provide an incentive'.[62] Is it really necessary, they ask, for copyright to last for seventy years after the death of the author? 'I am a great admirer of Ms. Rowling's work,' says Boyle, 'but my guess is that little extra incentive was provided by the thought that her copyright will endure seventy rather than merely fifty years after her death.' [63] Commonsists insist that creators and producers are motivated by a variety of incentives as well as money: fame, altruism, and creative and communal instincts.[64] Hence the success of open-source projects like Wikipedia and Linux.

At the heart of the commonsist argument is the belief that instead of fostering innovation and progress, excessive property rights blunt and stifle our creative powers. Locking inventions and creations up in patents and copyrights means that the next generation of producers must pay to build on them, which in practice may prevent them from doing so at all.[65] A commons of shared cultural resources, by contrast, would allow for creative adaptation, editing, remixing, parody, co-option, correction, criticism, commentary, and customization.[66]

Debate over the merits of the commons will continue to rage as long as intangible assets like ideas, inventions, designs, and software grow in economic importance. Recall McAfee and Brynjolffson's prediction that those capable of generating 'new ideas and innovations' will reap 'huge rewards'.[67] But justice requires us to ask: what about those who are not so capable? Or who never had the opportunity to make their ideas known? As with all types of property, intellectual property rights primarily benefit those who already own information, knowledge, and ideas. A commons-based system might even unleash a great hunger to learn: more people signed up for Harvard's free online courses in a single year than have attended the actual university in the nearly 400 years since it was founded.[68]

It might be objected that creators *deserve* great riches for their genius even if that leads to inequality. Why should my invention be owned by anyone other than me? It's not an unreasonable objection, but it'll become harder to sustain in the digital lifeworld, where access to ideas, knowledge, and information could mean the difference between stratospheric wealth and destitution. Surely, at the very least, we should make sure that property rights go no further than necessary to foster innovation.

'By and large,' argue Susskind squared, 'it would be better to live in a society in which most medical help, spiritual guidance, legal advice, the latest news, business assistance, accounting insight, and architectural know-how is widely available, at low or no cost.'[69] The same is true of great ideas and inventions. The question is where the balance should be struck between private property and the commons. That's a question of justice as well as economics. It's certainly not just a matter for tech firms and their lawyers.

Sharing

The 'sharing economy' offers another possible model of ownership for the digital lifeworld. These days, the term is used loosely to describe most forms of 'peer-to-peer' online transactions conducted online. The best example is Airbnb, which allows people to rent out their vacant residential properties to strangers. Sharing, with or without strings attached, is not new. What's new is the scale and extent enabled by digital technology.[70] At a glance, the ethos of the sharing economy resembles that of the commons. But there are a couple of key differences. First, in a commons no one strictly owns any of the stuff held in common, while in a sharing economy individuals retain the title of their property while letting others use it. Second, in the sharing economy, people tend to pay owners for their goods and services, while this would be antithetical to the ethos of the commons.

The sharing model is interesting because it encourages sellers to monetize assets that would otherwise be economically useless. And it offers purchasers the benefits of possession without the associated burdens of ownership. Thinking about the digital lifeworld, one might imagine a system that enabled people to share day-to-day items using a fleet of airborne drones that shuttle goods from one place to another:[71]

> [Let's say] you could just snap your fingers and have something magically appear in your hand whenever you wanted it [at] no cost, and it was instantaneous. You have a hammer in your home. You probably have a power drill. You use it one-10,000th of the time, maybe one-100,000th of the time . . . it could be shared by thousands of people, really safely, making everybody wealthier functionally because they would get the hammer when they need it without having to pay for the hammer and drain the world's resources by making all of these hammers that go almost entirely unused.

How does the sharing economy fit with our notions of social justice? Because it's essentially just another type of market, it will always favour those with something to 'share' and money to spend. If, theoretically, the whole economy were a sharing economy then the poor would do pretty badly. They'd be constantly 'asset stripping' their own homes in order to find things to rent out.[72] This would be a particular problem in the digital lifeworld, where we already expect an imbalance between asset-haves and asset-have-nots. Moreover, it increasingly looks like the people who benefit most from the sharing economy are not the participants but the platform-owners themselves: Uber, Airbnb, and the like. As Jonathan Allen explains, the early wealth effects of the sharing economy are hard to calculate but any wealth created 'will be highly concentrated in the hands of technology founders and early investors'.[73]

At its heart, the sharing economy doesn't really disrupt the Private Property Paradigm. The critical issue in the long-run, I suggest, will be *what* is shared. Sharing items of comfort, convenience, and amusement may help the asset-poor to make the most of what little they have, but it won't make them any wealthier over time. If, by contrast, what's shared is itself capital or productive technology then the sharing economy might just help to counteract the Wealth Cyclone.

The Data Deal

There are many things we could do with data. Data use could be subject to a tax. It could be held in a shared commons, free for general use or subject to constraints. It could be purchased and held by the state. It could be rented, 'shared', or loaned for charitable purposes. Evgeny Morozov suggests that it could accrue to a 'national data fund, co-owned by all citizens' with commercial access subject to heavy competition and regulation.[74]

Personal data—data about people—presents a special case because it seems to have its own logic of accumulation. I call it the *Data Deal*. The main method used by firms to gather personal data is to provide free or heavily discounted services, often personalized to the needs of each consumer, such as web search, access to platforms and networks, messaging and email, maps and guidance, videos and images, software and AI tools, and cloud storage. In exchange for these services, individuals surrender their data to the tech firms (knowingly or not).

From an economic perspective the Data Deal has the merit of efficiency. Our individual data is worth almost nothing, but by gathering together the data of millions, tech firms can create products, services, and platforms of value. From a political perspective, however, the bargain is a little more questionable. We know that increased *scrutability* is one of the consequences of a highly quantified society. Those with the data have a good deal of power (see chapter seven). There are also obvious implications for privacy and dignity. For now, let's focus on the Data Deal from the perspective of distributive justice.

At first glance, there's nothing overtly unjust about firms taking something for which we have little economic use (our data) and turning it into valuable social goods (for us) and profits (for them). We all benefit from free stuff, the poor arguably more than the rich. Plus, most of us don't see it as a sacrifice to get rewarded for data that would otherwise lie unused.

Nevertheless, the Data Deal is open to criticism. First, there is the *inequality objection*. Whatever its other merits, it leads to extraordinary disparities in wealth. It funnels vast riches toward tech firms while ordinary people get no financial compensation at all. Even a UBI skimmed from the profits made from personal data would barely scratch the surface of this inequality. The Data Deal, in short, contributes to the Wealth Cyclone.

Next there is the *exploitation objection*. On this view, the Data Deal is tantamount to exploitation because it involves the unjust extraction of economic value from 'uncompensated or low-wage labor'.[75] Just as the capitalist system is sometimes accused of paying workers less than the full value of their labour, the Data Deal is said to compensate people less than the full value of their data. On this view, Facebook's users 'have become *the largest unpaid workforce in history*'.[76] Tech guru Jaron Lanier argues that people should be compensated monetarily for the data they provide.[77] He proposes a system of 'micropayments' that allows people to earn 'royalties' on the 'tens of thousands of little contributions' made by their data over the course of their lives:'If observation of you yields data that makes it easier for a robot to seem like a natural conversationalist, or for a political campaign to target voters with its message, then you ought to be owed money for the use of that valuable data.'[78]

The exploitation objection is seductive but problematic. It assumes that firms would still be motivated to provide free services even if they had to pay for our data. It also glosses over the value of the social goods we *do* receive. It's hardly exploitation if we get something that we value in exchange for our data—and it's wrong to assume that hard cash is the only thing we value. Social networking platforms, global maps, search engines: these are not trivial things. They improve our lives. Andreas Weigend calculates that if Facebook had paid no dividends to shareholders and instead shared its 2015 profits between all its users, each user would have got about £3.50 for the year. As he puts it: 'Is having unlimited access to a communication platform for a year worth more to you than a cappuccino?'[79] As time goes on, the more data that's taken from us, the more we'll rightly expect in return. Moreover, some people's data is always likely to be more valuable than others', perhaps because they are from a group of particular interest to advertisers. Under Lanier's system, these lucky folk would enjoy greater data windfalls than others. Is that just? Why should some receive more for their data than others? Despite these difficulties, however, the exploitation objection is important, particularly when levelled at systems that provide no direct value to those whose data they extract. Where's the bargain in that?

A third objection to the Data Deal holds that it's unjust for other people to make economic use of our data without our informed consent. Call this the *knowledge/consent objection*. It's wrong for a burglar to steal your TV even if he leaves a wad of cash equal to the value of what was taken. By the same logic, it's wrong for a digital system to take people's data without properly informing them about it. 'Aha!' a lawyer might respond, 'but you clicked "accept" on the Terms and Conditions when you signed up to the service! *Didn't you?*' As we'll see in chapter nineteen, it's dubious whether signing a long and legalistic document ten years ago is an adequate way to ensure ongoing knowledge and consent on the part of ordinary people. Tech firms should inform people more clearly, concisely, and regularly which data is being extracted and for what purposes it's going to be used. This is partly a matter of ethics: if you knew that your data was going to be used for nasty purposes then you might conscientiously refrain from trading it in. But it's also a matter of economics: the more we understand the true *value* of our data, the easier it will be to assess whether the bargain in question is a fair one.

There's a final and more profound *structural objection* to be levelled at the Data Deal. As time goes on, it seems, society will increasingly divide itself into two classes: those who are able to process data usefully and amass wealth as a result, and those who are only able to sell their data in exchange. As well as leading to inequality, this will contribute to a serious imbalance in economic power. The direction of economic life will increasingly be shaped by a small group of entrepreneurs and industrialists who decide what data is needed and which services should be offered in return. Must the majority sell its personal data and take what it's given without complaint? 'Data unions' are one possible response to this structural imbalance, according to Pedro Domingos:[80]

> The twentieth century needed labor unions to balance the power of workers and bosses. The twenty-first needs data unions for a similar reason. Corporations have a vastly greater ability to gather and use data than individuals. This leads to an asymmetry in power... A data union lets its members bargain on equal terms with companies about the use of their data.

Another response might be to make sure that the technologies used to *process* data (both hardware and software) do not become the sole preserve of a tiny group of owners.

Future Justice

We've seen that, in the future, questions of distribution and recognition will increasingly be decided by code. Those who write the algorithms will wield awesome economic power. The line between software engineers and social engineers will grow blurry. At the same time, wealth will increasingly flow to those who own capital, particularly productive technologies, and data. Those who own productive digital technologies will be the prime economic beneficiaries of the transition into the digital lifeworld. These changes cannot go unnoticed by political theorists. The old ways of thinking will no longer do.

The economic structure of the digital lifeworld is significant for reasons other than social justice. The means of force, scrutiny, and perception give the tech firms that own them a degree of power; but that power is magnified many times over for firms that become the single or dominant provider of a particular digital system.

Recall from chapter eleven the idea of Digital Confederalism: that the best way to preserve liberty is to ensure that people can move between systems according to whose code they prefer. Digital Confederalism requires that for any important liberty—communication, news-gathering, search, transport—there must be a variety of available digital systems through which to exercise that liberty, and it must be possible to move between these systems without adverse consequences. As we see in chapter nineteen, a world of big tech monopolies could make this impossible.

PART VI

Future Politics

'Come, my friends,
'Tis not too late to seek a newer world.'

Alfred, Lord Tennyson, *Ulysses* (1842)

NINETEEN

Transparency and the New Separation of Powers

'Thinking too has a time for ploughing and a time for gathering the harvest.'

Ludwig Wittgenstein, *Culture and Value* (1970)

It has been a long journey through new and sometimes strange conceptual terrain. Let's take a moment to reflect on what we've seen. If you've leapt to this chapter from the beginning, be warned: it draws on concepts and arguments from throughout the book.

We stand at the edge of the digital lifeworld, a world populated by digital systems that will come to rival and surpass humans across a wide range of functions. Over time they'll grow more integrated, permeating the world around us in structures and objects that we never previously regarded as technology. Human life will be increasingly quantified, with ever more of our actions, utterances, movements, relationships, and emotions captured and recorded as data, then sorted, stored, and processed by digital systems.

Although we will enjoy dazzling new opportunities in the digital lifeworld, certain technologies will emerge as formidable instruments of power. Some will be able to force us to do things. Others will achieve the same result by scrutinizing us. Still others will exert power by controlling our perception of the world. The code animating those technologies will be highly adaptable and sophisticated, capable of shaping our lives in a dynamic and granular way. Those who control these technologies will be powerful in a broad sense, meaning they'll possess a stable and wide-ranging capacity to get others to do things of significance that they would not otherwise do, or not to do things they might otherwise have done. Any entity (public or private) that aspires to power will seek to control these technologies.

The most immediate political beneficiaries will be the state and tech firms, who will vie for control of the new technologies of power. The state will gain a supercharged ability to enforce the law, and certain powerful tech firms will be able to define the limits of our liberty, determine the health of our democracy, and decide vital questions of social justice. The economic structure of the digital lifeworld, left to itself, could cause wealth to become concentrated in the hands of a few mighty entities.

Almost a century ago, Max Weber wrote in *The Profession and Vocation of Politics* (1919) that the task of politics is to 'look at the realities of life with an unsparing gaze, to bear these realities and be a match for them inwardly'.[1] In this penultimate chapter I offer some unsparing thoughts on what may lie ahead.

Individual Responsibility

At its heart, this is not a book about technology, or even about political theory. It is a book about people.

Many of the problems referred to in these pages, past and future, can be attributed to the choices of individuals. The Airbnb host who refuses to host black guests; the troll who posts defamatory statements on social media; the anti-Semite who 'games' Google so that articles denying the Holocaust get more prominence;[2] the joker who teaches Microsoft's Tay to say 'Fuck my robot pussy daddy'; the hacker who holds people's medical data for ransom; the engineer who fails to train his systems to recognize women— these aren't problems with technology. They're problems with us.

The digital lifeworld will demand more of us all. From the shiniest CEO to the most junior programmer, those who work in tech will assume a role of particular influence. 'In the long history of the world,' said John F. Kennedy in 1961, 'only a few generations have been granted the role of defending freedom in its hour of maximum danger.' The future of politics will depend, in large part, on how the current generation of technologists approaches its work. That's their burden whether they like it or not. Indeed, whether they know it or not. Plato wrote in *The Republic* that 'there will be no end to the troubles of states . . . till philosophers become kings in this world, or till those we now call kings and rulers really and truly become philosophers.'[3] Now it is technologists who must become philosophers, if we are to preserve our liberty, enhance democracy, and keep the arc of history bent toward justice. We need *philosophical engineers* worthy of the name.

Those of us who don't work in tech can perhaps be guided by the principle of Digital Republicanism, which holds that a freedom that depends on the restraint of the powerful is no kind of freedom at all. We must be able to understand and shape the powers that govern our lives. It falls to us to design, theorize, and critique the world that's emerging. Vigilance, prudence, curiosity, persistence, assertiveness, and public-spiritedness: without them, the digital lifeworld will slip into darkness.

The Power of the Supercharged State

We have to develop systems and structures that encourage our best instincts while constraining our worst ones. In doing so, we'll face a number of systemic challenges. The first is what to do about the power of the supercharged state. In chapter ten I asked rhetorically whether we might be devising systems of power that are too complete and too effective for the flawed and damaged human beings they govern. In truth, I believe we are. The impending threats to liberty are unprecedented. We have to get working on the 'wise restraints' that could mitigate the oppressive consequences. In chapter eleven I sketched out a range of alternative perspectives on this issue, which I won't rehearse here, but it suffices to say that this is an area that *aches* for further theory, not just from political theorists but also lawyers, sociologists, psychologists, criminologists, technologists, and more.

Against the grain of recent history, I also suggest that democracy will be even more important in the future than it was in the past. Growth in the state's power will demand a corresponding growth in the people's capacity to hold it to account. Fortunately, as we've seen, new and interesting forms of democracy are emerging just at the time that they are needed most. '[H]istory', wrote Rosa Luxemburg, 'has the fine habit of always producing along with any real social need the means to its satisfaction, along with the task simultaneously the solution.'[4] The social need is clear: to protect people from servitude. The task is also clear: to keep the power of the supercharged state in check. The solution, I hope, will be a new and more robust form of democracy: not the tired competitive elitist model, but one that combines the most promising elements of Deliberative Democracy, Direct Democracy, Wiki Democracy, Data Democracy, and AI Democracy.

The Power of Tech Firms

The second systemic challenge lies with tech firms. As we've seen, the issue isn't (just) one of big companies getting richer and the rest of us getting poorer. It's about the power they'll wield through the technologies they control.

We don't typically let others dominate us without good reason, or at least without our permission. If tech firms are to acquire such power, then that power ought to be legitimate. For some this will sound odd. If an economic entity creates a product that consumers engage with, then why shouldn't it enjoy the power that comes with its success? This type of thinking is sensible up to a point, but ultimately it confuses the logic of economics with the logic of politics. In the marketplace, investment, risk-taking, and hard work will often lead to the legitimate acquisition of wealth. But to say that the legitimacy of a tech firm's political *power* should be judged by market standards because it originated in the market is like saying that the legitimacy of a military junta should be judged by military standards because it originated in a coup d'état. On any faintly liberal or democratic view, what matters for the purpose of legitimacy is the perspective of the people subject to the power in question. The political realm is not a marketplace. The principles that justify political power are different from those that justify commercial success. Riches, fame, and even celebrity can legitimately be won in the marketplace. But not great power. The acquisition of power must be justified by reference to political principles, not economic ones.

That being said, some of the principles historically used by the strong to justify their power have been unconvincing. *Divine Rule:* the king was appointed by God; we must obey him. *Original Sin:* our despicable urges must be restrained. *The Great Chain of Being:* we must all submit to our place in the natural hierarchy. *Tradition*: it's always been this way. *Patriarchy:* as the father justly rules his family, so too does the prince rule his subjects. *Might is right:* in politics there's no right and wrong, only strong and weak. *Pragmatism*: it's better not to rock the boat. *Charisma:* he's extraordinary, let's follow him! *Apathy*: no one really cares anyway.

These are all justifications, or excuses, made in the name of power. You may doubt whether they are good ones.

The citizen in a modern democracy has a solid idea of what makes the power of the state legitimate. It's that with the consent of the governed, the state guarantees a certain amount of liberty and enacts laws that are consistent (as far as possible) with the shared beliefs and interests of the governed. Call this *the liberal-democratic principle of legitimacy*. Because the power of tech firms will differ in scope and extent from that enjoyed by the state, the liberal-democratic principle may be too high a standard by which to judge the legitimacy of their power. You don't need to hold elections for the Board of Facebook to make its power legitimate.

Some will say that we should simply nationalize these new tech behemoths and be done with it. This would be a mistake, and not just because the market is a vital engine of innovation. The lesson of liberalism is that the state is itself a massive concentration of power that must be kept in check. We know that even democratic states can abuse their power to the detriment of citizens. Already our governments are moving to control the technologies of power without the hassle of owning them, by co-opting the help of tech firms, legislating to bring them under their control, and hacking their systems when other methods fail. Every increase in the state's power must be justifiable, and it's naïve to suppose that handing the technologies of power to the state wouldn't generate its own threats to liberty, democracy, and social justice.

A more sensible approach would be to regulate private tech firms in a way that gives their power an authentic stamp of legitimacy.[5] We already regulate private companies in various ways: consumer protection rules, employee rights, food standards, and so forth. The state aggressively regulates utilities to ensure that their power is exercised responsibly. Tech firms too are subject to regulation, but a growing number of scholars and commentators argue that we need to go further. Think of Google. As we've seen, its search results can be unjust. If you search for an African-American sounding name, you're more likely to see an advertisement for websites offering criminal background checks than if you enter a non-African-American sounding name. If you type 'Why do gay guys...', Google offers the completed question, 'Why do gay guys have weird voices?' (See chapter sixteen.) Legal scholar Frank Pasquale proposes that Google should face regulation to counter this sort of injustice. Among other measures, he would force Google to 'label, monitor, and explain hate-driven search results'; allow for 'limited outside annotations' on defamatory posts; 'hire more humans to judge complaints', and possibly even 'ban certain content'.[6] Elsewhere, Pasquale and Danielle Keats Citron propose a system of procedural safeguards, overseen by regulators, to allow people to challenge the decisions of algorithms that have made important decisions about their lives (like their credit scores).[7] This is sensible stuff. But the very idea of regulation will make some people queasy, precisely because it means handing even more power to the state. That queasiness will have to be suppressed, at least in part. Some of the power accruing to tech firms—the power, for instance, to control our perception of the world—is so extraordinary that it rivals or exceeds the power of any corporate entity of the past.

This is a book about principles and ideas. It isn't intended to offer specific regulatory proposals. But political theorists should be able to say, in broad terms, what would be needed to make tech firms legitimate in their exercise of power. I offer three suggestions.

The Consent Principle

The most obvious source of legitimacy is *consent*: the power of tech firms can be justified by the fact that people consent to what they do. The idea of consent is useful in situations where power is exerted in a small and ad hoc way.[8] For instance, if you type a single search query into Google, you tacitly accept Google's right, on that occasion, to filter the world for you as it sees fit. The same goes for *inferred consent*, where a system determines what a person 'would have' consented to in a particular instance.[9] To adapt Gerald Dworkin's example, if a robotic surgeon is repairing a broken bone and discovers a cancerous tumour, it may legitimately infer that the patient *would have consented* to it removing the tumour immediately.[10]

The use of consent to govern longer-term power relationships with tech firms, however, is more problematic. It's sometimes said in defence of big tech firms (usually by their lawyers) that when a person expressly agrees to a set of terms, he or she gives the tech firm permission to do whatever those terms provide.[11] But who in their right mind actually reads the terms and conditions before they use an app or a device? Scholars estimate it would take seventy-six working days to read all the privacy policies that we encounter in a year.[12] The timeworn practice of thrusting a huge legal document under someone's nose and saying 'sign here' is from another age. And even if we *could* and *did* read the terms and conditions, ticking a box once a decade is not a satisfactory means of surrendering important rights and freedoms.

Another difficulty with the consent principle is that often we won't have a choice whether or not to engage with a particular technology. This might be because of *ubiquity* (like public surveillance systems) or equally *necessity* (in a cashless economy, we'll have no choice but to use the standard payment technology). To confer legitimacy, consent must be free and informed. Necessary consent is not really consent at all. Finally, to return to the Google example, the consent principle doesn't protect us at a *systemic* level from abuses, even if it can justify an ad hoc exercise of power over a single search. Consent alone—at least consent of the kind given in the market economy—may not be enough.

The Fairness Principle

An alternative source of legitimacy comes from the principle of fairness: if you accept benefits from a digital platform, it is said, then you have a duty of fairness to accept the reasonable burdens too.[13] This is similar to the consent principle but differs in one important respect. Whereas the consent principle derives from the idea of *freedom* (if you freely consent to power, the loss of liberty you suffer is legitimate), the fairness principle comes from the idea of *justice* (if you receive X, it is just that you surrender Y in return).

The fairness principle is intuitively sensible but, as with the consent principle, if a tech firm extracts our data without making it clear that it is doing so, then it can scarcely be said that fairness requires us to let it do as it pleases. It's only legitimate if we know about it. Similarly, it's not exactly fair if people have no choice *but* to engage the services of a digital platform. When I walk through the streets of a 'smart' city I am watched by surveillance cameras. But I didn't *choose* to be the subject of surveillance. It's something that was done *to* me. First I'm given no choice but to engage with the technology, then I'm told that *because* I engaged with it I somehow have a duty of fairness to accept its rules and let its controllers use the data gathered about me for other commercial ends. That doesn't seem fair at all.

The Shared Values Principle

Finally, it might be said that a tech firm's power is legitimate to the extent that it reflects or embodies the shared values of its users.[14] According to this view, for instance, what would give a news platform the legitimacy to filter the news in a certain way is that its algorithms reflect or embody the shared values of its users as to how that filtering should be done. This means, however, that the power in question must always be sufficiently transparent that users can see whether it does indeed reflect or embody those values. If the news gathering, editing, and ranking algorithms are hidden from sight, its users can't possibly say whether or not they are engineered in a way that reflects their shared beliefs. They might *seem* to be, but it would be pretty difficult to know otherwise until something went wrong, as it did for Facebook in 2016 when Russian operatives purchased more than 3,000 targeted messages to interfere in the US presidential election.[15] Without transparency, the shared values principle is an article of faith and not a principle of legitimacy.

Regulation

It follows from these principles that at least two types of regulation are likely to be necessary in the digital lifeworld. The first is regulation to ensure transparency; the second is regulation to break up massive concentrations of power.

Transparency Regulation

It would be unacceptable for humans to become subject to significant powers that are utterly opaque to them. For this reason there must be transparency, not only in relation to algorithms but in relation to data use and the values (if any) that are consciously coded into technology. There should also be requirements of *simplicity*, compelling tech firms to explain their systems in basic terms that the rest of us can understand. (EU authorities already plan a legal *right to explanation*, although only in relation to fully automated decisions.)[16]

The case for transparency has secure philosophical foundations, the first of which is the concept of *legitimacy* just discussed. If we consider that people have effectively consented to be the subject of a tech firm's power (the consent principle), or that those who accept its benefits have a duty to accept the burdens (the fairness principle), or that its exercise of power reflects or embodies the shared values of its users (the shared values principle), then yes, the power of that tech firm may be said to be legitimate. But so long as tech firms keep their algorithms hidden under lock and key, their data policies obscure, and their values undefined, they cannot possibly claim a single one of these forms of legitimacy. How can we freely consent when we don't know to what we're consenting? How can we be said to have accepted burdens if we don't know what they are? How can we know if a tech firm shares our values if we don't know how its algorithms work, or what the firm does with our data? The case for transparency is reinforced by the principle of liberty. We are not truly free if we are subject to rules we cannot see, and that are determined by the whim and fancy of people other than ourselves.

Progress is being made toward developing principles of algorithmic audit.[17] One scholar suggests that there should be 'an FDA for algorithms' with a 'broad mandate to ensure that unacceptably dangerous algorithms are not released onto the market'.[18] This is an interesting idea. But tech firms will object that revealing their commercially sensitive algorithms would do irreparable harm to their business or enable malevolent parties to game and abuse the system (an argument favoured by Google). One way to address these concerns would be to entrust the work of algorithmic audit to discrete third-party professionals who 'would take a vow of impartiality and confidentiality, much as accountants and certain other professionals do now. They would evaluate the selection of data sources, the choice of analytical and predictive tools, and the interpretation of results' before issuing a certificate of good health.[19] On this model, rather than citizens holding tech firms to account, we'd leave it to regulatory authorities and independent auditors who would be better-placed, financially and technically, to serve as a check and balance. This idea is attractive but also a little regressive. It leaves us with yet another class of rulers—those who write the code and those who audit it—who understand the workings of power while the rest of us remain in the dark and rely on their benevolence and competence. A system of algorithmic audit would shift power away from tech firms, to be sure, but only to another technocratic élite of auditors, albeit one that has pledged to serve the public interest.

Structural Regulation

We'll also, I believe, need *structural* regulation. By this I mean political intervention to ensure that the technologies of power don't become too concentrated in the hands of a small number of firms and individuals. A thriving market economy might make this unnecessary, but if the analysis in Part V is correct then the trend will be toward greater concentration. Big tech firms may need to be broken up.

Again, there are sound philosophical reasons for structural intervention. The simplest is that it would prevent the accrual of dangerous levels of power. Imagine, for instance, that all or most political speech was channelled through a single digital platform. Users would have no choice but to accept its rules: they couldn't go elsewhere. And that platform would provide a single chokepoint for the stifling of speech altogether. That's too much power for one entity. Similarly, if a single platform (like the imaginary Delphi described in chapter eight) controlled all the apparatus of perception-control within an area, its capacity to shape the behaviour of the population would be unacceptably large.

Structural regulation could also be justified on grounds of *legitimacy*. You can't be said to owe a duty of fairness to a tech firm, or to have consented to its power, in circumstances where there is only one provider (or a few similar ones) and you have no other choice. A system with one tech firm that operated transparently might be able to satisfy the shared values principle—but even then we'd still be relying on its benevolence to keep things that way. From the Digital Republican perspective, this would be unacceptable. Digital Confederalists would go further, saying we need, at all times, a plurality of available digital systems and the opportunity to move between them without adverse consequences.

We already have some legal machinery to deal with concentrations of economic power. In the US it's called antitrust law and in Europe it's known as competition law. Both are aimed at the promotion of economic competition, that is, restricting cartels, preventing collusion, overseeing mergers and acquisitions, and preventing large economic entities from abusing their dominant position in the market. There have been some recent antitrust victories against big tech firms, like the €2.4 billion fine imposed by the EU on Google in 2017 for manipulating search results to promote its own shopping-comparison service over others'. (The US system is less robust. Google was not fined for the same conduct there.)

Important though it is, antitrust regulation was designed for a different set of problems from those we will confront in the digital lifeworld. There will be plenty of tech firms in the digital lifeworld, for example, that wield considerable power but are not unlawful monopolies for the purposes of antitrust law. Moreover, tech firms could routinely exercise power in ways that are wrongheaded, foolish, or unprincipled, but which don't properly fall into the category of 'abuses'. For now at least, the core aim of antitrust regulation is to prevent economic abuses in the form of price discrimination, predatory pricing, and the like, rather than to shape and constrain *political* power. But as we saw in chapter eighteen, the Data Deal means that many services are provided for *free*. No issue of economic abuse necessarily arises. The risk, therefore, is that the power of tech firms will fall outside the antitrust regulatory framework altogether. And even if an antitrust regulator *was* able to break up a tech monopoly, it still wouldn't necessarily guarantee the *plurality* of options that would be needed for true Digital Confederalism.

What we need from structural regulation is to provide citizens with some choice over the powers that govern them, not (just) fair prices for consumers. The philosopher Baron de Montesquieu is credited with developing the idea of a *separation of powers* to distribute political power across three branches of the state: the legislature, the executive, and the judiciary.[20] Montesquieu believed that the best way to prevent any one person or entity from assuming too much power was to keep the three branches independent from each other (together with a system of 'checks and balances', another term he invented). We can learn from Montesquieu. Arguably the most serious difficulty with antitrust regulation is that its regulatory domain is structured by reference to *markets*, not forms of power. What I mean by this is that antitrust regulators begin by identifying a particular market or markets—telecoms, road transportation, and so forth—within which a firm might be said to be abusing its dominance. But for political purposes, market dominance is not what matters. What ultimately matters are the forms of power themselves—force, scrutiny, perception-control—not the economic arenas in which they originate.

The idea should be to create a political system in which two conditions obtain. The first is that no firm is allowed a monopoly over *each* of the means of force, scrutiny, and perception-control. The second is that no firm is allowed significant control over *more than one* of the means of force, scrutiny, and perception-control together. Structurally, that's the best (and possibly only) way to ensure liberty and legitimacy. Instead of trying to bend antitrust law to serve a political function—hammering a legal peg into a political hole—we need to think in terms of a *new separation of powers*.

Technology and Democracy

A final way for citizens to hold on to control would be to give them a direct say in the rules that govern them. Make power accountable, not just transparent. Bring the values of democracy to the private sector. This could mean more 'free software'. Free software, as Lessig explains, is not free in the sense that 'code writers don't get paid' but rather free in the sense that anyone has the right to 'modify' the code 'as he or she sees fit'.[21] It also could mean more flexible operating systems, that is, fewer Apple-style systems that allow for little or no meaningful customization, and more Android-style systems that can be customized much more readily. Contrast a self-driving car that is programmed to kill the child rather than the trucker with one that lets *you* decide, in advance, what it ought to do in situations of that kind (see chapter five). It could mean the right to challenge or appeal a particular decision, prediction, or action of a digital system.[22] Engineers working on important new technologies could face public scrutiny before they unleash the consequences of their inventions.[23] And tech firms should be more answerable to their users. According to this view, Facebook shouldn't be able to change functionalities that affect liberty, democracy, and justice without the permission of the people affected by that change.

These are radical ideas but in philosophical terms they're not new. Many distinguished thinkers have rejected the idea that democracy begins and ends with the ballot box. Instead, they say, democratic principles should be present in as much of day-to-day life as possible, and at the very least should inform all decisions 'that concern the common activities which are among the conditions for self-development'.[24]

The Limits of the Lifeworld

Think for a moment about the finest political speech you've ever heard. It probably inspired you in some way or convinced you of a cause. Perhaps it moved you to tears. For many, Martin Luther King Jr's 'I Have a Dream' speech from August 1963 represents the pinnacle of rhetorical force:

> I have a dream that one day this nation will rise up and live out the true meaning of its creed: 'We hold these truths to be self-evident, that all men are created equal.'

> I have a dream that one day on the red hills of Georgia, the sons of former slaves and the sons of former slave owners will be able to sit down together at the table of brotherhood.

> I have a dream that one day even the state of Mississippi, a state sweltering with the heat of injustice, sweltering with the heat of oppression, will be transformed into an oasis of freedom and justice.

> I have a dream that my four little children will one day live in a nation where they will not be judged by the color of their skin but by the content of their character.

> I have a dream today!

In the very first chapter of this book I noted that engineers have now built an AI system capable of writing political speeches. How would you feel if you found out that the speech you so admire had been generated by that system? That it was the product of an algorithm that 'learned' to compose great speeches by analysing thousands of them from the past? Would knowledge of this fact diminish the value of that speech in your eyes?

For me, it would.

When we hear a great political speech, we hear words but we also hear the authentic voice of the speaker. We bear witness to their moral courage. We glimpse their soul. It's why we like to know if politicians write their own speeches or get someone else to draft them on their behalf. Once you know that a speech was written by a machine, its words sound a little more empty and formulaic.

At this point in the book (well done on making it this far, by the way) you've probably had quite enough of principles, but there's room for one more. It's this: there are some things that digital systems shouldn't do, even if they can (technically) do them better than humans. Such a principle might apply to things that have value precisely *because* they are the product of the human mind, the human hand, or the human heart. To borrow from Michael Sandel, these might be inherently 'corrupted or degraded', or at the very least, diminished, by the fact that they are the product of machines.[25] It might even be said that *we* are degraded in some way by having an algorithm for a boss, a VR system for a lover, a robot for a carer, or an AI system for a government.

As I said in the Introduction, the great debate of the last century was about how much of our collective life should be determined by the state and what should be left to market forces and civil society. In the future, the question will be how much of our collective life should be directed and controlled by powerful digital systems—and on what terms. We cannot, through inactivity, allow ourselves to become 'the plaything of alien powers', subject always to decisions made for us by entities and systems beyond our control and understanding.[26] That is the challenge of *Future Politics*.

TWENTY
Post-Politics

'All that is solid melts into air, all that is holy is profaned'
Karl Marx and Friedrich Engels, *Communist Manifesto* (1848)

The future stalks us more closely with each passing day. Staggering new achievements in science and technology are transforming the way we live together.

We aren't ready.

Most of our political ideas were dreamed up to describe a world that no longer exists. We need new ones, and I have tried to supply some in these pages. There may come a time, however, when the world is so utterly transformed by tech that almost none of our ideas make sense any more—even new-fangled ones like those in this book. In these closing pages, and with some caution, it's worth considering the fate of politics *after* the digital lifeworld.

Advances in genetic engineering, medicine, robotics, nano-technology, and AI are set to alter our biology in ways that the theorists of the past could scarcely have imagined.[1] The ability to edit our DNA,[2] to join together with machines (using implants, prosthetics, and interfaces) and adopt their unhuman power as our own, to manufacture fresh new organs and tissues, to personalize medical treatment for the genetic makeup of each patient—such faculties could forever change what it means to be human. Human enhancement is no longer the stuff of fiction.[3] It's expected that our descendants will be able to enhance their strength and resilience with bionic limbs, organs, and exoskeletons. They'll radically improve their minds, mood, and memory.[4] They'll be able to diminish the effects of pain and the need for sleep.[5] They'll sharpen their senses, giving them superhuman vision and hearing. They'll access entirely new worlds of emotion, sensation, and desire.[6] They'll have the capacity to dictate the characteristics of their unborn children.[7] They'll slow or reverse the ageing process, postponing death itself.[8]

Many of the ethical questions arising out of human enhancement are well-studied. Should enhancement be permitted? On what terms? Should it be a right?[9] Should self-enhancement even—as some argue—be seen as a *duty* to unborn generations and the gene pool itself?[10] Perhaps the most immediate political risk with human enhancement, as several authors have pointed out, is that access to it (like access to world-class healthcare today) may only be available to the rich.[11] Jaron Lanier imagines the morning on which we discover that the rich neighbours have undergone 'procedures that would extend their life spans by decades'. 'That's the kind of morning', says Lanier, 'that could turn almost anyone into a Marxist.'[12]

Whether we consider ourselves to be *more free* in such a world will depend, in part, on what we count as a constraint on freedom. Hobbes believed that the only constraints that mattered for the purposes of liberty were 'externall Impediments of motion'.[13] But to stretch the language of liberty a little, our own biological limitations could also be said to be constraints on our freedom. I'm not free to run a marathon because my scrawny legs constrain me. I'm not free to write great poetry because my feeble mind won't let me. If it's right that gene editing, human enhancement, and physical augmentation will mean that some people will enjoy vastly greater physical and cognitive abilities than others, then some will say that they're more 'free' than those still constrained by the limits of the human body.

If politics is about the collective life of human beings—why we live together, how we order and bind our collective life, and the ways in which we could or should order and bind that collective life differently—then any change in what it means to *be* a human being is likely to have profound political consequences. A world in which a class of 'new godlings'[14] emerges to live alongside the old *homo sapiens* is one in which the term *politics* itself ceases to have a clear or fixed meaning. As David Hume conjectured nearly three centuries ago, a society of more than one unequal species would be almost impossible to sustain:[15]

> Were there a species of creatures, intermingled with men, which . . . were possessed of such inferior strength, both of body and mind, that they were incapable of all resistance . . . the necessary consequence, I think, is, that we should be bound, by the laws of humanity, to give gentle usage to these creatures, but should not, properly speaking, lie under any restraint of justice with regard to them, nor could they possess any right or property, exclusive of such arbitrary lords. Our intercourse with them could not be called society, which supposes a degree of equality; but absolute command on the one side, and servile obedience on the other.

When we think about politics today, we assume a form of collective life in which the participants are all recognizably human. In the world to come, this assumption may no longer be safe. The difference between classes could look more like the difference between species.

This points to a deeper lesson. Our conception of politics is based on a set of implicit assumptions: that human beings are mortal, that they are vulnerable to pain and disease, that they are divided into at least two biological sexes, and so forth.[16] In *The Concept of Law* (1961) H. L. A. Hart argues that certain 'elementary truths' about human nature mean that 'certain rules of conduct' will always need to be enshrined in human law or morality. The first of Hart's 'truisms' is *human vulnerability*: that we are all susceptible to harm. The second is *approximate equality*: that although we differ in size or strength or intellect, no single person is so naturally powerful that he or she can dominate others alone. Next is *limited altruism*, the idea that 'if men are not devils, neither are they angels'. Then there's *limited resources*: there's not always enough to go around. Finally, *limited understanding and strength of will*.[17] Although Hart is talking about law and morality, his argument applies equally to politics. We take these truisms for granted when we speak and think about politics.

Yet if technology eventually allows humans to regenerate their organs, or live forever, then the precept of *human vulnerability* will plainly lose some of its force. Likewise, cognitive enhancement could diminish our *limited understanding and strength of will* while expanding the limits of our *altruism*. And if only the rich can access the technologies of enhancement, then *approximate equality* can't be taken for granted. As we saw in chapter eighteen, future generations could enjoy extraordinary economic abundance.[18] So much, then, for *limited resources*.

What would be the meaning of politics in such a world? Would we still seek to live together in the same way?

Now imagine a world in which AI systems achieve *superintelligence*, that is, an intellect 'that greatly exceeds the cognitive performance of humans in virtually all domains of interest'.[19] Serious students of AI consider this a real possibility. Nick Bostrom, a scholar at the University of Oxford, offers the 'weak conclusion' that:[20]

> it may be reasonable to believe that human-level machine intelligence has a fairly sizeable chance of being developed by mid-century, and that it has a non-trivial chance of being developed considerably sooner or much later; that it might perhaps fairly soon thereafter result in superintelligence.

It's hard for us to conceive of such a superintelligence. Bostrom cheerfully suggests it would not be smart in the sense that 'a scientific genius is smart compared with the average human being,' but rather 'smart in the sense that an average human being is smart compared with a beetle or a worm.' [21] He adds, not very reassuringly, that the advent of a superintelligent AI system could lead to a 'wide range of outcomes' including 'extremely good' ones but also 'outcomes that are as bad as human extinction'.[22] In such a world, politics would revert to its primordial purpose: to ensure survival in a harsh world.

Futurists such as Ray Kurzweil contend that in the long run, we're heading toward a *technological singularity*, that is, a point at which machine intelligence comes to saturate the universe, absorbing all matter and life in its path.[23] There would be no place for *homo sapiens* in such a world, let alone for politics.

I've deliberately avoided devoting many pages in this book to how all-powerful AI systems might come to destroy the world—not because such a scenario is impossible, but because it's already a popular topic of writing and one which can (unhelpfully) obscure the more immediate problems that we'll have to face in the digital lifeworld. As I stressed in chapter nineteen, many of the political issues awaiting us in the digital lifeworld will stem from the ideas and choices of people like you and me. Before politics disappears or becomes something different altogether, the fates of liberty, democracy, and social justice are in our hands.

I wrote this book because I do not believe that we are destined to the fate of the 'sorcerer' described by Marx who was 'no longer able to control the powers of the nether world' that he 'called up by his spells'.[24] The future stalks us, but we have more control over it than we realize. At the start of this book, I called for a fundamental change in the way we think about politics. This is truly the work of a generation, but it must begin now, and it will continue even after we are gone. For as Thomas Jefferson wrote in 1823, 'the generation which commences a revolution rarely completes it'.[25]

NOTES

Introduction

1. John Stuart Mill, *The Autobiography of John Stuart Mill* (US: Seven Treasures Publications, 2009), 93.

2. David Remnick, 'Obama Reckons With a Trump Presidency', *New Yorker*, 28 November 2016 <http://www.newyorker.com/magazine/2016/11/28/obama-reckons-with-a-trump-presidency> (accessed 30 November 2017).

3. Ronald Wright, *A Short History of Progress* (London: Canongate Books, 2006), 14, 55.

4. Wright, *Progress*, 14.

5. Jaron Lanier, *Who Owns the Future?* (London: Allen Lane, 2014), 17.

6. Karl Marx, *Theses on Feuerbach*, in *Karl Marx and Frederick Engels: Collected Works Vol. 5* (London: Lawrence & Wishart, 1976), 5.

7. Tim Berners-Lee, cited in H. Halpin, 'Philosophical Engineering. Towards a Philosophy of the Web', APA Newsletters, Newsletter on Philosophy and Computers 7, no. 2 (2008): 5–11, quoted in Mireille Hildebrandt, 'The Public(s) Onlife: A Call for Legal Protection by Design', in *The Onlife Manifesto: Being Human in a Hyperconnected Era*, ed. Luciano Floridi (Cham: Springer, 2015), 188.

8. Ibid.

9. Sheelah Kolhatkar, 'The Tech Industry's Gender-Discrimination Problem', *New Yorker*, 20 November 2017 <https://www.newyorker.com/magazine/2017/11/20/the-tech-industrys-gender-discrimination-problem> (accessed 12 December 2017).

10. Julia Wong, 'Segregated Valley: The Ugly Truth about Google and Diversity in Tech', *The Guardian*, 7 August 2017 <https://www.theguardian.com/technology/2017/aug/07/silicon-valley-google-diversity-black-women-workers> (accessed 28 November 2017).

11. Don Tapscott and Alex Tapscott, *Blockchain Revolution: How the Technology behind Bitcoin is Changing Money, Business, and the World* (London: Portfolio Penguin, 2016), 199.

12. Isaiah Berlin, 'The Purpose of Philosophy', in Isaiah Berlin, *The Power of Ideas*, ed. Henry Hardy (London: Pimlico, 2001), 35.

13. John S. Dryzek, Bonnie Honig, and Anne Phillips, 'Introduction', in *The Oxford Handbook of Political Theory*, eds. John S. Dryzek, Bonnie Honig, and Anne Phillips (New York: Oxford University Press, 2008), 4.

14. Onlife Initiative, 'Background Document: Rethinking Public Spaces in the Digital Transition', in *Onlife Manifesto*, 41.

15. 'Editors' Introduction', in *Political Innovation and Conceptual Change*, eds. Terence Ball, James Farr, and Russell L. Hanson (New York: Cambridge University Press, 1995), 1.

16. Ludwig Wittgenstein, *Tractatus Logico-Philosophicus* (Abingdon: Routledge, 2001), 68 (5.6).

17. C.f. Yochai Benkler, *The Wealth of Networks: How Social Production Transforms Markets and Freedom* (New Haven and London: Yale University Press, 2006), 17: 'Different technologies make different kinds of human action and interaction easier or harder to perform.'

18. Emmanuel G. Mesthene, 'The Social Impact of Technological Change', in *Philosophy of Technology: The Technological Condition: An Anthology* (Second Edition), eds. Robert C. Scharff and Val Dusek (Oxford: Wiley-Blackwell, 2014), 689.

19. Langdon Winner, 'Do Artifacts Have Politics?' in *Philosophy of Technology*, 669.

20. Langdon Winner, 'Do Artifacts Have Politics?'

21. Otto Mayr, *Authority, Liberty and Automatic Machinery in Early Modern Europe* (Baltimore: Johns Hopkins University Press, 1989), 102; Aristotle, *The Politics*, translated by T. A. Sinclair (London: Penguin, 1992), 1253a18, 60.

22. Mayr, *Authority,* 102.

23. Mayr, *Authority*, 27.

24. Mayr, *Authority*, 112.

25. Mayr, *Authority*, 119.

26. Mayr, *Authority*, 121.

27. E. M. Forster, *The Machine Stops* (London: Penguin, 2011).

28. Evgeny Morozov, *The Net Delusion: How Not to Liberate the World* (London: Penguin, 2011), xiii.

29. Ibid.

30. Evgeny Morozov, *To Save Everything, Click Here: Technology, Solutionism, and the Urge to Fix Problems That Don't Exist* (London: Penguin, 2014), 5.

31. See generally Andrew J. Beniger, *Control Revolution: Technological and Economic Origins of the Information Society* (Cambridge, Mass: Harvard University Press, 1986).

32. James Farr, 'Understanding Conceptual Change Politically', in *Political Innovation*, 25.

33. Yuval Noah Harari, *Sapiens: A Brief History of Humankind* (London: Vintage Books, 2011), 24–7.

34. Yuval Noah Harari, *Homo Deus: A Brief History of Tomorrow* (London: Harvill Secker, 2015), 167.

35. 'Domesday Book', *Wikipedia*, last modified 26 November 2017 <https://en.wikipedia.org/wiki/Domesday_Book> (accessed 28 November 2017).

36. Harari, *Homo Deus*, 167.

37. Paraphrasing Alain Desrosières, *The Politics of Large Numbers: A History of Statistical Reasoning*, translated by Camille Naish (Cambridge, Mass: Harvard University Press, 1998), 16.

38. Desrosières, *Politics of Large Numbers*, 9.

39. Alexander Hamilton, 'The Federalist No. 23', 18 December 1787, in Alexander Hamilton, James Madison, and John Jay, *The Federalist Papers* (New York: Penguin, 2012), 45; see Bruce Bimber, *Information and American Democracy: Technology in the Evolution of Political Power* (New York: Cambridge University Press, 2011), 45.

40. Desrosières, *Politics of Large Numbers*, 236.

41. Ibid.

42. Thomas Richards, *The Imperial Archive: Knowledge and the Fantasy of Empire* (London: Verso, 1993), 6.

43. Beniger, *Control Revolution*, 8.

44. Max Weber, *Economy and Society: An Outline of Interpretive Sociology* (Vol. 2), eds. Guenther Roth and Claus Wittich (Berkley: University of California Press, 2013), 990.

45. Weber, *Economy and Society* (Vol. 2), 973.

46. Desrosières, *Politics of Large Numbers*, 330.

47. James Gleick, *The Information: A History, A Theory, A Flood* (London: Fourth Estate, 2012), 42.

48. Harold Innis, *Empire and Communications* (Lanham: Rowman & Littlefield, 2007), 30.

49. Anthony M. Townsend, *Smart Cities: Big Data, Civic Hackers, and the Quest for a New Utopia* (New York: W. W. Norton & Company, 2014), 59–60.

50. Benkler, *Wealth of Networks*, 30.

51. Kevin Kelly, *What Technology Wants* (New York: Penguin, 2010), 191–2.

52. Vladimir Ilyich Lenin, 'Notes on Electrification', February 1921, reprinted (1977) in *Collected Works*, Vol. 42 (Moscow: Progress Publishers): 280–1, cited in Sally Wyatt, 'Technological Determinism is Dead; Long Live Technological Determinism', in *Philosophy of Technology*, 458.

53. Leon Trotsky, 'What is National Socialism?' *Marxists*, last modified 25 April 2007 <https://www.marxists.org/archive/trotsky/germany/1933/330610.htm> (accessed 28 November 2017).

54. Nadia Judith Enchassi and CNN Wire, 'New Zealand Passport Robot Thinks This Asian Man's Eyes Are Closed', *KFOR*, 11 December 2016 <http://kfor.com/2016/12/11/new-zealand-passport-robot-thinks-this-asian-mans-eyes-are-closed/> (accessed 2 December 2017).

55. Selena Larson, 'Research Shows Gender Bias in Google's Voice Recognition', *Daily Dot*, 15 July 2016 <https://www.dailydot.com/debug/google-voice-recognition-gender-bias/> (accessed 2 December 2017).

56. Alex Hern, 'Flickr Faces Complaints Over "Offensive" Auto-tagging for Photos', *The Guardian*, 20 May 2015 <https://www.theguardian.com/technology/2015/may/20/flickr-complaints-offensive-auto-tagging-photos> (accessed 2 December 2017).

57. Wittgenstein, *Tractatus*, Preface.

58. G. K. Chesterton, *Orthodoxy* (Cavalier Classics, 2015), 2.

Chapter 1

1. The term is from Richard Susskind and Daniel Susskind, *The Future of the Professions: How Technology Will Transform the Work of Human Experts* (Oxford: Oxford University Press, 2015).

2. I am grateful to Richard Susskind for his assistance in formulating this definition, although his preferred definition would be wider than mine (including manual and emotional tasks as well).

3. Yonghui Wu et al. 'Google's Neural Machine Translation System: Bridging the Gap between Human and Machine Translation', *arXiv*, 8 October 2016 <https://arxiv.org/abs/1609.08144> (accessed 6 December 2017); Yaniv Taigman et al., 'DeepFace: Closing the Gap to Human-Level Performance in Face Verification', 2014 IEEE Conference on Computer Vision and Pattern Recognition (CVPR) 2014 <https://www.cs.toronto.edu/~ranzato/publications/taigman_cvpr14.pdf> (accessed 11 December 2017); Aäron van den Oord et al., 'WaveNet: A Generative Model for Raw Audio', *arXiv*, 19 September 2016 <https://arxiv.org/abs/1609.03499> (accessed 6 December 2017).

4. Peter Campbell, 'Ford Plans Mass-market Self-driving Car by 2021', *Financial Times*, 16 August 2016 <https://www.ft.com/content/d2cfc64e-63c0-11e6-a08a-c7ac04ef00aa#axzz4HOGiWvHT> (accessed 28 November 2017); David Millward, 'How Ford Will Create a New Generation of Driverless Cars', *Telegraph*, 27 February 2017 <http://www.telegraph.co.uk/business/2017/02/27/ford-seeks-pioneer-new-generation-driverless-cars/> (accessed 28 November 2017).

5. Wei Xiong et al., 'Achieving Human Parity in Conversational Speech Recognition', *arXiv*, 17 February 2017 <https://arxiv.org/abs/1610.05256> (accessed 28 November 2017).

6. Yannis M. Assael et al., 'LipNet: End-to-End Sentence-level Lipreading', *arXiv*, 16 December 2016 <https://arxiv.org/abs/1611.01599> (accessed 6 December 2017).

7. Laura Hudson, 'Some Like it Bot', *FiveThirtyEight*, 29 September 2016 <http://fivethirtyeight.com/features/some-like-it-bot/> (accessed 28 November 2017).

8. Susskind and Susskind, *Future of the Professions*, 77.

9. Rory Cellan-Jones, '"Cut!"—the AI Director', *BBC News*, 23 June 2016, <http://www.bbc.co.uk/news/technology-36608933> (accessed 28 November 2017).

10. Cory Edwards, 'Why and How Chatbots Will Dominate Social Media', *TechCrunch*, 20 July 2016 <https://techcrunch.com/2016/07/20/why-and-how-chatbots-will-dominate-social-media/?ncid=rss&utm_

source=feedburner&utm_medium=feed&utm_campaign=Feed%3A+Techcrunch+%28TechCrunch%29&sr_share=twitter> (accessed 28 November 2017).

11. Valentin Kassarnig, 'Political Speech Generation', *arXiv,* 20 January 2016 <https://arxiv.org/abs/1601.03313> (accessed 28 November 2017).

12. Rob Wile, 'A Venture Capital Firm Just Named an Algorithm to its Board of Directors—Here's What it Actually Does', *Business Insider,* 13 May 2014 <http://www.businessinsider.com/vital-named-to-board-2014-5> (accessed 28 November 2017).

13. Krista Conger, 'Computers Trounce Pathologists in Predicting Lung Cancer Type, Severity', *Stanford Medicine News Center,* 16 August 2016 <http://med.stanford.edu/news/all-news/2016/08/computers-trounce-pathologists-in-predicting-lung-cancer-severity.html> (accessed 28 November 2017). See also Andre Esteva et al., 'Dermatologist-level Classification of Skin Cancer with Deep Neural Networks', *Nature* 542 (2 February 2017): 115–18.

14. Nikolaos Aletras, Dimitrios Tsarapatsanis, Daniel Preotiuc, and Vasileios Lampos, 'Predicting Judicial Decisions of the European Court of Human Rights: A Natural Language Processing Perspective' *Peer J Computer Science* 2, e93 (24 October 2016).

15. Sarah A. Topol, 'Attack of the Killer Robots', *BuzzFeed News,* 26 August 2016 <https://www.buzzfeed.com/sarahatopol/how-to-save-mankind-from-the-new-breed-of-killer-robots?utm_term=.nm1GdWDBZ#.vaJzgW6va>) (accessed 28 November 2017).

16. Cade Metz, 'Google's AI Wins Fifth and Final Game Against Go', *Wired,* 15 March 2016 <https://www.wired.com/2016/03/googles-ai-wins-fifth-final-game-go-genius-lee-sedol/> (accessed 28 November 2017); Nick Bostrom, *Superintelligence: Paths, Dangers, Strategies* (Oxford: Oxford University Press, 2014), 12–13.

17. Sam Byford, 'AlphaGo beats Ke Jie Again to Wrap Up Three-part March', *The Verge,* 25 May 2017 <https://www.theverge.com/2017/5/25/15689462/alphago-ke-jie-game-2-result-google-deepmind-china> (accessed 28 November 2017).

18. David Silver et al., 'Mastering the Game of Go Without Human Knowledge', *Nature* 550 (19 October 2017): 354–9.

19. David Silver et al., "A general reinforcement learning algorithm thatmasters chess, shogi, and Go through self-play' *Science*, 362 (7 Dec.2018): 1140–1144.

20. Susskind and Susskind, *Future of the Professions*, 165.

21. Ibid.

22. Ibid.

23. Kevin Kelly, *The Inevitable: Understanding the 12 Technological Forces that Will Shape Our Future* (New York: Viking, 2016), 31.

24. Emma Hinchliffe, 'IBM's Watson Supercomputer Discovers 5 New Genes Linked to ALS', *Mashable UK*, 14 December 2016 <http://mashable.com/2016/12/14/ibm-watson-als-research/?utm_cid=mash-com-Tw-tech-link%23sd613jsnjlqd#HJziN5r0aGq5> (accessed 28 November 2017).

25. Murray Shanahan, *The Technological Singularity* (Cambridge, Mass: MIT Press, 2015), 12.

26. BBC, 'Google Working on "Common-Sense" AI Engine at New Zurich Base', *BBC News*, 17 June 2016 <http://www.bbc.co.uk/news/technology-36558829> (accessed 30 November 2017); Blue Brain Project <https://bluebrain.epfl.ch/page-56882-en.html> (accessed 6 December 2017).

27. Bostrom, *Superintelligence*, 30.

28. Shanahan, *Technological Singularity*, 47.

29. Garry Kasparov, 'The Chess Master and the Computer', *New York Review of Books*, 11 February 2010, cited in Susskind and Susskind, *Future of the Professions*, 276.

30. Pedro Domingos, *The Master Algorithm: How The Quest for the Ultimate Learning Machine Will Remake Our World* (London: Allen Lane, 2015), xi.

31. Domingos, *Master Algorithm*, 8.

32. Domingos, *Master Algorithm*, xi.

33. Domingos, *Master Algorithm*, xvi.

34. Domingos, *Master Algorithm*, xiv.

35. Domingos, *Master Algorithm*, 8–9.

36. Cade Metz, 'Building AI is Hard—So Facebook is Building AI that Builds AI', *Wired*, 6 May 2016 <https://www.wired.com/2016/05/facebook-trying-create-ai-can-create-ai/> (accessed 28 November 2017).

37. Margaret A. Boden, *AI: Its Nature and Future* (Oxford: Oxford University Press, 2016), 47 (original emphasis).

38. Boden, *AI*, 40.

39. Cade Metz, 'Google's Dueling Neural Networks Spar to Get Smarter, No Humans Required', *Wired*, 11 April 2017 <https://www.wired.com/2017/04/googles-dueling-neural-networks-spar-get-smarter-no-humans-required/> (accessed 28 November 2017).

40. Silver et al., 'Mastering'.

41. Domingos, *Master Algorithm*, 7.

42. Neil Lawrence, quoted in Alex Hern, 'Why Data is the New Coal', *The Guardian*, 27 September 2016 <https://www.theguardian.com/technology/2016/sep/27/data-efficiency-deep-learning> (accessed 28 November 2017).

43. Ray Kurzweil, *The Singularity is Near* (New York: Viking, 2005), 127, cited in Susskind and Susskind, *Future of the Professions*, 157; Peter H. Diamandis and Steven Kotler, *Abundance: The Future is Better Than You Think* (New York: Free Press, 2014), 55.

44. Paul Mason, *Postcapitalism: A Guide to Our Future* (London: Allen Lane, 2015), 121.

45. Luciano Floridi, *The 4th Revolution: How the Infosphere is Reshaping Human Reality* (Oxford: Oxford University Press, 2015), 7.

46. Samuel Greengard, *The Internet of Things* (Cambridge, Mass: MIT Press, 2015), 28.

47. Domingos, *Master Algorithm*, 73.

48. Jaron Lanier, *Who Owns the Future?* (London: Allen Lane, 2014), 6.

49. Gordon Moore, 'Cramming More Components onto Integrated Circuits', *Proceedings of the IEEE* 86, no. 1 (January 1998), 83.

50. Walter Isaacson, *The Innovators: How a Group of Hackers, Geniuses and Geeks Created the Digital Revolution* (London: Simon & Schuster, 2014), 184.

51. Susskind and Susskind, *Future of the Professions*, 157; Kevin Kelly, *What Technology Wants* (New York: Penguin, 2010), 166–7; Eric Schmidt and Jared Cohen, *The New Digital Age: Reshaping the Future of People, Nations and Business* (London: John Murray, 2014), 5; Shanahan, *Technological Singularity*, xviii; Erik Brynjolfsson and Andrew McAfee, *The Second Machine Age: Work, Progress, and Prosperity in a Time of Brilliant Technologies* (New York: W. W. Norton & Company, 2014), 49; Wendell Wallach, *A Dangerous Master: How to Keep Technology from Slipping Beyond Our Control* (New York: Basic Books, 2015), 67.

52. Jamie Condliffe, 'Chip Makers Admit Transistors Are About to Stop Shrinking', *MIT Technology Review*, 25 July 2016 <https://www.technologyreview.com/s/601962/chip-makers-admit-transistors-are-about-to-stop-shrinking/?> (accessed 28 November 2017); Tom Simonite, 'Moore's Law is Dead. Now What?' *MIT Technology Review*, 13 May 2016 <https://www.technologyreview.com/s/601441/moores-law-is-dead-now-what/> (accessed 28 November 2017); Thomas L. Friedman, *Thank You for Being Late: An Optimist's Guide to Thriving in the Age of Accelerations* (New York: Farrar, Straus, and Giroux, 2016), 43; Tim Cross, 'Beyond Moore's Law', in *Megatech: Technology in 2050*, ed. Daniel Franklin (New York: Profile Books, 2017), 56–7.

53. Shanahan, *Technological Singularity*, 160; Kelly, *What Technology Wants*, 166.

54. Friedman, *Thank You for Being Late*, 21.

55. Kristian Vättö, 'Samsung SSD 850 Pro (128GB, 256GB & 1TB) Review: Enter the 3D Era', *AnandTech*, 1 July 2014 <http://www.anandtech.com/show/8216/samsung-ssd-850-pro-128gb-256gb-1tb-review-enter-the-3d-era> (accessed 28 November 2017); Intel, 'New Technology Delivers an Unprecedented Combination of Performance and Power Efficiency', *Intel 22 NM Technology* <http://www.intel.com/content/www/us/en/silicon-innovations/intel-22nm-technology.html> (accessed 28 November 2017).

56. Shanahan, *Technological Singularity*, 35.

57. Norm Jouppi, 'Google Supercharges Machine Learning Tasks with TPU Custom Chip', *Google Cloud Platform Blog*, 18 May 2016 <https://cloud-platform.googleblog.com/2016/05/Google-supercharges-machine-learning-tasks-with-custom-chip.html> (accessed 28 November 2017).

58. Cade Metz, 'Microsoft Bets its Future on a Reprogrammable Computer Chip', *Wired*, 25 August 2016 <https://www.wired.com/2016/09/microsoft-bets-future-chip-reprogram-fly/?mbid=social_twitter> (accessed 28 November 2017).

59. M. Mitchell Waldrop, 'The Chips are Down for Moore's Law', *Nature* 530, no. 7589 (9 February 2016): 144–7.

60. M. Mitchell Waldrop, 'Neuroelectronics: Smart Connections', *Nature* 503, no. 7474 (6 November 2013): 22–44.

61. Shanahan, *Technological Singularity*, 34.

Chapter 2

1. Cited in William J. Mitchell, *Me ++: The Cyborg Self and the Networked City* (Cambridge, Mass: MIT Press, 2003), 3.

2. Eric Schmidt and Jared Cohen, *The New Digital Age: Reshaping the Future of People, Nations and Business* (London: John Murray, 2014), 172.

3. Marc Goodman, *Future Crimes: A Journey to the Dark Side of Technology—and How to Survive it* (London: Bantam Press, 2015), 59.

4. See Rob Kitchin, *The Data Revolution: Big Data, Open Data, Data Infrastructures and their Consequences* (London: Sage Publications Ltd, 2014), 83.

5. David Rose, *Enchanted Objects: Design, Human Desire, and the Internet of Things* (New York: Scribner, 2014), 7.

6. Adam Greenfield, *Everyware: The Dawning Age of Ubiquitous Computing* (Berkley: New Riders, 2006).

7. Statista, 'Internet of Things 'IoT connected Devices installed base worldwide from 2015 to 2025 (in billions)', <https://www.statista.com/statistics/471264/iot-number-of-connected-devices-world-wide/> (accessed 5 Jul. 2019).

8. Samuel Greengard, *The Internet of Things* (Cambridge, Mass: MIT Press, 2015), 13.

9. Greenfield, *Everyware*, 1.

10. Greengard, *Internet of Things*; Greenfield, *Everyware*; Kitchin, *Data Revolution*.

11. Mat Smith, 'Ralph Lauren Made a Great Fitness Shirt that Also Happens to Be "Smart"', *Engadget*, 18 March 2016 <https://www.engadget.com/2016/03/18/ralph-lauren-polotech-review/> (accessed 6 December 2017).

12. Casey Newton, 'Here's How Snapchat's New Spectacles Will Work', *The Verge*, 24 September 2016 <http://www.theverge.com/2016/9/24/13042640/snapchat-spectacles-how-to-use> (accessed 28 November 2017).

13. Katherine Bourzac, 'A Health-Monitoring Sticker Powered by Your Cell Phone', *MIT Technology Review*, 3 August 2016 <https://www.technologyreview.com/s/602067/a-health-monitoring-sticker-powered-by-your-cell-phone/?utm_campaign=socialflow&utm_source=twitter&utm_medium=post> (accessed 29 November 2017).

14. Brian Heater, 'Wilson's Connected Football is a $200 Piece of Smart Pigskin', *TechCrunch*, 8 August 2016 <https://techcrunch.com/2016/08/08/wilson-x-football/?ncid=rss> (accessed 29 November 2017).

15. See Greengard, *Internet of Things*; Greenfield, *Everyware*; Kitchin, *Data Revolution*.

16. Tanvi Misra, '3 Cities Using Open Data in Creative Ways to Solve Problems', *CityLab*, 22 April 2015 <http://www.citylab.com/cityfixer/2015/04/3-cities-using-open-data-in-creative-ways-to-solve-problems/391035/> (accessed 29 November 2017).

17. Internet Live Stats, 'Internet Users' <http://www.internetlivestats.com/internet-users/> (accessed 30 November 2017).

18. Cisco, 'VNI Global Fixed and Mobile Internet Traffic Forecasts, Complete Visual Networking Index (VNI) Forecast', 2016 <https://www.cisco.com/c/en/us/solutions/service-provider/visual-networking-index-vni/index.html#~mobile-forecast> (accessed 30 November 2017).

19. Statista, 'Number of Monthly Active Facebook Users Worldwide as of 3rd Quarter 2017 (in Millions)' <https://www.statista.com/statistics/264810/number-of-monthly-active-facebook-users-worldwide/> (accessed 11 December 2017).

20. Twitter.com <https://about.twitter.com/company> (accessed 30 November 2017).

21. YouTube for Press <https://www.youtube.com/intl/en-GB/yt/about/press/> (accessed 30 November 2017).

22. See Yochai Benkler, *The Wealth of Networks: How Social Production Transforms Markets and Freedom* (New Haven and London: Yale University Press, 2006) and *The Penguin and the Leviathan: How Cooperation Triumphs over Self-Interest* (New York: Crown Publishing, 2011).

23. Don Tapscott and Alex Tapscott, *Blockchain Revolution: How the Technology Behind Bitcoin is Changing Money, Business and the World* (London: Portfolio Penguin, 2016), 7.

24. Tapscott and Tapscott, *Blockchain Revolution*, 16.

25. Tapscott and Tapscott, *Blockchain Revolution*, 153–4; Stan Higgins, 'IBM Invests $200 Million in Blockchain-Powered IoT', *CoinDesk,* 4 October 2016 <https://www.coindesk.com/ibm-blockchain-iot-office/> (accessed 30 November 2017).

26. Melanie Swan, *Blockchain: Blueprint for a New Economy* (Sebastopol, CA: O'Reilly, 2015), 14.

27. *Economist*, 'Not-so-clever Contracts', 28 July 2016 <http://www.economist.com/news/business/21702758-time-being-least-human-judgment-still-better-bet-cold-hearted?frsc=dg%7Cd> (accessed 30 November 2017).

28. Tapscott and Tapscott, *Blockchain Revolution*, 18.

29. Tapscott and Tapscott, *Blockchain Revolution*, 253–9; Benjamin Loveluck and Primavera De Filippi, 'The Invisible Politics of Bitcoin: Governance Crisis of a Decentralized Infrastructure', *Internet Policy Review* 5, no. 3 (30 September 2016) <http://policyreview.info/articles/analysis/invisible-politics-bitcoin-governance-crisis-decentralised-infrastructure> (accessed 30 November 2017); Erik Brynjolfsson and Andrew McAfee, *Machine Platform Crowd: Harnessing Our Digital Future* (New York: W. W. Norton & Company, 2017), 306–7. *Economist*, 'Not-so-clever Contracts'; BBC, 'Hack Attack Drains Start-up Investment Fund', *BBC News*, 21 June 2016 <http://www.bbc.co.uk/news/technology-36585930> (accessed 30 November 2017).

30. Schwab, Klaus, *The Fourth Industrial Revolution* (Geneva: World Economic Forum, 2016), 19; Laura Shin, 'The First Government to

Secure Land Titles on the Bitcoin Blockchain Expands Project', *Forbes*, 7 February 2017 <https://www.forbes.com/sites/laurashin/2017/02/07/the-first-government-to-secure-land-titles-on-the-bitcoin-blockchain-expands-project/#432b8b494dcd> (accessed 30 November 2017); Joon Ian Wong, 'Sweden's Blockchain-powered Land Registry is Inching Towards Reality', *Quartz Media*, 3 April 2017 <https://qz.com/947064/sweden-is-turning-a-blockchain-powered-land-registry-into-a-reality/> (accessed 30 November 2017).

31. Daniel Palmer, 'Blockchain Startup to Secure 1 Million e-Health Records in Estonia', *CoinDesk*, 3 March 2016 <http://www.coindesk.com/blockchain-startup-aims-to-secure-1-million-estonian-health-records/> (accessed 30 November 2017).

32. Harriet Green, 'Govcoin's Co-founder Robert Kay Explains Why His Firm is Using Blockchain to Change the Lives of Benefits Claimants', *City AM*, 10 October 2016 <http://www.cityam.com/250993/govcoins-co-founder-robert-kay-explains-why-his-firm-using> (accessed 30 November 2017).

33. Kyle Mizokami, 'The Pentagon Wants to Use Bitcoin Technology to Protect Nuclear Weapons', *Popular Mechanics*, 11 October 2016 <http://www.popularmechanics.com/military/research/a23336/the-pentagon-wants-to-use-bitcoin-technology-to-guard-nuclearweapons/?utm_content=buffer98698&utm_medium=social&utm_source=twitter.com&utm_campaign=buffer> (accessed 30 November 2017).

34. Nick Bostrom, *Superintelligence: Paths, Dangers, Strategies* (Oxford: Oxford University Press, 2014), ch. 10.

35. Murray Shanahan, *The Technological Singularity* (Cambridge, Mass: MIT Press, 2015), 153.

36. Matt Burgess, 'Samsung is Working on Putting AI Voice Assistant Bixby in Your TV and Fridge', *Wired*, 27 June 2017 <https://www.wired.co.uk/article/samsung-bixby-television-refrigerator> (accessed 30 November 2017).

37. James O'Malley, 'Bluetooth Mesh Is Going to Be a Big Deal: Here Are 6 Reasons Why You Should Care', *Gizmodo*, 18 July 2017 <http://www.gizmodo.co.uk/2017/07/bluetooth-mesh-is-

going-to-be-a-big-deal-here-are-6-reasons-why-you-should-care/> (accessed 30 November 2017).

38. John Palfrey and Urs Gasser, *Interop: The Promise and Perils of Highly Interconnected Systems* (New York: Basic Books, 2012), 249–50.

39. *Telegraph*, 'Brain-to-brain "Telepathic" Communication Achieved for First Time', 5 September 2014 <http://www.telegraph.co.uk/news/worldnews/northamerica/usa/11077094/Brain-to-brain-telepathic-communication-achieved-for-first-time.html> (accessed 30 November 2017).

40. Muse.com <http://www.choosemuse.com/> (accessed 30 November 2017).

41. Zoltan Istvan, 'Will Brain Wave Technology Eliminate the Need for a Second Language?' in *Visions of the Future*, ed. J. Daniel Batt (Reno: Lifeboat Foundation, 2015), 641.

42. Cade Metz, 'Elon Musk isn't the Only One Trying to Computerize Your Brain', *Wired*, 31 March 2017 <https://www.wired.com/2017/03/elon-musks-neural-lace-really-look-like/?mbid=social_twitter> (accessed 30 November 17).

43. Tim Berners-Lee with Mark Fischetti, *Weaving the Web: The Original Design and Ultimate Destiny of the World Wide Web* (New York: HarperCollins, 2000), 1.

44. Daniel Kellmereit and Daniel Obodovski, *The Silent Intelligence: The Internet of Things* (DND Ventures LLC, 2013), 3.

45. Kitchin, *Data Revolution*, 91.

46. Kitchin, *Data Revolution*, 89.

47. Kitchin, *Data Revolution*, 91.

48. Kitchin, *Data Revolution*, 89.

49. Kitchin, *Data Revolution*, 91.

50. Tucker, Patrick, *The Naked Future: What Happens in a World that Anticipates Your Every Move?* (London: Current, 2015), 8.

51. *Economist*, 'How Cities Score', 23 May 2016 <https://www.economist.com/news/special-report/21695194-better-use-data-could-make-cities-more-efficientand-more-democratic-how-cities-score> (accessed 30 November 2017).

52. Kitchin, *Data Revolution*, 92; Margarita Angelidou, 'Smart City Strategy: PlanIT Valley (Portugal)', *Urenio*, 26 January 2015 <http://

www.urenio.org/2015/01/26/smart-city-strategy-planlt-valley-portugal/> (accessed 30 November 2017).

53. *Economist*, 'How Cities Score'.

54. Greengard, *Internet of Things*, 48.

55. Jane Wakefield, 'Google, Facebook, Amazon Join Forces on Future of AI', *BBC News*, 28 September 2016 <http://www.bbc.com/news/technology-37494863> (accessed 30 November 2017).

56. Margaret A. Boden, *AI: Its Nature and Future* (Oxford: Oxford University Press, 2016), 41.

57 Bostrom, *Superintelligence*, 15.

58. BBC, 'Beijing Park Dispenses Loo Roll Using Facial Recognition', *BBC News*, 20 March 2017 <http://www.bbc.com/news/world-asia-china-39324431> (accessed 30 November 2017).

59. Rose, *Enchanted Objects*, 17.

60. Robert Scoble and Israel Shel, *The Fourth Transformation: How Augmented Reality and Artificial Intelligence Change Everything* (CreateSpace Independent Publishing Platform, 2017), 61.

61. Lisa Fischer, 'Control Your Phone with these Temporary Tattoos', *CNN Tech* (undated) <http://money.cnn.com/video/technology/2016/08/15/phone-control-tattoos.cnnmoney/index.html?sr=twCNN091216phone-control-tattoos.cnnmoney1112PMVideo Video&linkId=28654785> (accessed 30 November 2017).

62. Ben Popper, 'Electrick Lets You Spray Touch Controls Onto Any Object or Surface', *The Verge*, 8 May 2017 <https://www.theverge.com/2017/5/8/15577390/electrick-spray-on-touch-controls-future-interfaces-group> (accessed 30 November 2017).

63. Yuval Noah Harari. *Homo Deus: A Brief History of Tomorrow* (London: Harvill Secker, 2015), 45.

64. Schwab, *Fourth Industrial Revolution*, 122.

65. Wendell Wallach, *A Dangerous Master: How to Keep Technology from Slipping Beyond Our Control* (New York: Basic Books, 2015), 181–2.

66. Schwab, *Fourth Industrial Revolution*, 120.

67. *Riley v. California* 134 S. Ct. 2473 Supreme Court 2014, *per* Chief Justice Roberts at III.

68. Affectiva.com <http://www.affectiva.com/> (accessed 30 November 2017).

69. Raffi Khatchadourian, 'We Know How You Feel', *New Yorker*, 19 January 2015 <http://www.newyorker.com/magazine/2015/01/19/know-feel> (accessed 30 November 2017).

70. Ludwig Wittgenstein, *Culture and Value*, translated by Peter Winch (Chicago: University of Chicago Press, 1980), 23e.

71. Richard Susskind and Daniel Susskind, *The Future of the Professions: How Technology Will Transform the Work of Human Experts* (Oxford: Oxford University Press, 2015), 171.

72. Khatchadourian, 'We Know How You Feel'.

73. Robby Berman, 'New Tech Uses WiFi to Read Your Inner Emotions—Accurately, and From Afar', *Big Think*, 2016 <http://bigthink.com/robby-berman/new-tech-can-accurately-read-the-emotions-you-may-be-hiding> (accessed 30 November 2017).

74. L. R. Sudha, and R. Bhavani, 'Biometric Authorization System Using Gait Biometry', *arXiv*, 2011 <https://arxiv.org/pdf/1108.6294.pdf%3b%20Boden/39-40.pdf> (accessed 30 November 2017).

75. Khatchadourian, 'We Know How You Feel'.

76. Boden, *AI*, 74.

77. Boden, *AI*, 162.

78. Alan Winfield, *Robotics: A Very Short Introduction* (Oxford: Oxford University Press, 2012), 16.

79. 'Moravec's Paradox', *Wikipedia*, last modified 9 May 2017. <https://en.wikipedia.org/wiki/Moravec%27s_paradox> (accessed 6 December 2017).

80. Bostrom, *Superintelligence*, 15.

81. Schwab, *Fourth Industrial Revolution*, 153.

82. Susskind and Susskind, *Future of the Professions*, 168; *Time*, 'Meet the Robots Shipping Your Amazon Orders', *Time Robotics*, 1 December 2014 <http://time.com/3605924/amazon-robots/> (accessed 30 November 2017).

83. Brynjolfsson and McAfee, *Machine Platform Crowd*, 101.

84. IFR, 'World Robotics Report 2016', *IFR Press Release* <https://ifr.org/ifr-press-releases/news/world-robotics-report-2016> (accessed 30 November 2017).

85. Alison Sander and Meldon Wolfgang, 'The Rise of Robotics', *BCG Perspectives*, 27 August 2014 <https://www.bcgperspectives.com/

content/articles/business_unit_strategy_innovation_rise_of_robotics/>
(accessed 30 November 2017).

86 Susskind and Susskind, *Future of the Professions*, 50.

87. Waymo, *Google* <https://www.google.com/selfdrivingcar/> (accessed 30 November 2017).

88. Danielle Muoio, 'Here's Everything We Know About Google's Driverless Cars', *Business Insider*, 25 July 2016 <http://uk.businessinsider.com/google-driverless-car-facts-2016-7?r=US&IR=T/#the-cars-have-been-in-a-few-minor-accidents-only-one-of-which-could-be-argued-to-have-been-the-google-cars-fault-11> (accessed 30 November 2017).

89. Wallach, *Dangerous Master*, 220; Bryant Walker Smith, 'Human Error as a Cause of Vehicle Crashes', *Stanford Center for Internet and Society*, 18 December 2013 <http://cyberlaw.stanford.edu/blog/2013/12/human-error-cause-vehicle-crashes> (accessed 30 November 2017).

90. Boden, *AI*, 102.

91. Wyss Institute <http://wyss.harvard.edu/viewpage/457> (accessed 30 November 2017).

92. CBC, 'Cockroach-inspired Robots Designed for Disaster Search and Rescue', *CBC The Associated Press*, 8 February 2016 <http://www.cbc.ca/beta/news/technology/robot-roach-1.3439138> (accessed 30 November 2017).

93. Greengard, *Internet of Things*, 162.

94. Paul Ratner, 'Harvard Scientists Create a Revolutionary Robot Octopus', *Big Think*, 2016 <http://bigthink.com/paul-ratner/harvard-team-creates-octobot-the-worlds-first-autonomous-soft-robot> (accessed 30 November 2017).

95. Zoe Kleinman, 'Toyota Launches "Baby" Robot for Companionship', *BBC News*, 3 October 2016 <http://www.bbc.co.uk/news/technology-37541035> (accessed 30 November 2017).

96. Boden, *AI*, 74.

97. Jack Lynch, 'For the Price of a Smartphone You Could Bring a Robot Home', *World Economic Forum*, 7 June 2016 <https://www.weforum.org/agenda/2016/06/for-the-price-of-a-smartphone-you-could-bring-a-robot-home?utm_content=bufferafeb1&utm_

medium=social&utm_source=twitter.com&utm_campaign=buffer> (accessed 30 November 2017).

98. Robby Berman, 'So the Russians Just Arrested a Robot at a Rally', *Big Think*, 2016 <http://bigthink.com/robby-berman/so-the-russians-just-arrested-a-robot-at-a-rally> (accessed 30 November 2017).

99. Wallach, *Dangerous Master*, 82.

100. Susskind and Susskind, *Future of the Professions*, 54.

101. Tom Whipple, 'Nanorobots Could Deliver Drugs by Power of Thought', *Times*, 27 August 2016 <http://www.thetimes.co.uk/article/226da2de-6baf-11e6-998d-9617c077f056> (accessed 30 November 2017).

102. George Dvorsky, 'Record-Setting Hard Drive Writes Information One Atom at a Time', *Gizmodo*, 18 July 2016 <http://gizmodo.com/record-setting-hard-drive-writes-information-one-atom-a-1783740015 > (accessed 30 November 2017).

103. Wallach, *Dangerous Master*, 59; Rick Kelly, 'The Next Battle for Internet Freedom Could Be Over 3D Printing', *TechCrunch*, 26 August 2012 <https://techcrunch.com/2012/08/26/the-next-battle-for-internet-freedom-could-be-over-3d-printing/> (accessed 30 November 2017).

104. Jaron Lanier, *Who Owns the Future?* (London: Allen Lane, 2014), 79.

105. Wallach, *Dangerous Master*, 59.

106. Stuart Dredge, '30 Things Being 3D Printed Right Now (and None of them are Guns)', *The Guardian*, 29 January 2014 <https://www.theguardian.com/technology/2014/jan/29/3d-printing-limbs-cars-selfies> (accessed 30 November 2017).

107. Jerome Groopman, 'Print Thyself', *New Yorker*, 24 November 2014 <https://www.newyorker.com/magazine/2014/11/24/print-thyself> (accessed 30 November 2017).

108. Greengard, *Internet of Things*, 100.

109. Ibid.

110. Groopman, 'Print Thyself'.

111. Schmidt and Cohen, *New Digital Age*, 16.

112. Dredge, '30 things being 3D printed right now'.

113. BBC, 'Flipped 3D Printer Makes Giant Objects', *BBC News*, 24 August 2016 <http://www.bbc.co.uk/news/technology-37176662?ocid=socialflow_twitter> (accessed 30 November 2017).

114. Clare Scott, 'Chinese Construction Company 3D Prints an Entire Two-Story House On-Site in 45 Days', 16 June 2016 <https://3dprint.com/138664/huashang-tengda-3d-print-house/> (accessed 30 November 2017).

115. Kelly, 'Next Battle for Internet Freedom'.

116. Ariel Bogle, 'Good News: Replicas of 16th Century Sculptures Are Not Off-Limits for 3-D Printers', *Slate*, 26 January 2015 <http://www.slate.com/blogs/future_tense/2015/01/26/_3_d_printing_and_copyright_replicas_of_16th_century_sculptures_are_not.html?wpisrc=obnetwork> (accessed 30 November 2017).

117. Dredge, '30 Things Being 3D Printed Right Now'.

118. Schwab and Cohen, *New Digital Age*, 161.

119. Skylar Tibbits, *TED*, 2013 <https://www.ted.com/talks/skylar_tibbits_the_emergence_of_4d_printing?language=en> (accessed 30 November 2017).

120. Luciano Floridi, *The 4th Revolution: How the Infosphere is Reshaping Human Reality* (Oxford: Oxford University Press, 2015), 145.

121. Tim Wu, *The Master Switch: The Rise and Fall of Information Empires* (London: Atlantic, 2010), 171.

122. Rose, *Enchanted Objects*, 17.

123. Dave Gershgorn, 'Google Has Built Earbuds that Translate 40 Languages in Real Time', *Quartz*, 4 October 2017 <https://qz.com/1094638/google-goog-built-earbuds-that-translate-40-languages-in-real-time-like-the-hitchhikers-guides-babel-fish/> (accessed 7 December 2017).

124. Andrea Peterson, 'Holocaust Museum to Visitors: Please Stop Catching Pokémon Here', *Washington Post*, 12 July 2016 <https://www.washingtonpost.com/news/the-switch/wp/2016/07/12/holocaust-museum-to-visitors-please-stop-catching-pokemon-here/> (accessed 30 November 2017).

125. BBC, 'Pokemon Go: Is the Hugely Popular Game a Global Safety Risk?' *BBC News*, 21 July 2016 <http://www.bbc.co.uk/news/world-36854074> (accessed 30 November 2017).

126. Jamie Fullerton, 'Democracy Hunters Use Pokémon to Conceal Rallies', *The Times*, 3 August 2016 <http://www.thetimes.co.uk/article/democracy-hunters-use-pokemon-to-conceal-rallies-j6xrv59jl> (accessed 30 November 2017).

127. Aaron Frank, 'You Can Ban a Person, But What About Their Hologram?' *Singularity Hub*, 17 March 2017 <https://singularityhub.com/2017/03/17/you-can-ban-a-person-but-what-about-their-hologram/> (accessed 30 November 2017).

128. Dean Takahashi, 'Magic Leap Sheds Light on its Retina-based Augmented Reality 3D Displays', *VentureBeat*, 20 February 2015 <http://venturebeat.com/2015/02/20/magic-leap-sheds-light-on-its-retina-based-augmented-reality-3d-displays/> (accessed 30 November 2017).

129. Tom Simonite, 'Oculus Finally Delivers the Missing Piece for VR', *MIT Technology Review*, 6 October 2016 <https://www.technologyreview.com/s/602570/oculus-finally-delivers-the-missing-piece-for-vr/?utm_campaign=socialflow&utm_medium=post&utm_source=twitter&set=602564> (accessed 30 November 2017).

130. Richard Lai, 'bHaptics' TactSuit is VR Haptic Feedback Done Right', *Engadget*, 7 February 2017 <https://www.engadget.com/2017/07/02/bhaptics-tactsuit-vr-haptic-feedback-htc-vive-x-demo-day/?sr_source=Twitter> (accessed 30 November 2017).

131. Jordan Belamaire, 'My First Virtual Reality Groping', *Medium*, 20 October 2016 <https://medium.com/athena-talks/my-first-virtual-reality-sexual-assault-2330410b62ee#.i1o6j1vjy> (accessed 30 November 2017).

Chapter 3

1. Desjardins, Jeff. 'How much data is generated each day?' <https://www.weforum.org/agenda/2019/04/how-much-data-is-generated-each-day-cf4bddf29f/> (accessed 5 Jul. 2019).

2. Richard Susskind and Daniel Susskind, *The Future of the Professions: How Technology Will Transform the Work of Human Experts* (Oxford: Oxford University Press, 2015), 161.

3. Marc Goodman, *Future Crimes: A Journey to the Dark Side of Technology—and How to Survive It* (London: Bantam Press, 2015), 85.

4. Rob Kitchin, *The Data Revolution: Big Data, Open Data, Data Infrastructures and their Consequences* (London: Sage Publications Ltd, 2014), 69.

5. Viktor Mayer-Schönberger and Kenneth Cukier, *Big Data: A Revolution That Will Transform How We Live, Work and Think* (London: John Murray, 2013), 78.

6. Kenneth Cukier and Viktor Mayer-Schönberger, 'The Rise of Big Data', *Foreign Affairs*, May/June 2013 <https://www.foreignaffairs.com/articles/2013-04-03/rise-big-data> (accessed 30 November 2017).

7. Mayer-Schönberger and Cukier, *Big Data*, 101.

8. Elizabeth Eisenstein, *The Printing Press as an Agent of Change: Communications and Cultural Transformations in Early-modern Europe, Volumes I and II* (Cambridge: Cambridge University Press, 2009), 45. See Mayer-Schönberger and Cukier, *Big Data*, 10.

9. EMC, 'The Digital Universe of Opportunities'.

10. Radicati Group Inc, 'Email Statistics Report, 2015-2019' <http://www.radicati.com/wp/wp-content/uploads/2015/02/Email-Statistics-Report-2015-2019-Executive-Summary.pdf> (accessed 30 November 2017).

11. Cooper Smith, 'Facebook Users Are Uploading 350 Million New Photos Each Day', *Business Insider*, 18 September 2013 <http://www.businessinsider.com/facebook-350-million-photos-each-day-2013-9?IR=T> (accessed 30 November 2017); Internet Live Stats, 'Twitter Users.' <http://www.internetlivestats.com/twitter-statistics/> (accessed 30 November 2017).

12. Mayer-Schönberger and Cukier, *Big Data*, 93.

13. Kitchin, *Big Data*, 96.

14. Mayer-Schönberger and Cukier, *Big Data*, 7.

15. Mayer-Schönberger and Cukier, *Big Data*, 113.

16. Goodman, *Future Crimes*, 62.

17. Bruce Schneier, *Data and Goliath: The Hidden Battles to Collect Your Data and Control Your World* (New York: W. W. Norton & Company, 2016), 2.

18. Goodman, *Future Crimes*, 62.

19. Danny Sullivan, 'Google Now Handles at Least 2 trillion Searches Per Year', *Search Engine Land*, 24 May 2016 <http://searchengineland.com/google-now-handles-2-999-trillion-searches-per-year-250247> (accessed 30 November 2017).

20. Goodman, *Future Crimes*, 50.

21. Rob Crossley, 'Where in the World is My Data and How Secure is it?' *BBC News*, 9 August 2016 <http://www.bbc.com/news/business-36854292> (accessed 30 November 2017).

22. Kitchin, *Big Data*, 72.

23. Mayer-Schönberger and Cukier, *Big Data*, 7.

24. Mayer-Schönberger and Cukier, *Big Data*, 133.

25. Kitchin, *Big Data*, 10.

26. Mayer-Schönberger and Cukier, *Big Data*, 19.

27. Mayer-Schönberger and Cukier, *Big Data*, 38–9.

28. Goodman, *Future Crimes*, 55; Mayer-Schönberger and Cukier, *Big Data*, 119.

29. Mayer-Schönberger and Cukier, *Big Data*, 5; Steve Jones, 'Why "Big Data" is the Fourth Factor of Production', *Financial Times*, 27 December 2012 <https://www.ft.com/content/5086d700-504a-11e2-9b66-00144feab49a> (accessed 9 December 2017); Neil Lawrence, quoted in Alex Hern, 'Why Data is the New Coal', *The Guardian*, 27 September 2016 <https://www.theguardian.com/technology/2016/sep/27/data-efficiency-deep-learning> (accessed 9 December 2017).

30. Jamie Bartlett, *The Dark Net: Inside the Digital Underworld* (London: William Heinemann, 2014), 169.

31. Susskind and Susskind, *Future of the Professions*, 1.

32. Schneier, *Data and Goliath*, 4.

Chapter 4

1. Arthur C. Clarke, *Profiles of the Future: An Inquiry into the Limits of the Possible* (London: Victor Gollancz, 1999), 2.

2. See J. G. A. Pocock, *Politics, Language, and Time: Essays on Political Thought and History* (Chicago: University of Chicago Press, 1989).

3. Peter P. Nicholson, 'Politics and the Exercise of Force', in *What is Politics?* ed. Adrian Leftwich (Cambridge: Polity Press, 2015), 42.

4. See, e.g. Judith Squires, 'Politics Beyond Boundaries: A Feminist Perspective', in *What is Politics?*

5. See Bernard Crick, 'Politics as Form of Rule: Politics, Citizenship, and Democracy', in *What is Politics?*

6. Crick, 'Politics as Form of Rule', esp. 67–70.

7. Squires, 'Politics Beyond Boundaries'.

8. See Peter Bachrach and Morton S. Baratz, 'Two Faces of Power', *American Political Science Review* 56, no. 4 (December 1962): 947–52.

9. Adrian Leftwich, 'Thinking Politically: On the Politics of Politics', in *What is Politics?*

10. See, e.g. Nicholson, 'Politics and the Exercise of Force'.

11. See generally, *Political Innovation and Conceptual Change*, eds. Terence Ball, James Farr, and Russell L. Hanson (New York: Cambridge University Press, 1995); Michael Freeden, *Ideologies and Political Theory: A Conceptual Approach* (Oxford: Oxford University Press, 1996).

12. On the concept/conception distinction, see John Rawls, *A Theory of Justice* (Cambridge, Mass: Harvard University Press, 2003), 5.

13. See Freeden, *Ideologies*.

14. Freeden, *Ideologies*, 53.

15. Yuval Noah Harari, *Sapiens: A Brief History of Humankind* (London: Vintage Books, 2011), 121.

16. Larry Siedentop, *Inventing the Individual: The Origins of Western Liberalism* (London: Allen Lane, 2014), 16–17.

17. Plato, *The Laws*, translated by Tom Griffith (Cambridge: Cambridge University Press, 2016), XI 414–5; see Siedentop, *Inventing the Individual*, 16–17.

18. Andrew J. Beniger, *Control Revolution: Technological and Economic Origins of the Information Society* (Cambridge, Mass: Harvard University Press, 1990), 7.

19. Sandra Braman, *Change of State: Information, Policy, and Power* (Cambridge, Mass: MIT Press, 2009), 2.

20. James Gleick, *The Information: A History, A Theory, A Flood* (London: Fourth Estate, 2012), 7–8.

21. Erik Brynjolfsson and Andrew McAfee, *The Second Machine Age: Work, Progress, and Prosperity in a Time of Brilliant Technologies* (New York: W. W. Norton & Company, 2014), 16.

22. Karl Marx, *The German Ideology*, in *Karl Marx and Frederick Engels: Collected Works Vol. 5* (London: Lawrence & Wishart, 1976), 36.

23. Karl Mannheim, *Ideology and Utopia: An Introduction to the Sociology of Knowledge*, translated by Louis Wirth and Edward Shils (Connecticut: Martino Publishing, 2015), 3.

24. Marx, *German Ideology*, 59.

25. Cited in Gleick, *Information*, 51.

26. Eric Hobsbawm, *The Age of Revolution: 1789–1848* (New York: Vintage Books, 1996), 1.

27. William Blake, *London*, Poetry Foundation <https://www.poetryfoundation.org/poems/43673/london-56d222777e969> (accessed 7 December 2017).

28. Adam Swift, 'Political Philosophy and Politics', in *What is Politics?* 141.

29. Swift, 'Political Philosophy and Politics', 140.

30. George Orwell, 'Politics and the English Language', in *Essays* (London: Penguin, 2000), 359.

31. Winston Churchill, *My Early Life: A Roving Commission* (London: Reprint Society, 1944), 66.

32. Ludwig Wittgenstein, *Tractatus Logico-Philosophicus* (Abingdon: Routledge, 2001), 3.

33. See William E. Connolly, *The Terms of Political Discourse* (Third Edition) (Oxford: Basil Blackwell, 1994).

34. Daniel McDermott, 'Analytical Political Philosophy', in *Political Theory: Methods and Approaches*, eds. David Leopold and Marc Stears (Oxford: Oxford University Press, 2010), 11.

35. Adam Swift and Stuart White, 'Political Theory, Social Science, and Real Politics', in *Political Theory*, 52.

36. Marx, *German Ideology*, 36.

Chapter 5

1. Steven Lukes, *Power: A Radical View* (Second Edition) (Basingstoke: Palgrave Macmillan, 2005), 34.

2. Robert Dahl, 'The Concept of Power', *Behavioral Science* 2, 201–15, cited in Lukes, *PRV*, 16.

3. Lukes, *PRV*, 5.

4. See Robert Dahl, 'Power as the Control of Behaviour' in *Power*, ed. Steven Lukes (New York: New York University Press, 1986), 41; Lukes, *PRV,* 74–5.

5. Lukes, *PRV*, 21–2; Peter Bachrach and Morton S. Baratz, *Power and Poverty: Theory and Practice* (New York: Oxford University Press, 1970).

6. Rob Kitchin and Martin Dodge, *Code/Space: Software and Everyday Life* (Cambridge, Mass: MIT Press, 2014), 3–5.

7. James Grimmelmann, 'Regulation by Software' *Yale Law Journal* 114, no. 7 (May 2005), 1729.

8. Steiner, Christopher, *Automate This: How Algorithms Came to Rule Our World* (London: Portfolio, 2012), 55; Ed Finn, *What Algorithms Want: Imagination in the Age of Computing* (Cambridge, Mass: MIT Press, 2017), 17.

9. See generally Lawrence Lessig, *Code Version 2.0* (New York: Basic Books, 2006).

10. Grimmelmann, 'Regulation by Software', 1729.

11. Gordon Brown, My Life, Our Times (London: Bodley Head, 2017), 326.

12. Julie E. Cohen, *Configuring the Networked Self: Law, Code, and the Play of Everyday Practice* (New Haven and London: Yale University Press, 2012), 155.

13. Lessig, *Code 2.0*, 298.

14. Finn, *What Algorithms Want*, 6.

Chapter 6

1. Michel Foucault, *Discipline and Punish: The Birth of the Prison*, translated by Alan Sheridan (New York: Vintage Books, 1995), 74.

2. Christopher Dandeker, *Surveillance, Power and Modernity* (Cambridge: Polity, 1990), 119.

3. Foucault, *Discipline and Punish*.

4. See Tim O'Reilly, 'Open Data and Algorithmic Regulation', in *Beyond Transparency: Open Data and the Future of Civic Innovation*, eds. Brett Goldstein and Lauren Dyson (San Francisco: Code for America Press, 2013), 195.

5. Richard Susskind and Daniel Susskind, *The Future of the Professions: How Technology Will Transform the Work of Human Experts* (Oxford: Oxford University Press, 2015), 70.

6. Nanette Byrnes, 'As Goldman Embraces Automation, Even the Masters of the Universe Are Threatened', *MIT Technology Review*, 7 February 2017 <https://www.technologyreview.com/s/603431/as-goldman-embraces-automation-even-the-masters-of-the-universe-are-threatened/?set=603585&utm_content=bufferd5a8f&utm_

medium=social&utm_source=twitter.com&utm_campaign=buffer> (accessed 1 December 2017).

7. O'Reilly, 'Open Data and Algorithmic Regulation', 291.

8. Steve Rosenbush, 'The Morning Download: China's Facial Recognition ID's Citizens and Soon May Score Their Behaviour', *Wall Street Journal*, 27 July 2017 <https://blogs.wsj.com/cio/2017/06/27/the-morning-download-chinas-facial-recognition-ids-citizens-and-soon-may-score-their-behavior/> (accessed 1 December 2017).

9. Hans Kelsen, *Pure Theory of Law*, translated from the Second (Revised and Enlarged) German Edition by Max Knight (New Jersey: Law Book Exchange, 2009).

10. H. L. A. Hart, *The Concept of Law* (Second Edition) (Oxford: Oxford University Press, 1997), 35–6.

11. Foucault, *Discipline and Punish*, 82.

12. Foucault, *Discipline and Punish*, ch.1.

13. Foucault, *Discipline and Punish*, 9.

14. Lessig, *Code 2.0*, 82.

15. *New York Times*, 'Why Not Smart Guns in This High-Tech Era?' Editorial, 26 November 2016 <http://mobile.nytimes.com/2016/11/26/opinion/sunday/why-not-smart-guns-in-this-high-tech-era.html?smid=tw-nytopinion&smtyp=cur&referer=> (accessed 1 December 2017).

16. Hart, *Concept of Law*, 27–8, 48.

17. Melanie Swan, *Blockchain: Blueprint for a New Economy* (Sebastopol, CA: O'Reilly, 2015), 14.

18. See Primavera De Filippi and Aaron Wright, *Blockchain and the Law: The Rule of Code* (Cambridge, Mass: Harvard University Press, 2018), ch. 12; on computable contracts see also Harry Surden, 'Computable Contracts', *UC Davis Law Review* 46, (2012): 629–700.

19. See De Filippi and Wright, *Blockchain and the Law*, ch. 12.

20. O'Reilly, 'Open Data and Algorithmic Regulation', 295.

21. Grimmelmann, 'Regulation by Software', 1732.

22. Mark Bridge, 'AI Can Identify Alzheimer's Disease a Decade before Symptoms Appear', *The Times*, 20 September 2017 <https://www.thetimes.co.uk/article/ai-can-identify-alzheimer-s-a-decade-before-symptoms-appear-9b3qdrrf7> (accessed 1 December 2017).

23. Wendell Wallach and Colin Allen, *Moral Machines: Teaching Robots Right from Wrong* (Oxford: Oxford University Press, 2009), 27.

24. Nikolaos Aletras, Dimitrios Tsarapatsanis, Daniel Preotiuc, and Vasileios Lampos. 'Predicting Judicial Decisions of the European Court of Human Rights: A Natural Language Processing Perspective'. *Peer J Computer Science* 2, e93 (24 October 2016). See further Harry Surden, 'Machine Learning and Law', *Washington Law Review* 89, no. 1 (2014): 87–115.

25. Erik Brynjolfsson and Andrew McAfee *Machine Platform Crowd: Harnessing Our Digital Future* (New York: W.W. Norton & Company, 2017), 41.

26. See Anthony J. Casey and Anthony Niblett, 'The Death of Rules and Standards', *Indiana Law Journal* 92, no. 4 (2017); Anthony J. Casey and Anthony Niblett, 'Self-Driving Laws' *University of Toronto Law Journal* 429 (Fall 2016) 66: 428–42.

27. Casey and Niblett, 'Death of Rules and Standards'; 'Self-Driving Laws'.

28. Oliver Wendell Holmes, 'The Path of the Law' *Harvard Law Review* 10, no. 457 (1897); Casey and Niblett, 'Death of Rules and Standards', 1422.

29. Primavera De Filippi and Samer Hassan, 'Blockchain Technology as a Regulatory Technology: From Code is Law to Law is Code', *First Monday* 21, no. 12, 5 December 2016.

30. Walter Ong, *Orality and Literacy* (Abingdon: Routledge, 2012), 31.

31. Richard Susskind, *The Future of Law: Facing the Challenges of Information Technology* (Oxford: Oxford University Press, 1998), 92–4.

32. Eric A. Havelock, *The Greek Concept of Justice: From its Shadow in Homer to its Substance in Plato.* (Cambridge, Mass: Harvard University Press, 1978), 135, 14.

33. Havelock, *Greek Concept of Justice*, 23–36; Eric A. Havelock, *The Muse Learns to Write: Reflections on Orality and Literacy from Antiquity to the Present* (New Haven and London: Yale University Press, 1986), 4.

34. Susskind, *Future of Law*, 92–94.

35. Alexis de Tocqueville, *Democracy in America*, translated by George Lawrence (New York: HarperCollins, 2006), 49.

36. Casey and Niblett, 'Death of Rules and Standards'; 'Self-Driving Laws'.

37. Ibid.

38. Ibid.

39. Max Weber, 'The Profession and Vocation of Politics', in *Political Writings*, eds. Peter Lassman and Ronald Speirs (Cambridge: Cambridge University Press, 2010), 310–11.

40. Thomas Hobbes, *Leviathan* (Cambridge: Cambridge University Press, 2007), [62], 88.

41. Jean-Jacques Rousseau, *The Social Contract*, translated by Maurice Cranston (London: Penguin, 1968), 371.

42. David Hume, 'Of the Original Contract', in *Selected Essays* (Oxford: Oxford University Press, 2008), 283.

43. See Manuel Castells, *Communication Power* (Oxford: Oxford University Press, 2013), 43–5.

44. Aaron Perzanowksi and Jason Schultz, *The End of Ownership: Personal Property in the Digital Economy* (Cambridge, Mass: MIT Press, 2016), 4.

45. See Philippa Foot, 'The Problem of Abortion and the Doctrine of the Double Effect', in *Virtues and Vices and Other Essays in Moral Philosophy* (Oxford: Oxford University Press, 2009).

46. Lessig, *Code 2.0*, 78.

47. Elizabeth Anderson, *Private Government: How Employers Rule Our Lives (and Why We Don't Talk About It)* (Princeton and Oxford: Princeton University Press, 2017), 55.

48. Wallach and Allen, *Moral Machines*, 26–7.

49. De Filippi and Wright, *Blockchain and the Law*, 159.

50. De Filippi and Wright, *Blockchain and the Law*, 22.

51. De Filippi and Wright, *Blockchain and the Law*, 165.

52. The example of an airport security system is from Wallach and Allen, *Moral Machines*, 15.

Chapter 7

1. Michel Foucault, *Power/Knowledge: Selected Interviews and Other Writings, 1972–1977* (New York: Vintage Books, 1980), 152.

2. John Milton, *Paradise Lost* (London: Penguin, 2003) Book IX, 203–04.

3. On chilling effects, see Jon Penney, 'Internet Surveillance, Regulation, and Chilling Effects Online: A Comparative Case Study', *Internet Policy Review* 6, no. 2 (2017): 1–38.

4. Sandra Bartky, 'Foucault, Femininity, and the Modernization of Patriarchal Power', cited in Steven Lukes, *Power: A Radical View* (Second Edition) (Basingstoke: Palgrave Macmillan, 2005), 99.

5. Foucault, *Power/Knowledge*, 39.

6. Foucault, *Power/Knowledge*, 158.

7. Foucault, *Power/Knowledge*, 155.

8. Michel Foucault, *Discipline and Punish: The Birth of the Prison*, translated by Alan Sheridan (New York: Vintage Books, 1995), 173.

9. James C. Scott, *Seeing Like a State* (New Haven and London: Yale University Press, 1998).

10. Scott, *Seeing Like a State*, 77.

11. Benjamin Constant, *De l'esprite de conquête*, cited in Scott, *Seeing Like a State*, 30 (original emphasis).

12. Scott, *Seeing Like a State*, 54–7.

13. Scott, *Seeing Like a State*, 66.

14. Scott, *Seeing Like a State*, 67.

15. Scott, *Seeing Like a State*, 65.

16. Scott, *Seeing Like a State*, 71.

17. Viktor Mayer-Schönberger and Kenneth Cukier, *Big Data: A Revolution That Will Transform How We Live, Work and Think* (London: John Murray, 2013), 152.

18. Zeynep Tufekci, *Twitter and Tear Gas: The Power and Fragility of Networked Protest* (New Haven: Yale University Press, 2017), 6.

19. John Cheney-Lippold, *We Are Data: Algorithms and the Making of Our Digital Selves* (New York: New York University Press, 2017); see also Gilles Deleuze, 'Postscript on the Societies of Control', *October* 59 (Winter, 1992): 3–7.

20. See generally Cheney-Lippold, *We Are Data*.

21. Cheney-Lippold, *We Are Data*, 6.

22. Cheney-Lippold, *We Are Data*, 10.

23. Friedrich Hayek, 'The Use of Knowledge in Society', *The American Economic Review* 35, no. 4 (September 1945), 521–4.

24. Scott, *Seeing Like a State*, 87.

25. Jake Swearingen, 'Can an Amazon Echo Testify Against You?' *NY Mag*, 27 December 2016 <http://nymag.com/selectall/2016/12/can-an-amazon-echo-testify-against-you.html> (accessed 1 December

2017); Billy Steele, 'Police Seek Amazon Echo Data in Murder Case', *Engadget*, 27 December 2016 <https://www.engadget.com/2016/12/27/amazon-echo-audio-data-murder-case/> (accessed 1 December 2017).

26. Christine Hauser, 'In Connecticut Murder Case, a Fitbit Is a Silent Witness', *New York Times*, 27 April 2017 <https://www.nytimes.com/2017/04/27/nyregion/in-connecticut-murder-case-a-fitbit-is-a-silent-witness.html?smid=tw-nytimes&smtyp=cur> (accessed 1 December 2017).

27. Sam Machkovech, 'Marathon Runner's Tracked Data Exposes Phony Time, Cover-up Attempt', *Ars Technica UK*, 22 February 2017 <https://arstechnica.com/gadgets/2017/02/suspicious-fitness-tracker-data-busted-a-phony-marathon-run/> (accessed 1 December 2017).

28. Cleve R. Wootson Jr, 'A Man Detailed His Escape from a Burning House. His Pacemaker Told Police a Different Story', *Washington Post*, 8 February 2017 <https://www.washingtonpost.com/news/to-your-health/wp/2017/02/08/a-man-detailed-his-escape-from-a-burning-house-his-pacemaker-told-police-a-different-story/?tid=sm_tw&utm_term=.531d8fabc6d2> (accessed 1 December 2017).

29. *Semayne's Case* (1604) 5 Coke Reports 91a 77 E.R. 194.

30. David Rose, *Enchanted Objects: Design, Human Desire, and the Internet of Things* (New York: Scribner, 2014), 7.

31. Leo Mirani, 'Personal Technology Gets Truly Personal', in *Megatech: Technology in 2050*, ed. Daniel Franklin (New York: Profile Books, 2017), 150.

32. Alex Hern, 'Vibrator Maker Ordered to Pay Out C$4m for Tracking Users' Sexual Activity', *The Guardian*, 14 March 2017 <https://www.theguardian.com/technology/2017/mar/14/we-vibe-vibrator-tracking-users-sexual-habits?CMP=Share_iOSApp_Other> (accessed 1 December 2017).

33. Spencer Ackerman and Sam Thielman, 'US Intelligence Chief: We Might Use the Internet of Things to Spy on You', *The Guardian*, 9 February 2016 <https://www.theguardian.com/technology/2016/feb/09/internet-of-things-smart-home-devices-government-surveillance-james-clapper> (accessed 1 December 2017).

34. Plato, *Phaedrus*, translated by Christopher Rowe (London: Penguin, 2005), 62.

35. See David Rieff, *In Praise of Forgetting: Historical Memory and its Ironies* (New Haven and London: Yale University Press, 2017).

36. Viktor Mayer-Schönberger and Kenneth Cukier, *Delete: The Virtue of Forgetting in the Digital Age* (Princeton: Princeton University Press, 2009), 2.

37. Mayer-Schönberger and Cukier, *Delete*, 6.

38. Mayer-Schönberger and Cukier, *Delete*, 104.

39. Nadia Khomami, 'Ministers Back Campaign to Give Under-18s Right to Delete Social Media Posts', *The Guardian*, 28 July 2015 <https://www.theguardian.com/media/2015/jul/28/ministers-back-campaign-under-18s-right-delete-social-media-posts> (accessed 1 December 2017).

40. Meg Leta Jones, *Ctrl + Z: The Right to Be Forgotten* (New York: New York University Press, 2016), 1.

41. Leta Jones, *Ctrl + Z*, 9–11.

42. Eric Siegel, *Predictive Analytics: The Power to Predict Who Will Click, Buy, Lie, or Die* (New Jersey: John Wiley & Sons, Inc., 2016), 11.

43. Walter Perry et al., *Predictive Policing: The Role of Crime Forecasting in Law Enforcement Operations* (Santa Monica: RAND Corporation, 2013).

44. Siegel, *Predictive Analytics*, centrefold (table 5).

45. Frank Pasquale, *The Black Box Society: The Secret Algorithms that Control Money and Information* (Cambridge, Mass: Harvard University Press, 2015), 23–6.

46. Josh Chin and Gillian Wong, 'China's New Tool for Social Control: A Credit Rating for Everything', *Wall Street Journal*, 28 November 2016 <http://www.wsj.com/articles/chinas-new-tool-for-social-control-a-credit-rating-for-everything-1480351590> (accessed 1 December 2017); *Economist*, 'China Invents the Digital Totalitarian State', 17 December 2016 <http://www.economist.com/news/briefing/21711902-worrying-implications-its-social-credit-project-china-invents-digital-totalitarian> (accessed 1 December 2017).

47. See Mara Hvistendahl, 'Inside China's Vast New Experiment in Social Ranking', *Wired*, 14 December 2017 <https://www.wired.com/story/age-of-social-credit/> (accessed 21 January 2018).

Chapter 8

1. See Steven Lukes, *Power: A Radical View* (Second Edition) (Basingstoke: Palgrave Macmillan, 2005).

2. See Peter Bachrach and Morton S. Baratz, *Power and Poverty: Theory and Practice* (New York: Oxford University Press, 1970).

3. See E. E. Schattschneider, *The Semisovereign People: A Realist's View of Democracy in America* (South Melbourne, Victoria: Wadsworth Thomson Learning, 1975).

4. Lukes, *PRV*, 20–5.

5. See Lukes, *PRV*, 27–8.

6. Manuel Castells, *Communication Power* (Oxford: Oxford University Press, 2013), 3.

7. Karl Marx, *The German Ideology*, in *Karl Marx and Frederick Engels: Collected Works Vol. 5* (London: Lawrence & Wishart, 1976), 41.

8. See Antonio Gramsci, *Selections from the Prison Notebooks* (London: Lawrence & Wishart, 2007).

9. Karl Marx, *Contribution to Critique of Hegel's Philosophy of Law. Introduction*, in *Karl Marx and Frederick Engels: Collected Works Vol. 3.* (London: Lawrence & Wishart, 1975), 187.

10. Yochai Benkler, *The Wealth of Networks: How Social Production Transforms Markets and Freedom* (New Haven & London: Yale University Press, 2006), 130.

11. Benkler, *Wealth of Networks*, 168.

12. Yochai Benkler, 'Degrees of Freedom, Dimensions of Power', *Daedalus* 145, no. 1 (Winter 2016), 21.

13. Eric Schmidt and Jared Cohen, *The New Digital Age: Reshaping the Future of People, Nations and Business.* (London: John Murray, 2014), 82.

14. Benkler, 'Degrees of Freedom', 21.

15. Eric Siegel, *Predictive Analytics: The Power to Predict Who Will Click, Buy, Lie, or Die* (New Jersey: John Wiley & Sons, Inc, 2016), centrefold (table 1).

16. Robert Epstein, 'The New Censorship', *US News*, 22 July 2016 <http://www.usnews.com/opinion/articles/2016-06-22/google-is-the-worlds-biggest-censor-and-its-power-must-be-regulated> (accessed 1 December 2017).

17. See e.g. Allison Linn, 'Microsoft Creates AI that Can Read a Document and Answer Questions About it As Well As a Person', *The AI Blog*,

15 January 2018 <https://blogs.microsoft.com/ai/microsoft-creates-ai-can-read-document-answer-questions-well-person/> (accessed 21 January 2018).

18. See Jonathan Zittrain, 'Apple's Emoji Gun Control', *New York Times*, 16 August 2016 <https://mobile.nytimes.com/2016/08/16/opinion/get-out-of-gun-control-apple.html?_r=0&referer=https://www.google.com/> (accessed 1 December 2017).

19. Lotus Ruan, Jeffrey Knockel, Jason Q. Ng, and Masashi Crete-Nishihata, 'One App, Two Systems', *The Citizen Lab*, 30 November 2016 <https://citizenlab.ca/2016/11/wechat-china-censorship-one-app-two-systems/> (accessed 1 December 2017).

20. Zittrain, 'Apple's Emoji Gun Control'.

21. Robert Booth, 'Facebook Reveals News Feed Experiment to Control Emotions', *The Guardian*, 30 June 2004 <https://www.theguardian.com/technology/2014/jun/29/facebook-users-emotions-news-feeds> (accessed 11 December 2017).

22. Halting Problem, 'Tech Bro Creates Augmented Reality App to Filter Out Homeless People', *Medium*, 23 February 2016 <https://medium.com/halting-problem/tech-bro-creates-augmented-reality-app-to-filter-out-homeless-people-3bf8d827b0df> (accessed 7 December 2017).

23. Frank Pasquale, *The Black Box Society: The Secret Algorithms that Control Money and Information* (Cambridge, Mass: Harvard University Press, 2015), 63; Benkler, 'Degrees of Freedom', 18.

24. Pasquale, *Black Box Society*, 60.

25. Bobby Johnson, 'Amazon Kindle Users Surprised by "Big Brother" Move', *The Guardian*, 17 July 2009 <https://www.theguardian.com/technology/2009/jul/17/amazon-kindle-1984> (accessed 8 December 2017).

26. Jonathan Zittrain, 'Engineering an Election', *Harvard Law Review Forum*, 20 June 2014 <https://harvardlawreview.org/2014/06/engineering-an-election/> (accessed 1 December 2017).

Chapter 9

1. 'Who? Whom?' *Wikipedia*, last modified 3 June 2017 <https://en.wikipedia.org/wiki/Who,_whom%3F> (accessed 7 December 2017).

2. Michael Walzer, *Spheres of Justice: A Defense of Pluralism and Equality* (New York: Basic Books, 1983), xiii.

3. Walzer, *Spheres*, 11.

4. Carol Gould, *Rethinking Democracy: Freedom and Social Cooperation in Politics, Economy, and Society* (Cambridge: Cambridge University Press, 1990), 271.

5. Gould, *Rethinking Democracy*, 272.

6. See Joshua A. T. Fairfield, *Owned: Property, Privacy, and the New Digital Serfdom* (Cambridge: Cambridge University Press, 2017).

7. Sheila Jasanoff, *The Ethics of Invention: Technology and the Human Future* (New York: W. W. Norton & Company, 2016), 169.

8. Frank Pasquale, *The Black Box Society: The Secret Algorithms That Control Money and Information* (Cambridge, Mass: Harvard University Press, 2015), 21.

9. John Markoff, *Machines of Loving Grace: the Quest for Common Ground Between Humans and Robots* (New York: HarperCollins, 2015), xvi.

10. Robert W. McChesney, *Digital Disconnect: How Capitalism is Turning The Internet Against Democracy* (New York: The New Press, 2014), 166.

11. McChesney, *Digital Disconnect*, 162.

12. Philip N. Howard, *Pax Technica: How the Internet of Things May Set Us Free or Lock Us Up* (New Haven and London: Yale University Press, 2015), xix–xx.

13. Lodewijk F. Asscher, '"Code" as Law: Using Fuller to Assess Code Rules', in *Coding Regulation: Essays on the Normative Role of Information Technology*, eds. E. J. Dommering and Lodewijk F. Asscher (The Hague: TMC Asser, 2006), 69.

14. Jasanoff, *Ethics of Invention*, 171.

15. Yochai Benkler, 'Degrees of Freedom, Dimensions of Power', *Daedalus* 145, no. 1 (Winter 2016), 23.

16. Pasquale, *Black Box Society*, 94.

17. See Jaron Lanier, *Who Owns the Future?* (London: Allen Lane, 2014), 240.

18. John Nichols, 'If Trump's FCC Repeals Net Neutrality, Elites Will Rule the Internet—and the Future', *Nation*, 24 November 2017 <https://www.thenation.com/article/if-trumps-fcc-repeals-net-neutrality-elites-will-rule-the-internet-and-the-future/> (accessed 1 December 2017).

19. Walzer, *Spheres*, 294.

20. Elizabeth Anderson, *Private Government: How Employers Rule Our Lives (and Why We Don't Talk About It)* (Princeton and Oxford: Princeton University Press, 2017), 9.

21. Ibid.

22. Alexis de Tocqueville, *Democracy in America*, translated by George Lawrence (New York: HarperCollins, 2006), 692.

Chapter 10

1. Friedrich Hayek, *The Constitution of Liberty* (Abingdon: Routledge, 2009), 17.

2. Erich Fromm, *The Fear of Freedom* (Abingdon: Routledge, 2009), 3.

3. Inaugural Address, 20 January 1961.

4. Gerald Dworkin, *The Theory and Practice of Autonomy* (Cambridge: Cambridge University Press, 1989) 15–20.

5. Jean-Jacques Rousseau, *The Social Contract*, translated by Maurice Cranston (London: Penguin, 1968), 65.

6. Thomas Scanlon, 'A Theory of Freedom of Expression', *Philosophy and Public Affairs* 1, no. 2 (1972), 215.

7. Isaiah Berlin, 'Two Concepts of Liberty', in *Four Essays on Liberty* (Oxford: Oxford University Press, 1969), 134.

8. Freedom.to <https://freedom.to/> (accessed 7 December 2017).

9. Fromm, *Fear of Freedom*, 208 (emphasis removed).

10. See, e.g., Quentin Skinner, *Liberty Before Liberalism* (Cambridge: Cambridge University Press, 2012) and Philip Pettit, 'The Republican Ideal of Freedom', in *The Liberty Reader*, ed. David Miller (Edinburgh: Edinburgh University Press, 2006).

11. Skinner, *Liberty Before Liberalism*, 23.

12. Dworkin, *Autonomy*, 13.

13. Quentin Skinner, 'The Republican Ideal of Political Liberty', in *Machiavelli and Republicanism*, eds. Gisela Bock, Quentin Skinner, and Maurizio Viroli (Cambridge: Cambridge University Press, 1993), 303.

14. Pettit, 'The Republican Ideal of Freedom', 226; Skinner, 'A Third Concept of Liberty', in *Liberty Reader*, 250.

15. Skinner, 'A Third Concept of Liberty', 250.

16. Skinner, 'A Third Concept of Liberty', 254.

17. Donald A. Norman, *The Design of Future Things* (New York: Basic Books, 2007), 68.

18. Daniel Cooper, 'These Subtle Smart Gloves Turn Sign Language into Text', *Engadget,* 31 May 2017 <https://www.engadget.com/2017/05/31/these-subtle-smart-gloves-turn-sign-language-into-words/?sr_source=Twitter> (accessed 1 December 2017).

19. Brian D. Wassom, *Augmented Reality Law, Privacy, and Ethics: Law, Society, and Emerging AR Technologies.* (Rockland: Syngress, 2015), 250.

20. Bruce Goldman, 'Typing With Your Mind: How Technology is Helping the Paralyzed Communicate', *World Economic Forum*, 1 March 2017 <https://www.weforum.org/agenda/2017/03/this-technology-allows-paralysed-people-to-type-using-their-mind?utm_content=buffer8a986&utm_medium=social&utm_source=twitter.com&utm_campaign=buffer)%20%20(Ref)%20ch.21%20of%20leviathan?> (accessed 1 December 2017).

21. See, e.g., Francis Fukuyama, *Our Posthuman Future: Consequences of the Biotechnology Revolution* (London: Profile Books, 2002); Max More and Natasha Vita-More, eds., *The Transhumanist Reader: Classical and Contemporary Essays on the Science, Technology, and Philosophy of the Human Future* (Chichester: John Wiley & Sons, Inc, 2013); Julian Savulescu, Ruud ter Meulen, and Guy Kahane, eds., *Enhancing Human Capacities* (Chichester: Wiley-Blackwell, 2011); Justin Nelson et al., 'The Effects of Transcranial Direct Current Stimulation (tDCS) on Multitasking Throughput Capacity' *Frontiers in Human Neuroscience*, 29 November 2016 <https://www.frontiersin.org/articles/10.3389/fnhum.2016.00589/full> (accessed 8 December 2017); Michael Bess, 'Why Humankind Isn't Ready for the Bionic Revolution', *Ozy*, 24 October 2016. <http://www.ozy.com/opinion/why-humankind-isnt-ready-for-the-bionic-revolution/72555?utm_source=dd&utm_medium=email&utm_campaign=10242016&variable=af3d1702308a23693509dd3317fe68e7> (accessed 8 December 2017).

22. *The Correspondence of John Stuart Mill and Auguste Comte*, ed. Oscar A. Haac (London: Transaction, 1995), Foreword and Introduction.

23. Helen Nissenbaum, *Privacy in Context: Technology, Policy, and the Integrity of Social Life* (Stanford: Stanford University Press, 2010), 83.

24. Cass R. Sunstein, *The Ethics of Influence: Government in the Age of Behavioral Science* (New York: Cambridge University Press, 2016), 82.

25. Dworkin, *Autonomy*, 18.

26. Sarah Dean, 'A Nation of "Micro-Criminals": The 11 Sneaky Crimes We Are Commonly Committing', *iNews*, 22 October 2016 <https://

inews.co.uk/essentials/news/uk/nation-micro-criminals-11-sneaky-crimes-commonly-committing/> (accessed 1 December 2017).

27. Blaise Agüera y Arcas, Margaret Mitchell, and Alexander Todorov, 'Physiognomy's New Clothes', *Medium*, 6 May 2017 <https://medium.com/@blaisea/physiognomys-new-clothes-f2d4b59fdd6a> (accessed 1 December 2017).

28. Bernard E. Harcourt, *Against Prediction: Profiling, Policing, and Punishing in an Actuarial Age* (Chicago: University of Chicago Press, 2007), 179.

29. Harcourt, *Against Prediction*, 174.

30. Agüera y Arcas et al., 'Physiognomy's New Clothes'.

31. Ibid.

32. John Stuart Mill, *On Liberty and Other Writings* (Cambridge: Cambridge University Press, 2008), 15.

33. Kenneth Cukier, 'The Data-driven World', in *Megatech: Technology in 2050*, ed. Daniel Franklin (New York: Profile Books, 2017), 171.

34. Jason Tashea, 'Courts are Using AI to Sentence Criminals: That Must Stop Now', *Wired*, 17 April 2017 <https://www.wired.com/2017/04/courts-using-ai-sentence-criminals-must-stop-now/> (accessed 1 December 2017).

35. Julia Anwin, Jeff Larson, Surya Mattu, and Lauren Kirchner, 'Machine Bias', *ProPublica*, 23 May 2016 <https://www.propublica.org/article/machine-bias-risk-assessments-in-criminal-sentencing> (accessed 1 December 2017).

36. Isaiah Berlin, 'Historical Inevitability', in *Four Essays on Liberty*, 63.

37. Berlin, 'Historical Inevitability', 57.

38. Auguste Comte, 'Plan of the Scientific Work Necessary for the Reorganization of Society', in *Early Political Writings* (Cambridge: Cambridge University Press, 1998), 100.

39. Comte, 'Plan', 81–121.

40. Aristotle, *The Politics*, translated by T. A. Sinclair (London: Penguin, 1992), 1281a2, 198.

41. Aristotle, *Nichomachean Ethics*, translated by Terence Irwin (Second Edition) (Indianapolis: Hackett Publishing Company, 1999), II, i, 18–9.

42. Roger Brownsword and Morag Goodwin, *Law and the Technologies of the Twenty-First Century: Texts and Materials* (Cambridge: Cambridge University Press, 2012), 447.

43. Alfred North Whitehead, *An Introduction to Mathematics* (Milton Keynes: Watchmaker, 2011), 61.

44. Jathan Sadowski and Frank Pasquale, 'The Spectrum of Control: A Social Theory of the Smart City', *First Monday* 20, no. 7, 6 July 2015 <http://digitalcommons.law.umaryland.edu/cgi/viewcontent.cgi?article=2545&context=fac_pubs> (accessed 1 December 2017).

45. Rob Kitchin, *The Data Revolution: Big Data, Open Data, Data Infrastructures and their Consequences*. (London: Sage Publications Ltd, 2014), 71.

46. See Sadowski and Pasquale, 'Spectrum of Control'; Cory Doctorow, 'Riot Control Drone that Fires Paintballs, Pepper-spray and Rubber Bullets at Protesters', *Boing Boing*, 17 June 2014 <https://boingboing.net/2014/06/17/riot-control-drone-that-paintb.html> (accessed 7 December 2017); Desert Wolf, 'Skunk Riot Control Copter', <http://www.desert-wolf.com/dw/products/unmanned-aerial-systems/skunk-riot-control-copter.html> (accessed 1 December 2017).

47. See Richard Yonck, *Heart of the Machine: Our Future in a World of Artificial Intelligence* (New York: Arcade Publishing, 2017), 137.

48. Henry David Thoreau, *On the Duty of Civil Disobedience* (1854) in *Political Thought*, eds. Michael Rosen and Jonathan Wolff (Oxford: Oxford University Press, 1999), 81.

49. John Rawls, *A Theory of Justice* (Cambridge, Mass: Harvard University Press, 2003), 319–23.

50. Martin Luther King, *Letter from Birmingham City Jail* (1963) in *Political Thought*, 85.

51. E. Gabriella Coleman, *Coding Freedom: The Ethics and Aesthetics of Hacking* (Princeton: Princeton University Press, 2013), 19.

52. Ibid.

53. Tom Simonite, 'Pentagon Bot Battle Shows How Computers Can Fix Their Own Flaws', *MIT Technology Review*, 4 August 2016 <https://www.technologyreview.com/s/602071/pentagon-bot-battle-shows-how-computers-can-fix-their-own-flaws/?utm_campaign=socialflow&utm_source=twitter&utm_medium=post> (accessed 1 December 2017).

54. Rawls, *Theory of Justice*, 326–31.

55. Steven Levy, *Crypto: How the Code Rebels Beat the Government—Saving Privacy in the Digital Age* (New York: Penguin, 2002), 1.

56. Robert Scoble and Israel Shel, *The Fourth Transformation: How Augmented Reality and Artificial Intelligence Change Everything* (CreateSpace Independent Publishing Platform, 2017), 124.

57. BBC, 'German Parents Told to Destroy Cayla Dolls Over Hacking Fears', *BBC News*, 17 February 2017 <http://www.bbc.co.uk/news/world-europe-39002142> (accessed 1 December 2017).

58. Scoble and Shel, *Fourth Transformation*, 124.

59. Marc Goodman, *Future Crimes: A Journey to the Dark Side of Technology—and How to Survive It* (London: Bantam Press, 2015), 22–3.

60. Goodman, *Future Crimes*, 249.

61. William J. Mitchell, *Me ++: The Cyborg Self and the Networked City* (Cambridge, Mass: MIT Press, 2003), 5.

62. Justin Clark et al., 'The Shifting Landscape of Global Internet Censorship', *Internet Monitor*, 29 June 2017 <https://thenetmonitor.org/research/2017-global-internet-censorship> (accessed 1 December 2017).

63. Reuters, 'Turkey Blocks Wikipedia Under Law Designed to Protect National Security', *The Guardian*, 30 April 2017 <https://www.theguardian.com/world/2017/apr/29/turkey-blocks-wikipedia-under-law-designed-to-protect-national-security> (accessed 8 December 2017); Dahir, Abdi Latif. 'Egypt Has Blocked Over 100 Local and International Websites Including HuffPost and Medium'. *Quartz*, 29 June 2017 <https://qz.com/1017939/egypt-has-blocked-huffington-post-al-jazeera-medium-in-growing-censorship-crackdown/> (accessed 8 December 2017).

64. Clark et al., 'Shifting Landscape'.

65. Berkman Center for Internet and Society, 'DON'T PANIC', 1 February 2016 <https://cyber.harvard.edu/pubrelease/dont-panic/Dont_Panic_Making_Progress_on_Going_Dark_Debate.pdf> (accessed 1 December 2017).

66. BBC, 'WhatsApp Must Not Be "Place For Terrorists to Hide"', *BBC News*, 26 March 2017 <http://www.bbc.co.uk/news/uk-39396578> (accessed 1 December 2017); Tom Pritchard, 'The EU Wants to Enforce Encryption, and Ban Backdoor Access', *Gizmodo*, 19 June 2017 <http://www.gizmodo.co.uk/2017/06/the-eu-wants-to-enforce-encryption-and-ban-backdoor-access/> (accessed 1 December 2017).

67. Thomas Hobbes, *Leviathan* (Cambridge: Cambridge University Press, 2007), [62], 88.

68. Immanuel Kant, 'Idea for a Universal History from a Cosmopolitan Point of View', in *Philosophy of Technology: The Technological Condition: An Anthology* (Second Edition), eds. Robert C. Scharff and Val Dusek (Oxford: Wiley-Blackwell, 2014), 49–50.

Chapter 11

1. Niccolò Machiavelli, *Discourses on Livy*, translated by Julia Conaway Bondanella and Peter Bondanella (Oxford: Oxford University Press, 2008), 158.

2. Tim Wu, *The Master Switch: The Rise and Fall of Information Empires* (London: Atlantic, 2010), 292. See also Jonathan Zittrain, *The Future of the Internet (and How to Stop It)* (London: Allen Lane, 2008).

3. John Stuart Mill, *On Liberty and other writings* (Cambridge: Cambridge University Press, 2008), 54.

4. Nick Hopkins, 'Revealed: Facebook's Internal Rulebook on Sex, Terrorism and Violence', *The Guardian*, 21 May 2017 <https://amp.theguardian.com/news/2017/may/21/revealed-facebook-internal-rulebook-sex-terrorism-violence> (accessed 1 December 2017).

5. Electronic Frontier Foundation, 'Free Speech', <https://www.eff.org/free-speech-weak-link/> (accessed 1 December 2017).

6. Facebook Newsroom, 'Facebook, Microsoft, Twitter and YouTube Announce Formation of the Global Internet Forum to Counter Terrorism', 26 June 2017 <https://newsroom.fb.com/news/2017/06/global-internet-forum-to-counter-terrorism/> (accessed 1 December 2017).

7. Samuel Arbesman, *Overcomplicated: Technology at the Limits of Comprehension* (New York: Current, 2016), 34.

8. Arbesman, *Overcomplicated*, 4.

9. Arbesman, *Overcomplicated*, 21–2.

10. Frank Pasquale, *The Black Box Society: The Secret Algorithms that Control Money and Information* (Cambridge, Mass: Harvard University Press, 2015), 4–6.

11. Daniel J. Solove, *The Digital Person: Technology and Privacy in the Information Age* (New York: New York University Press, 2004), 38.

12. John Stuart Mill, *The Autobiography of John Stuart Mill* (US: Seven Treasures Publications, 2009), 6.

13. Mill, *Autobiography*, 6.

14. Mill, *Autobiography*, 16.

15. Isaiah Berlin, 'John Stuart Mill and the Ends of Life', in *Four Essays on Liberty* (Oxford: Oxford University Press, 1969), 177.

16. Stefan Collini, 'Introduction', *On Liberty*, xi.

17. Stefan Collini, 'Introduction', *On Liberty*, xiii.

18. Mill, *On Liberty*, 67.

19. Mill, *On Liberty*, 13 (emphasis added).

20. Joel Feinberg, *Harm to Others: The Moral Limits of the Criminal Law* (Oxford: Oxford University Press, 1984), 12.

21. See, e.g., Moley <http://www.moley.com/> (accessed 1 December 2017).

22. Sensifall <http://www.sensifall.com/> (accessed 12 December 2017).

23. Mill, *On Liberty*, 13.

24. Patrick Devlin, 'Morals and the Criminal Law', in *The Enforcement of Morals* (London: Oxford University Press, 1965), 6.

25. Devlin, 'Morals', 7.

26. 'Teledildonics', *Wikipedia*, last modified 29 November 2017 <https://en.wikipedia.org/wiki/Teledildonics> (accessed 8 December 2017).

27. Rachel Metz, 'Controlling VR With Your Mind', *MIT Technology Review*, 22 March 2017 <https://www.technologyreview.com/s/603896/controlling-vr-with-your-mind/> (accessed 1 December 2017).

28. James Fitzjames Stephen, *Liberty, Equality, Fraternity and Three Brief Essays* (Chicago: University of Chicago Press, 1991), 139.

29. Cited in J. W. Harris, *Legal Philosophies* (Second Edition) (New York: Oxford University Press, 2004), 133.

30. Devlin, 'Morality', 9.

31. Devlin, 'Morality', 10.

32. Cited in Tim Gray, *Freedom* (Basingstoke: Macmillan Education, 1991), 114.

33. H. L. A. Hart, *Law, Liberty, and Morality* (Oxford: Oxford University Press, 1991), 50.

34. Robin Rosenberg, Shawnee Baughman, and Jeremy Bailenson, 'Virtual Superheroes: Using Superpowers in Virtual Reality to Encourage Prosocial Behaviour', *PLoS ONE*, (8)1, 30 January 2013 <https://www.ncbi.nlm.nih.gov/pubmed?Db=pubmed&Cmd=ShowDetailView&TermToSearch=23383029> (accessed 1 December 2017).

35. Joel Feinberg, *Harmless Wrongdoing: The Moral Limits of the Criminal Law* (Oxford: Oxford University, 1990), 4.

36. Feinberg, *Harmless Wrongdoing*, 3.

37. Rosenberg et al., 'Virtual Superheroes'; Feinberg, *Harmless Wrongdoing*, 3.

38. Devlin, 'Morals', 10.

39. Lodewijk F. Asscher, '"Code" as Law: Using Fuller to Assess Code Rules', in *Coding Regulation: Essays on the Normative Role of Information Technology*, eds. E. J. Dommering, and Lodewijk F. Asscher (The Hague: TMC Asser, 2006), 80.

40. Douglas Rushkoff, *Program or Be Programmed: Ten Commands for a Digital Age* (New York: Soft Skull Press, 2011), 140.

41. Rushkoff, *Program*, 13.

42. Jean-Jacques Rousseau, *The Social Contract*, translated by Maurice Cranston (London: Penguin, 1968), 65.

Chapter 12

1. Bernard Crick, 'Politics as Form of Rule: Politics, Citizenship, and Democracy', in *What is Politics?* ed. Adrain Leftwich (Cambridge: Polity Press, 2015), 75.

2. Adam Swift, *Political Philosophy: A Beginners' Guide for Students and Politicians* (Second Edition) (Cambridge: Polity Press, 2007), 179.

3. Amartya Sen, 'Democracy as a Universal Value' *Journal of Democracy* 10, no. 3 (1999): 3–17.

4. Hélène Landemore, *Democratic Reason: Politics, Collective Intelligence, and the Rule of the Many* (Princeton: Princeton University Press, 2017), 1.

5. John Dunn, *Setting the People Free: The Story of Democracy* (London: Atlantic, 2005), 23.

6. David Held, *Models of Democracy* (Third Edition) (Cambridge: Polity, 2006), x.

7. David Van Reybrouck, *Against Elections: The Case for Democracy* (London: Bodley Head, 2016), 1.

8. Francis Fukuyama, *The Origins of Political Order: From Prehuman Times to the French Revolution* (London: Profile Books, 2012), 3.

9. Brian Klaas, *The Despot's Accomplice: How the West is Aiding and Abetting the Decline of Democracy* (Oxford: Oxford University Press, 2016), 1.

10. Reybrouck, *Against Elections*, 16.

11. Douglas Haven, 'The uncertain future of democracy', *BBC futurenow*, 30 March 2017 <http://www.bbc.com/future/story/20170330-the-uncertain-future-of-democracy?ocid=ww.social.link.twitter> (accessed 1 December 2017).

12. Plato, *The Republic*, translated by Desmond Lee (London: Penguin, 2003), [557a], 292.

13. See J. Lively, *Democracy* (1975), 30, cited in Held, *Models*, 2.

14. Alan Ryan, *On Politics: A History of Political thought from Herodotus to the Present* (London: Penguin, 2013), 11–13; Held, *Models*, 16–19; Dunn, *Setting the People Free*, 35.

15. Thucydides, *The Peloponnesian War*, translated by Martin Hammond (New York: Oxford University Press, 2009), Book II, §§37, 91.

16. Dunn, *Setting the People Free*, 35.

17. Dunn, *Setting the People Free*, 34.

18. Held, *Models*, 27–9.

19. Held, *Models*, 27–33.

20. Dunn, *Setting the People Free*, 55, 58.

21. Thomas Aquinas, *Political Writings*, translated by R. W. Dyson (Cambridge: Cambridge University Press, 2002), 9.

22. Held, *Models*, 1.

23. Dunn, *Setting the People Free*; Russell L. Hanson, 'Democracy', in *Political Innovation and Conceptual Change*, eds. Terence Ball, James Farr, and Russell L. Hanson (New York: Cambridge University Press, 1995), 75.

24. Dunn, *Setting the People Free*, 16; Hanson, 'Democracy', 72.

25. Hanson, 'Democracy', 76.

26. Giacomo Casanova, *The Story of My Life*, translated by Sophie Hawkes (London: Penguin, 2000), 373.

27. Cited in Niccolò Machiavelli, *Discourses on Livy*, translated by Julia Conaway Bondanella and Peter Bondanella (Oxford: Oxford University Press, 2008), 141.

28. Held, *Models*, 59–62.

29. Niccolò Machiavelli, *Discourses on Livy*, translated by Julia Conaway Bondanella and Peter Bondanella (Oxford: Oxford University Press, 2008), 142.

30. Joseph Schumpeter, *Capitalism, Socialism and Democracy* (Abingdon: Routledge, 2010), 220.

31. Cited in Carol Pateman, *Participation and Democratic Theory* (Cambridge: Cambridge University Press, 1999), 5.

32. Ryan, *On Politics*, 961.

33. See Sasha Issenberg, 'How Obama's Team Used Big Data to Rally Voters', *MIT Techology Review*, 19 December 2012 <https://www.technologyreview.com/s/509026/how-obamas-team-used-big-data-to-rally-voters/> (accessed 1 December 2017).

34. 'Joseph Schumpeter', Wikipedia, last edited 23 December 2017 <https://en.wikipedia.org/wiki/Joseph_Schumpeter> (accessed 21 January 2018).

35. Pedro Domingos, *The Master Algorithm: How the Quest for the Ultimate Learning Machine Will Remake Our World* (London: Allen Lane, 2015), 17.

36. Carole Cadwalladr, 'Robert Mercer: The Big Data Billionaire Waging War on Mainstream Media', *The Guardian*, 26 February 2017 <https://www.theguardian.com/politics/2017/feb/26/robert-mercer-breitbart-war-on-media-steve-bannon-donald-trump-nigel-farage> (accessed 1 December 2017).

37. Edward L. Bernays, 'The Engineering of Consent', *ANNALS of the American Academy of Political and Social Science* 250, no. 1 (1947), 113–20, cited in Zeynep Tufekci, 'Engineering the Public: Big Data, Surveillance and Computational Politics', *First Monday* 19, no. 7 (7 July 2014).

38. Berit Anderson and Brett Horvath, 'The Rise of the Weaponized AI Propaganda Machine', *Medium*, 12 February 2017 <https://medium.com/join-scout/the-rise-of-the-weaponized-ai-propaganda-machine-86dac61668b> (accessed 1 December 2017).

39. See Lauren Moxley, 'E-Rulemaking and Democracy' *Administrative Law Review* 68, no. 4 (2016): 661–99.

40. Julie Simon et al., 'Digital Democracy: The Tools Transforming Political Engagement', *Nesta*, February 2017 <http://www.nesta.org.uk/sites/default/files/digital_democracy.pdf> (accessed 1 December 2017).

41. Simon et al., 'Digital Democracy'.

42. Beth Simone Noveck, *Smart Citizens, Smarter State: The Technologies of Expertise and the Future of Governing* (Cambridge, Mass: Harvard University Press, 2015), 1–16; Simon et al., 'Digital Democracy'.

43. Noveck, *Smart Citizens*, 110.

44. Helen Margretts et al., *Political Turbulence: How Social Media Shape Collective Action* (Princeton: Princeton University Press, 2016), 211.

45. See e.g. Robert A. Dahl, *Who Governs? Democracy and Power in an American City* (New Haven and London: Yale University Press, 1961).

46. Alexis de Tocqueville, *Democracy in America*, translated by George Lawrence (New York: HarperCollins, 2006), 192.

47. Jean-Jacques Rousseau, *The Social Contract*, translated by Maurice Cranston (London: Penguin, 1968), 61.

48. Aristotle, *The Politics*, translated by T. A. Sinclair (London: Penguin, 1992), 1253a1, 59.

49. Aristotle, *Politics*, 1281a2, 198.

50. Thucydides, *Peloponnesian War*, Book II, §40, 92.

51. John Stuart Mill, 'Considerations on Representative Government' *Project Gutenberg* <https://www.gutenberg.org/files/5669/5669-h/5669-h.htm> (accessed 1 December 2017).

52. See Landemore, *Democratic Reason*.

53. Aristotle, *Politics*, 1281a39, 202.

54. Josiah Ober, *Democracy and Knowledge: Innovation and Learning in Classical Athens* (Princeton: Princeton University Press, 2008).

55. See Landemore, *Democratic Reason*; Philip E. Tetlock, *Expert Political Judgment: How Good Is It? How Can We Know?* (Princeton: Princeton University Press, 2006).

56. Baruch Spinoza, *Tractatus Theologico-Politicus* (1670), cited in Landemore, *Democratic Reason*, 67.

57. James Surowiecki, *The Wisdom of Crowds: Why the Many are Smarter than the Few* (London: Abacus, 2005).

58. Landemore, *Democratic Reason*, 157.

59. Jürgen Habermas, cited in Landemore, *Democratic Reason*, xvii; see also Landemore, *Democratic Reason*, 97.

60. Tocqueville, *Democracy in America*, 70.

61. Rousseau, *Social Contract*, 64.

62. Landemore, *Democratic Reason*, xv–xvii.

Chapter 13

1. See generally Jürgen Habermas, *Between Facts and Norms* (Cambridge: Polity Press in association with Oxford: Basil Blackwell, 2010).

2. David Held, *Models of Democracy* (Third Edition) (Cambridge: Polity, 2006), 237–42; Amy Gutmann and Dennis Thompson, *Why Deliberative Democracy?* (Princeton: Princeton University Press, 2004), 10–14.

3. See Yochai Benkler, *The Wealth of Networks: How Social Production Transforms Markets and Freedom* (New Haven and London: Yale University Press, 2006).

4. See e.g. Robert Faris et al., 'Partisanship, Propaganda, and Disinformation: Online Media and the 2016 U.S. Presidential Election', *Berkman Klein Center Research Paper* <https://papers.ssrn.com/sol3/papers.cfm?abstract_id=3019414> (accessed 8 December 2017).

5. See Cass R. Sunstein, *Republic.com 2.0* (Princeton: Princeton University Press, 2007); Cass R. Sunstein, *#Republic: Divided Democracy in the age of Social Media* (Princeton: Princeton University Press, 2017); Alex Krasodomski-Jones, 'Talking To Ourselves?' *Demos*, September 2016 <https://www.demos.co.uk/wp-content/uploads/2017/02/Echo-Chambers-final-version.pdf> (accessed 1 December 2017).

6. Bruce Bimber, *Information and American Democracy: Technology in the Evolution of Political Power* (New York: Cambridge University Press, 2011), 206–9.

7. Sunstein, *#Republic*, 121.

8. Timothy J. Penny, 'Facts Are Facts', *National Review*, 4 September 2003 <http://www.nationalreview.com/article/207925/facts-are-facts-timothy-j-penny> (accessed 9 December 2017).

9. David Remnick, 'Obama Reckons With a Trump Presidency', *New Yorker*, 28 November 2016 <http://www.newyorker.com/magazine/2016/11/28/obama-reckons-with-a-trump-presidency> (accessed 30 November 2017).

10. Craig Silverman, 'This Analysis Shows How Viral Fake Election News Stories Outperformed Real News on Facebook', *BuzzFeed News*, 16 November 2017 <https://www.buzzfeed.com/craigsilverman/viral-fake-election-news-outperformed-real-news-on-facebook?utm_term=.ufqYm8llgv#.sf9JbwppAm> (accessed 1 December 2017).

11. Matthew D'Ancona, *Post Truth: The New War on Truth and How to Fight Back* (London: Ebury Press, 2017), 54.

12. See discussion in Zeynep Tufekci, 'Engineering the Public: Big Data, Surveillance and Computational Politics' *First Monday* 19, no. 7 (7 July 2014).

13. Sunstein, *#Republic*, 71.

14. See Jamie Bartlett, *The Dark Net: Inside the Digital Underworld* (London: William Heinemann, 2014), 41.

15. Plato, *The Republic*, translated by Desmond Lee (London: Penguin, 2003).

16. I understand that the account '@imposterbusters' has itself been suspended by Twitter.

17. Peter Martinez, 'Study Reveals Whopping 48M Twitter Accounts Are Actually Bots', *CBS News*, 10 March 2017 <http://www.cbsnews.com/news/48-million-twitter-accounts-bots-university-of-southern-california-study/?ftag=CNM-00-10aab7e&linkId=35386687> (accessed 1 December 2017).

18. Carole Cadwalladr, 'Robert Mercer: The Big Data Billionaire Waging War on Mainstream Media', *The Guardian*, 26 February 2017 <https://www.theguardian.com/politics/2017/feb/26/robert-mercer-breitbart-war-on-media-steve-bannon-donald-trump-nigel-farage> (accessed 1 December 2017).

19. See Leo Kelion and Shiroma Silva, 'Pro-Clinton Bots "Fought Back but Outnumbered in Second Debate"', *BBC News*, 19 October 2016<http://www.bbc.com/news/technology-37703565> (accessed 1 December 2017); Amanda Hess, 'On Twitter, a Battle Among Political Bots', *New York Times*, 14 December 2016 <https://mobile.nytimes.com/2016/12/14/arts/on-twitter-a-battle-among-political-bots.html?contentCollection=weekendreads&referer=> (accessed 1 December 2017); Bence Kollanyi, Philip N. Howard, and Samuel C. Woolley, 'Bots and Automation over Twitter during the U.S. Election', *Computational Propaganda Project*, 2016 <http://comprop.oii.ox.ac.uk/2016/11/17/bots-and-automation-over-twitter-during-the-u-s-election/> (accessed 1 December 2017); John Markoff, 'Automated Pro-Trump Bots Overwhelmed Pro-Clinton Messages, Researchers Say', *New York Times*, 17 November 2016 http://www.nytimes.com/2016/11/18/technology/automated-pro-trump-bots-overwhelmed-pro-clinton-messages-researchers-say.html)> (accessed 1 December 2017).

20. Ian Sample, 'Study Reveals Bot-on-Bot Editing Wars Raging on Wikipedia's Pages', *The Guardian*, 23 February 2017 <https://www.theguardian.com/technology/2017/feb/23/wikipedia-bot-editing-war-study> (accessed 1 December 2017).

21. Julie Simon et al., 'Digital Democracy: The Tools Transforming Political Engagement', *Nesta*, February 2017 <http://www.nesta.org.uk/sites/default/files/digital_democracy.pdf> (accessed 1 December 2017).

22. Full Fact <https://fullfact.org/> (accessed 1 December 2017).

23. Evgeny Morozov, *To Save Everything Click Here: Technology, Solutionism, and the Urge to Fix Problems that Don't Exist* (London: Penguin, 2014), 119; Andy Greenberg, 'Now Anyone Can Deploy Google's Troll-Fighting AI', *Wired*, 23 February 2017 <https://www.wired.com/2017/02/googles-troll-fighting-ai-now-belongs-world/?mbid=social_twitter> (accessed 1 December 2017).

24. James Weinstein, 'An Overview of American Free Speech Doctrine and its Application to Extreme Speech', in *Extreme Speech and Democracy*, eds. Ivan Hare and James Weinstein (Oxford: Oxford University Press, 2010), 81–9.

25. Rebecca MacKinnon, *Consent of the Networked: The Worldwide Struggle for Internet Freedom* (New York: Basic Books, 2013), 127.

26. Matthew Prince, 'Why We Terminated Daily Stormer', *Cloudfare*, 16 August 2017 <https://blog.cloudflare.com/why-we-terminated-daily-stormer/> (accessed 1 December 2017).

27. Lizzie Plaugic, 'Spotify Pulls Several "Hate Bands" from its Service', *The Verge*, 16 August 2017 <https://www.theverge.com/2017/8/16/16158502/spotify-racist-bands-streaming-service-southern-poverty-law-center> (accessed 1 December 2017).

28. Rishabh Jain, 'Charlottesville Attack: Facebook, Reddit, Google and GoDaddy Shut Down Hate Groups', *IBT*, 16 August 2017 <http://www.ibtimes.com/charlottesville-attack-facebook-reddit-google-godaddy-shut-down-hate-groups-2579027> (accessed 1 December 2017).

29. Zeynep Tufekci, *Twitter and Tear Gas: The Power and Fragility of Networked Protest* (New Haven: Yale University Press, 2017), 149–50.

30. Tufekci, *Twitter*, 150.

31. John Stuart Mill, *On Liberty* in *On Liberty and other writings* (Cambridge: Cambridge University Press, 2008), 56.

32. See Martin Jay, *The Virtues of Mendacity: On Lying in Politics* (Charlottesville: University of Virginia Press, 2010).

33. Richard Hofstadter, *The Paranoid Style in American Politics* (New York: Vintage Books, 2008), 3.

34. Hannah Arendt, 'Truth and Politics', in *Between Past and Future* (London: Penguin, 2006), 223.

35. George Orwell, *Diaries* (London: Penguin, 2009), 24 April 1942, 335.

36. Alvin I. Goldman, *Knowledge in a Social World* (Oxford: Oxford University Press, 2003), 7–10.

37. Michel Foucault, *Power/Knowledge: Selected Interviews and other writings, 1972–1977* (New York: Vintage Books, 1980), 93.

38. Don Tapscott and Alex Tapscott, *Blockchain Revolution: How the Technology Behind Bitcoin is Changing Money, Business and the World* (London: Portfolio Penguin, 2016), 131.

39. D'Ancona, *Post-Truth*, 100–1.

40. Jean-Jacques Rousseau, *The Social Contract*, translated by Maurice Cranston (London: Penguin, 1968), 112.

41. Agoravoting.com <https://agoravoting.com/> (accessed 1 December 2017).

42. Danny Bradbury, 'How Block Chain Technology Could Usher in Digital Democracy', *CoinDesk*, 16 June 2014 <http://www.coindesk.com/block-chain-technology-digital-democracy/> (accessed 1 December 2017).

43. Karl Marx, *The Civil War in France*, in *Karl Marx and Frederick Engels: Collected Works Vol. 22.* (London: Lawrence & Wishart, 1986), 333.

44. Thomas Christiano, *The Rule of the Many: Fundamental Issues in Democratic Theory* (Westview Press: Colorado & London, 1996), 109.

45. Sunstein, *#Republic*, 48.

46. James Madison, 'Federalist No. 63', in *The Federalist Papers* (New York: Penguin, 2012), 114 (original emphasis). See Sunstein, *#Republic*.

47. DemocracyOS <http://democracyos.org/> (accessed 1 December 2017).

48. Tapscott and Tapscott, *Blockchain Revolution*, 218; Micah L. Sifry, *The Big Disconnect: Why the Internet Hasn't Transformed Politics (Yet)* (New York and London: OR Books, 2014), 212; Steven Johnson, *Future Perfect: The Case for Progress in a Networked Age* (London: Penguin, 2013), 152–76.

49. John Stuart Mill, 'Thoughts on Parliamentary Reform', *Collected Works of John Stuart Mill, Volume XIX—Essays on Politics and Society Part 2*, eds. John M. Robson (Toronto: University of Toronto Press,

London: Routledge and Kegan Paul, 1977) <http://oll.libertyfund. org/titles/mill-the-collected-works-of-john-stuart-mill-volume-xix-essays-on-politics-and-society-part-2#lf0223-19_head_002> (accessed 8 December 2017).

50. See generally Yochai Benkler, *The Wealth of Networks: How Social Production Transforms Markets and Freedom* (New Haven and London: Yale University Press, 2006).

51. See Beth Simone Noveck, *Wiki Government: How Technology Can Make Government Better, Democracy Stronger, and Citizens More Powerful* (Washington, DC: Brookings Institution Press), 2009; Alan Watkins and Iman Straitens, *Crowdocracy: The End of Politics* (Rochester: Urbane Publications, 2016).

52. Daren C. Brabham, *Crowdsourcing* (Cambridge, Mass: MIT Press, 2013), 34.

53. Julie Simon et al., 'Digital Democracy'.

54. Noveck, *Wiki Government*, 39.

55. Jürgen Habermas, 'Further Reflections on the Public Sphere', cited in Douglas Torgerson, 'Democracy Through Policy Discourse', in *Deliberative Policy Analysis: Understanding Governance in the Network Society*, eds. Maarten A. Hajer and Hendrik Wagenaar (New York: Cambridge University Press, 2003), 115.

56. Jaron Lanier, *Who Owns the Future?* (London: Allen Lane, 2014), 57.

57. Richard Susskind and Daniel Susskind, *The Future of the Professions: How Technology Will Transform the Work of Human Experts* (Oxford: Oxford University Press, 2015), 161.

58. Hiroki Azuma, *General Will 2.0: Rousseau, Freud, Google* (New York: Vertical, Inc, 2014); Yuval Noah Harari, *Homo Deus: A Brief History of Tomorrow* (London: Harvill Secker, 2015), 329–40.

59. John O. McGinnis, *Accelerating Democracy: Transforming Governance through Technology* (Princeton: Princeton University Press, 2013), 123–5; Hélène Landemore, *Democratic Reason: Politics, Collective Intelligence, and the Rule of the Many* (Princeton: Princeton University Press, 2017), 125; Watkins and Straitens, *Crowdocracy,* 116.

60. See, e.g., Johan Bollen, Huina Mao, and Xiao-Jun Zeng, 'Twitter Mood Predicts the Stock Market', *arXiv,* 14 October 2010 <https:// arxiv.org/pdf/1010.3003.pdf> (accessed 1 December 2017).

61. Harari, *Homo Deus*, 340.

62. Jamie Bartlett and Nathaniel Tkacz, 'Governance by Dashboard', *Demos*, March 2017 <https://www.demos.co.uk/wp-content/uploads/2017/04/Demos-Governance-by-Dashboard.pdf> (accessed 1 December 2017).

63. Auguste Comte, 'Plan of the Scientific Work Necessary for the Reorganization of Society', in *Early Political Writings*, translated by H. S. Jones (Cambridge: Cambridge University Press, 1998), 100.

64. See e.g. voteforpolicies.org.uk <https://voteforpolicies.org.uk/> (accessed 1 December 2017) and Crowdpac <https://www.crowd-pac.co.uk/> (accessed 1 December 2017).

65. Voter.xyz <http://www.voter.xyz/> (accessed 1 December 2017).

66. See Pedro Domingos, *The Master Algorithm: How The Quest for the Ultimate Learning Machine Will Remake Our World* (London: Allen Lane, 2015), 19.

67. Alan Ryan, *On Politics: A History of Political thought from Herodotus to the Present* (London: Penguin, 2013), 8.

Chapter 14

1. John Rawls, *A Theory of Justice* (Cambridge, Mass: Harvard University Press, 2003), 3.

2. Jonathan P. Allen, *Technology and Inequality: Concentrated Wealth in a Digital World* (Kindle Edition: Palgrave Macmillan, 2017), Kindle Locations 245–7.

3. Klaus Schwab, *The Fourth Industrial Revolution* (Geneva: World Economic Forum, 2016), 92–3.

4. Karl Marx, *Contribution to Critique of Hegel's Philosophy of Law. Introduction*, in *Karl Marx and Frederick Engels: Collected Works Vol. 3.* (London: Lawrence & Wishart, 1975), 185 (original emphasis). See Jamie Susskind, *Karl Marx and British Intellectuals in the 1930s* (Burford: Davenant Press, 2011), 1.

5. John Rawls, 'Reply to Alexander and Musgrave', in *The Ideal of Equality*, eds. Matthew Clayton and Andrew Williams (New York: Palgrave Macmillan, 2002), 22.

6. Iris Marion Young, *Justice and the Politics of Difference* (Princeton: Princeton University Press, 2011), 33.

7. Aristotle, *The Politics*, translated by T. A. Sinclair (London: Penguin, 1992), 1280a7, 195; Larry Siedentop, *Inventing the Individual: The Origins of Western Liberalism* (London: Allen Lane, 2014), 51.

8. Will Kymlicka, *Contemporary Political Philosophy: An Introduction* (Second Edition) (Oxford: Oxford University Press, 2002), 3–4; Adam Swift, *Political Philosophy: A Beginners' Guide for Students and Politicians* (Second Edition) (Cambridge: Polity Press, 2007), 93.

9. Swift, *Political Philosophy*, 121.

10. Harry Frankfurt, 'Equality as a Moral Ideal' *Ethics* 98, no. 1 (October 1987): 21–43.

11. Derek Parfit, 'Equality or Priority?' in *The Ideal of Equality*.

12. Swift, *Political Philosophy*, 99–100.

13. See Larry Temkin, 'Equality, Priority, and the Levelling Down Objection', in *The Ideal of Equality*.

14. Thomas Scanlon, 'The Diversity of Objections to Equality', in *The Ideal of Equality*.

15. Swift, *Political Philosophy*, 104.

16. Swift, *Political Philosophy*, 19.

17. David Hume, *An Enquiry Concerning the Principles of Morals* (Indianapolis: Hackett Publishing Company, 1983), 28.

18. Robert Nozick, *Anarchy, State, and Utopia* (Oxford: Blackwell Publishing, 2008), 149.

19. Elizabeth Anderson, *Private Government: How Employers Rule Our Lives (and Why We Don't Talk About It)* (Princeton and Oxford: Princeton University Press, 2017), 2.

20. Cathy O'Neil, *Weapons of Math Destruction: How Big Data Increases Inequality and Threatens Democracy* (New York: Crown, 2016), 114.

21. O'Neil, *Weapons*, 120.

22. Laurence Mills, 'Numbers, Data and Algorithms: Why HR Professionals and Employment Lawyers Should Take Data Science and Analytics Seriously', *Future of Work Hub*, 4 April 2017 <http://www.futureofworkhub.info/comment/2017/4/4/numbers-data-and-algorithms-why-hr-professionals-and-employment-lawyers-should-take-data-science-seriously> (accessed 1 December 2017); Ifeoma Ajunwa, Kate Crawford, and Jason Schultz, 'Limitless Worker Surveillance', *California Law Review* 105, no. 3, 13 March 2016 <https://

papers.ssrn.com/sol3/papers.cfm?abstract_id=2746211> (accessed 1 December 2017).

23. Olivia Solon, 'World's Largest Hedge Fund to Replace Managers with Artificial Intelligence', *The Guardian*, 22 December 2016 <https://www.theguardian.com/technology/2016/dec/22/bridgewater-associates-ai-artificial-intelligence-management> (accessed 1 December 2017).

24. Danielle Keats Citron and Frank Pasquale, 'The Scored Society: Due Process for Automated Predictions', *Washington Law Review* 89, no. 1 (26 March 2014) <https://digital.law.washington.edu/dspace-law/bitstream/handle/1773.1/1318/89WLR0001.pdf?sequence=1> (accessed 1 December 2017).

25. Eric Siegel, *Predictive Analytics: The Power to Predict Who Will Click, Buy, Lie, or Die* (New Jersey: John Wiley & Sons, Inc, 2016), 10.

26. Siegel, *Predictive Analytics*, 292–3; Citron and Pasquale, 'Scored Society'.

27. See Rawls, *Theory of Justice*, 79: 'primary social goods...are things which it is supposed a rational man wants whatever else he wants'.

28. Jaron Lanier, *Who Owns the Future?* (London: Allen Lane, 2014), xvi.

29. Allen, *Technology and Inequality*, Kindle Locations 968–70.

30. O'Neil, *Weapons*, 144.

31. Jennifer Valentino-DeVries, Jeremy Singer-Vine, and Ashkan Soltani, 'Websites Vary Prices, Deals Based on Users' Information', *Wall Street Journal*, 24 December 2012 <https://www.wsj.com/articles/SB10001424127887323777204578189391813881534> (accessed 1 December 2017).

32. Sam Schechner, 'Why Do Gas Station Prices Constantly Change? Blame the Algorithm', *Wall Street Journal*, 8 May 2017 <https://www.wsj.com/articles/why-do-gas-station-prices-constantly-change-blame-the-algorithm-1494262674?mod=e2tw> (accessed 1 December 2017).

33. Jeremy Useem, 'How Online Shopping Makes Suckers of Us All', *Atlantic*, May 2017 Issue <https://www.theatlantic.com/magazine/archive/2017/05/how-online-shopping-makes-suckers-of-us-all/521448/?utm_source=nextdraft&utm_medium=email> (accessed 1 December 2017).

34. Benjamin Reed Shiller, 'First-Degree Price Discrimination Using Big Data', *Brandeis University*, 19 January 2014 <http://benjaminshiller.

com/images/First_Degree_PD_Using_Big_Data_Jan_18,_2014.pdf> (accessed 1 December 2017).

35. Shiller, 'First-Degree Price Discrimination'.

36. See Lawrence Lessig, *Code Version 2.0* (New York: Basic Books, 2006).

Chapter 15

1. See Axel Honneth, *The Struggle for Recognition: The Moral Grammar of Social Conflicts*, translated by Joel Anderson (Cambridge: Polity Press, 2005).

2. Robert H. Frank, *Choosing the Right Pond: Human Behavior and the Quest for Status* (New York: Oxford University Press, 1985), 9.

3. Honneth, *Struggle*.

4. Translator's Note, Honneth, *Struggle*.

5. Elizabeth Anderson, 'Against Luck Egalitarianism: What is the Point of Equality?' in *Social Justice*, eds. Matthew Clayton, and Andrew Williams (Oxford: Blackwell Publishing, 2005), 155.

6. Anderson, 'What is the Point of Equality?'

7. Iris Marion Young, *Justice and the Politics of Difference* (Princeton: Princeton University Press, 2011), 53–61.

8. Erika Harrell, 'Crime Against Persons with Disabilities, 2009-2015: Statistical Tables', *Bureau of Justice Statistics*, July 2017 <https://www.bjs.gov/content/pub/pdf/capd0915st.pdf> (accessed 2 December 2017).

9. Judith Squires, 'Equality and Difference', in *The Oxford Handbook of Political Theory*, eds. John S. Dryzek, Bonnie Honig, and Anne Phillips (New York: Oxford University Press, 2008), 479.

10. Michael Walzer, *Spheres of Justice: A Defense of Pluralism and Equality* (New York: Basic Books, 1983), 249; Elizabeth Anderson, *Private Government: How Employers Rule Our Lives (and Why We Don't Talk About It)* (Princeton and Oxford: Princeton University Press, 2017), 3–4.

11. Nadia Judith Enchassi and CNN Wire, 'New Zealand Passport Robot Thinks This Asian Man's Eyes Are Closed', *KFOR*, 11 December 2016 <http://kfor.com/2016/12/11/new-zealand-passport-robot-thinks-this-asian-mans-eyes-are-closed/> (accessed 2 December 2017).

12. Richard Yonck, *Heart of the Machine: Our Future in a World of Artificial Intelligence* (New York: Arcade Publishing, 2017), 50.

13. Douglas Rushkoff, *Throwing Rocks at the Google Bus: How Growth Became the Enemy of Prosperity* (New York: Portfolio/Penguin, 2016), 31.

14. Nick Couldry, *Media, Society, World: Social Theory and Digital Media Practice* (Cambridge: Polity, 2012), 25.

15. Frank, *Choosing*, 7, 26.

16. Christopher Steiner, *Automate This: How Algorithms Came to Rule Our World* (London: Portfolio, 2012), 55.

Chapter 16

1. This example is from the Executive Office of the President, 'Big Data: A Report on Algorithmic Systems, Opportunity, and Civil Rights', *Obama White House Archives*, May 2016 <https://obamawhitehouse. archives.gov/sites/default/files/microsites/ostp/2016_0504_data_ discrimination.pdf> (accessed 2 December 2017).

2. Ian Tucker, '"A White Mask Worked Better": Why Algorithms Are Not Colour Blind', *The Guardian*, 28 May 2017 <https://www.theguardian. com/technology/2017/may/28/joy-buolamwini-when-algorithms- are-racist-facial-recognition-bias> (accessed 2 December 2017).

3. Selena Larson, 'Research Shows Gender Bias in Google's Voice Recognition', *Daily Dot*, 15 July 2016 <https://www.dailydot.com/debug/ google-voice-recognition-gender-bias/> (accessed 2 December 2017).

4. Jordan Pearson, 'Why an AI-Judged Beauty Contest Picked Nearly All White Winners', *Motherboard*, 5 September 2016 <https://motherboard. vice.com/en_us/article/78k7de/why-an-ai-judged-beauty-contest- picked-nearly-all-white-winners> (accessed 2 December 2017).

5. Alex Hern, 'Flickr Faces Complaints Over "Offensive" Auto-tagging for Photos', *The Guardian*, 20 May 2015 <https://www.theguardian. com/technology/2015/may/20/flickr-complaints-offensive-auto- tagging-photos> (accessed 2 December 2017).

6. Alistair Barr, 'Google Mistakenly Tags Black People as "Gorillas", Showing Limits of Algorithms', *Wall Street Journal*, 1 July 2015 <https://blogs.wsj.com/digits/2015/07/01/google-mistakenly- tags-black-people-as-gorillas-showing-limits-of-algorithms/> (accessed 2 December 2017).

7. Executive Office of the President, 'Big Data'; Cathy O'Neil, *Weapons of Math Destruction: How Big Data Increases Inequality and Threatens Democracy* (New York: Crown, 2016), 7, 156.

8. Executive Office of the President, 'Big Data', 18.

9. Executive Office of the President, 'Big Data', 15.

10. Christian Sandvig et al., 'When the Algorithm Itself is a Racist: Diagnosing Ethical Harm in the Basic Components of Software', *International Journal of Communications* 10 (2016): 4972–4990.

11. Julia Angwin and Jeff Larson, 'The Tiger Mom Tax: Asians Are Nearly Twice as Likely to Get a Higher Price from Princeton Review', *ProPublica*, 1 September 2015 <https://www.propublica.org/article/asians-nearly-twice-as-likely-to-get-higher-price-from-princeton-review> (accessed 3 December 2017).

12. Frank Pasquale, *The Black Box Society: The Secret Algorithms that Control Money and Information* (Cambridge, Mass: Harvard University Press, 2015), 39; Emerging Technology from the arXiv, 'Racism is Poisoning Online Ad Delivery, Says Harvard Professor', *MIT Technology Review*, 4 February 2013 <https://www.technologyreview.com/s/510646/racism-is-poisoning-online-ad-delivery-says-harvard-professor/> (accessed 3 December 2017).

13. Paul Baker and Amanda Potts, '"Why Do White People Have Thin Lips?" Google and the Perpetuation of Stereotypes via Auto-complete Search Forms', *Critical Discourse Studies* 10, no. 2 (2013) <http://www.tandfonline.com/doi/full/10.1080/17405904.2012.744320?scroll=top&needAccess=true> (accessed 3 December 2017).

14. Francesco Bonchi, Carlos Castillo, and Sara Hajian, 'Algorithmic Bias: From Discrimination Discovery to Fairness-aware Data Mining', *KDD 2016 Tutorial* <http://francescobonchi.com/tutorial-algorithmic-bias.pdf> (accessed 3 December 2017).

15. Tom Slee, *What's Yours is Mine: Against the Sharing Economy* (New York and London: OR Books, 2015), 94.

16. Slee, *What's Yours is Mine*, 95.

17. Josh Chin and Gillian Wong, 'China's New Tool for Social Control: A Credit Rating for Everything', *Wall Street Journal*, 28 November 2016 <http://www.wsj.com/articles/chinas-new-tool-for-social-control-a-credit-rating-for-everything-1480351590> (accessed 1 December 2017); *Economist*, 'China Invents the Digital Totalitarian State', 17 December 2016 <http://www.economist.com/news/briefing/21711902-worrying-implications-its-social-credit-project-china-invents-digital-totalitarian> (accessed 1 December 2017).

18. Andrew Whitby, Audun Jøsang, and Jadwiga Indulska, 'Filtering Out Unfair Ratings in Bayesian Reputation Systems', *Proceedings of the Workshop on Trust in Agent Societies, at the Autonomous Agents and Multi Agent Systems Conference*, July 2004 <https://www.csee. umbc.edu/~msmith27/readings/public/whitby-2004a.pdf> (accessed 3 December 2017).

19. Benjamin Edelman, Michael Luca, and Dan Svirsky, 'Racial Discrimination in the Sharing Economy: Evidence from a Field Experiment', *American Economic Journal: Applied Economics* 9, no. 2 (April 2017): 1–22.

20. Slee, *What's Yours is Mine*, 95.

21. Tolga Bolukbasi et al., 'Man is to Computer Programmer as Woman is to Homemaker? Debiasing Word Embeddings', *arXiv*, 21 July 2016 <https://arxiv.org/pdf/1607.06520.pdf> (accessed 3 December 2017).

22. Iris Marion Young, *Justice and the Politics of Difference* (Princeton: Princeton University Press, 2011), 97.

23. Young, *Justice*, 98.

24. See Thomas Nagel, *The View from Nowhere* (New York: Oxford University Press, 1986).

25. Richard Yonck, *Heart of the Machine: Our Future in a World of Artificial Intelligence* (New York: Arcade Publishing, 2017), 90.

26. Pasquale, *Black Box Society*.

27. Computerscience.org. 'Women in Computer Science: Getting Involved in STEM' <http://www.computerscience.org/resources/ women-in-computer-science/> (accessed 3 December 2017); Sheelah Kolhatkar, 'The Tech Industry's Gender-Discrimination Problem', *New Yorker*, 20 November 2017 <https://www.newyorker.com/ magazine/2017/11/20/the-tech-industrys-gender-discrimination- problem> (accessed 12 December 2017).

28. Julia Wong, 'Segregated Valley: The Ugly Truth About Google and Diversity in Tech', *The Guardian*, 7 August 2017 <https://www. theguardian.com/technology/2017/aug/07/silicon-valley-google- diversity-black-women-workers> (accessed 3 December 2017).

Chapter 17

1. Jeff Guo, 'We're So Unprepared for the Robot Apocalypse', *Washington Post*, 30 March 2017 <https://www.washingtonpost.com/news/

wonk/wp/2017/03/30/were-so-unprepared-for-the-robot-apocalypse/?utm_term=.caeece2d19b4> (accessed 8 December 2017).

2. Federica Cocco, 'Most US Manufacturing Jobs Lost to Technology, Not Trade', *Financial Times*, 2 December 2016 <https://www.ft.com/content/dec677c0-b7e6-11e6-ba85-95d1533d9a62> (accessed 8 December 2017).

3. Michael Chui, James Manyika, and Mehdi Miremadi, 'Where Machines Could Replace Humans—and Where They Can't (Yet)', *McKinsey Quarterly*, July 2016 <https://www.mckinsey.com/business-functions/digital-mckinsey/our-insights/where-machines-could-replace-humans-and-where-they-cant-yet> (accessed 8 December 2017).

4. Richard Susskind and Daniel Susskind, *The Future of the Professions: How Technology Will Transform the Work of Human Experts* (Oxford: Oxford University Press, 2015).

5. Susskind and Susskind, *Future of the Professions*.

6. See Daniel Susskind, 'Re-thinking the Capabilities of Machines in Economics', Oxford University Discussion Paper no. 825, version 1 May 2017 (May 2017); 'A Model of Technological Unemployment', Oxford University Discussion Paper no. 819, version 6 July 2017 (July 2017), both at <https://www.danielsusskind.com/research> (accessed 5 December 2017).

7. Karl Marx, *Economic and Philosophical Manuscripts of 1844*, in *Karl Marx and Frederick Engels: Collected Works Vol. 3*. (London: Lawrence & Wishart, 1975), 235.

8. Ryan Avent, *The Wealth of Humans: Work, Power, and Status in the Twenty-First Century* (New York: St. Martin's Press, 2016), 6.

9. Cocco, 'Most US Manufacturing Jobs Lost to Technology, Not Trade'.

10. Sam Shead, 'Amazon's Supermarket of the Future Could Operate With Just 3 Staff—and Lots of Robots', *Business Insider*, 6 February 2017 <http://www.businessinsider.com/amazons-go-supermarket-of-the-future-3-human-staff-2017-2?r=UK&IR=T> (accessed 8 December 2017); Yiting Sun, 'In China, a Store of the Future—No Checkout, No Staff', *MIT Technology Review*, 16 June 2017 <https://www.technologyreview.com/s/608104/in-china-a-store-of-the-future-no-checkout-no-staff/> (accessed 8 December 2017).

11. Martin Ford, *Rise of the Robots: Technology and the Threat of a Jobless Future* (New York: Basic Books, 2015), 12.

12. Laura Tyson and Michael Spence, 'Exploring the Effects of Technology on Income and Wealth Inequality', in *After Piketty: The Agenda for Economics and Inequality*, eds. Heather J. Boushey, Bradford DeLong, and Marshall Steinbaum (Cambridge, Mass: Harvard University Press, 2017), 177.

13. Chui et al., 'Where Machines Could Replace Humans'.

14. See Susskind and Susskind, *Future of the Professions*.

15. I am grateful to Richard Susskind for both the point and the examples.

16. Kory Schaff, 'Introduction', in *Philosophy and the Problems of Work: A Reader*, ed. Kory Schaff (Lanham, Maryland: Rowman & Littlefield Publishers, 2001), 6.

17. Schaff, 'Introduction', 9.

18. Marie Jahoda, *Employment and Unemployment: A Social-psychological Analysis* (Cambridge: Cambridge University Press, 1982), 24.

19. Jahoda, *Employment and Unemployment*, 22.

20. Jahoda, *Employment and Unemployment*, 60–1.

21. Sigmund Freud, *Civilization and its Discontents* (Oregon: Rough Draft Printing, 2013), 19 fn. 11; Jahoda, *Employment and Unemployment*, 60.

22. Jon Elster, 'Is There (or Should There Be) a Right to Work?' in *Philosophy and the Problems of Work*, 283.

23. Elster, 'Right to Work'.

24. James Livingston, 'Fuck Work', *Aeon*, 25 November 2016 <https://aeon.co/essays/what-if-jobs-are-not-the-solution-but-the-problem> (accessed 8 December 2017).

25. Kevin J. Delaney, 'The Robot that Takes Your Job Should Pay Taxes, Says Bill Gates', *Quartz*, 17 February 2017 <https://qz.com/911968/bill-gates-the-robot-that-takes-your-job-should-pay-taxes/> (accessed 8 December 2017).

26. Philippe van Parijs and Yannick Vanderborght, *Basic Income: A Radical Proposal for a Free Society and a Sane Economy* (Cambridge, Mass: Harvard, 2017), 4.

27. Van Parijs and Vanderborght, *Basic Income*, 8.

28. Avent, *Wealth of Humans*, 201.

29. Karl Marx, *Critique of the Gotha Programme*, in *Karl Marx and Frederick Engels: Collected Works Vol. 24* (London: Lawrence & Wishart, 1989), 87.

30. Richard Arneson, 'Is Work Special? Justice and the Distribution of Employment', in *Philosophy and the Problems of Work*, 208.

31. William Shakespeare, *Hamlet* (Oxford: Oxford University Press, 2008), 363.

32. Herbert Spencer, *The Man Versus the State* (London: Watts & Co, 1909) cited in Arneson, 'Is Work Special?', 201.

33. Van Parijs and Vanderborght, *Basic Income*, 101.

34. Friedrich Engels, *The Condition of the Working-Class in England* in *Karl Marx and Frederick Engels: Collected Works Vol. 4* (London: Lawrence & Wishart, 1975), 187.

35. Elizabeth Anderson, *Private Government: How Employers Rule Our Lives (and Why We Don't Talk About It)* (Princeton and Oxford: Princeton University Press, 2017), 129.

36. Oscar Wilde, 'The Soul of Man Under Socialism', cited in Michael Walzer, *Spheres of Justice: A Defense of Pluralism and Equality* (New York: Basic Books, 1983), 167.

37. Walzer, *Spheres*, 185.

38. William Shakespeare, *Henry IV, Part I*, cited in Walzer, *Spheres*, 195.

39. Jahoda, *Employment and Unemployment*, 59.

40. For further reading, see Nick Srnicek and Alex Williams, *Inventing the Future: Postcapitalism and a World Without Work* (London: Verso, 2015); David Frayne, *The Refusal of Work: The Theory and Practice of Resistance to Work* (London: Zed Books, 2015); André Gorz, *Reclaiming Work: Beyond the Wage-Based Society*, translated by Chris Turner (Cambridge: Polity Press, 2005); André Gorz, *Capitalism, Socialism, Ecology*, translated by Martin Chalmers (London and New York: Verso, 2012); and Bertrand Russell, *In Praise of Idleness* (Abingdon: Routledge, 2004).

Chapter 18

1. Robert Nozick, *Anarchy, State, and Utopia* (Oxford: Blackwell Publishing, 2008), 169.

2. Tim Wu, *The Master Switch: The Rise and Fall of Information Empires* (London: Atlantic, 2010), 276.

3. Ibid.

4. Cited in Wu, *Master Switch*, 276–7.

5. Thomas Piketty, *Capital in the Twenty-First Century* (Cambridge, Mass: The Belknapp Press of Harvard University Press, 2014), 18.

6. Piketty, *Capital*, 26.

7. Piketty, *Capital*, 22; Ryan Avent, *The Wealth of Humans: Work, Power, and Status in the Twenty-First Century* (New York: St. Martin's Press, 2016), 119–20.

8. Avent, *Wealth of Humans*, 119–20.

9. Erik Brynjolfsson and Andrew McAfee, *The Second Machine Age: Work, Progress, and Prosperity in a Time of Brilliant Technologies* (New York: W. W. Norton & Company, 2014), 118.

10. Erik Brynjolfsson, Andrew McAfee, and Michael Spence. 'New World Order: Labor, Capital, and Ideas in the Power Law Economy', *Foreign Affairs*, July/August 2014 <https://www.foreignaffairs.com/articles/united-states/2014-06-04/new-world-order> (accessed 8 December 2017).

11. Robert W. McChesney, *Digital Disconnect: How Capitalism is Turning The Internet Against Democracy* (New York: The New Press, 2014), 134.

12. Brynjolfsson et al., 'New World Order'.

13. MIT Technology Review Custom, in partnership with Oracle, 'The Rise of Data Capital', *MIT Technology Review*, 21 March 2016 <https://www.technologyreview.com/s/601081/the-rise-of-data-capital/> (accessed 8 December 2017).

14. Viktor Mayer-Schönberger and Kenneth Cukier, *Big Data: A Revolution That Will Transform How We Live, Work and Think* (London: John Murray, 2013), 5; Steve Jones, 'Why "Big Data" is the Fourth Factor of Production', *Financial Times*, 27 December 2012 <https://www.ft.com/content/5086d700-504a-11e2-9b66-00144feab49a> (accessed 9 December 2017); Neil Lawrence, quoted in Alex Hern, 'Why Data is the New Coal', *The Guardian*, 27 September 2016 <https://www.theguardian.com/technology/2016/sep/27/data-efficiency-deep-learning> (accessed 9 December 2017).

15. Gustavo Grullon, Yelena Larkin, and Roni Michaely, 'Are U.S. Industries Becoming More Concentrated?' *SSRN*, 2017 <https://papers.ssrn.com/sol3/papers.cfm?abstract_id=2612047> (accessed 8 December 2017).

16. David Dayen, 'This Budding Movement Wants to Smash Monopolies', *Nation*, 4 April 2017 https://www.thenation.com/article/this-budding-movement-wants-to-smash-monopolies/(accessed 8 December 2017).

17. David Autor et al., 'The Fall of the Labor Share and the Rise of Superstar Firms', 1 May 2017 <https://economics.mit.edu/files/12979> (accessed 8 December 2017).

18. Angelo Young, 'How to Break Up Alphabet, Amazon and Facebook', *Salon*, 31 May 2017 <https://www.salon.com/2017/05/31/how-to-break-up-alphabet-amazon-and-facebook/> (accessed 8 December 2017).

19. Paula Dwyer, 'Should America's Tech Giants Be Broken Up?' *Bloomberg Businessweek*, 20 July 2017 <https://www.bloomberg.com/news/articles/2017-07-20/should-america-s-tech-giants-be-broken-up> (accessed 8 December 2017).

20. James Ball, 'Let's Challenge Google While We Still Can', *The Guardian*, 16 April 2015 <https://www.theguardian.com/commentisfree/2015/apr/16/challenge-google-while-we-can-eu-anti-trust> (accessed 8 December 2017).

21. BI Intelligence, 'Amazon Accounts for 43% of US Online Retail Sales', *Business Insider E-Commerce Briefing*, 3 February 2017 <http://uk.businessinsider.com/amazon-accounts-for-43-of-us-online-retail-sales-2017-2?r=US&IR=T> (accessed 9 December 2017).

22. Dwyer, 'Should America's Tech Giants be Broken Up?'

23. Jonathan Taplin, *Move Fast and Break Things: How Facebook, Google, and Amazon Cornered Culture and Undermined Democracy* (New York: Little, Brown and Company, 2017), 8.

24. Connie Chan, cited in Frank Pasquale, 'Will Amazon Take Over the World?' *Boston Review*, 20 July 2017 <https://bostonreview.net/class-inequality/frank-pasquale-will-amazon-take-over-world> (accessed 8 December 2017).

25. Klaus Schwab, *The Fourth Industrial Revolution* (Geneva: World Economic Forum, 2016), 10.

26. Martin Ford, *Rise of the Robots: Technology and the Threat of a Jobless Future* (New York: Basic Books, 2015), 175.

27. See David Singh Grewal, *Network Power: The Social Dynamics of Globalization* (New Haven & London: Yale University Press, 2008).

28. See Michael Lewis, *Flash Boys: Cracking the Money Code*. London: Allen Lane, 2014.

29. Jaron Lanier, *Who Owns the Future?* (London: Allen Lane, 2014), xvi.

30. Jonathan P. Allen, *Technology and Inequality: Concentrated Wealth in a Digital World* (Kindle Edition: Palgrave Macmillan, 2017), Kindle Locations 596–601.

31. Cited in Elizabeth Anderson, *Private Government: How Employers Rule Our Lives (and Why We Don't Talk About It)* (Princeton and Oxford: Princeton University Press, 2017), 30.

32. Alan Ryan, *On Politics: A History of Political Thought from Herodotus to the Present* (London: Penguin, 2013), 212.

33. John Locke, *Second Treatise of Government*, in *Two Treatises of Government and A Letter Concerning Toleration*, eds. Ian Shapiro (New Haven and London: Yale University Press, 2003), 111.

34. Locke, *Second Treatise*, 112.

35. Jean-Jacques Rousseau, *Discourse on the Origins of Inequality*, translated by Donald A. Cress (Indianapolis: Hackett Publishing Company, 1992), 44.

36. Karl Marx, *Capital Vol. 1* in *Karl Marx and Frederick Engels: Collected Works Vol. 35* (London: Lawrence & Wishart, 1996), 705.

37. Cicero, *De Officiis*, translated by W. Miller (Cambridge, Mass: Harvard University Press, 1913), cited in Eric Nelson, 'Republican Visions', in *The Oxford Handbook of Political Theory*, eds. John S. Dryzek, Bonnie Honig, and Anne Phillips (New York: Oxford University Press, 2008), 197.

38. Larry Siedentop, *Inventing the Individual: The Origins of Western Liberalism* (London: Allen Lane, 2014), 16–17.

39. Aaron Perzanowksi and Jason Schultz, *The End of Ownership: Personal Property in the Digital Economy* (Cambridge, Mass: MIT Press, 2016), 17.

40. Adam Smith, *Wealth of Nations* (Ware: Wordsworth, 2012), Book III, ch. ii, 382.

41. Friedrich Hayek, *The Constitution of Liberty* (Abingdon: Routledge, 2009), 123.

42. G. W. F. Hegel, *Elements of the Philosophy of Right*, translated by H. B. Nisbet (Cambridge: Cambridge University Press, 2008), §§41, 73.

43. Smith, *Wealth of Nations*, Book V, ch. i, 709.

44. Karl Marx and Friedrich Engels, *Manifesto of the Communist Party*, in *Karl Marx and Frederick Engels: Collected Works Vol. 6* (London: Lawrence & Wishart, 1976), 500.

45. Letter from Thomas Jefferson to Isaac McPherson, 13 August 1813, cited in James Boyle, *The Public Domain: Enclosing the Commons of the Mind* (New Haven and London:Yale University Press, 2008), 19.

46. See discussion in Robert A. Dahl, *A Preface to Economic Democracy* (Cambridge: Polity Press in association with Oxford: Basil Blackwell, 1985), 67.

47. Piketty, *Capital*, 471.

48. Kevin J. Delaney, 'The Robot that Takes Your Job Should Pay Taxes, Says Bill Gates', *Quartz*, 17 February 2017 <https://qz.com/911968/bill-gates-the-robot-that-takes-your-job-should-pay-taxes/> (accessed 8 December 2017).

49. Allen, *Technology and Inequality*, Kindle Locations 638–44.

50. Allen, *Technology and Inequality*, Kindle Locations 379–81.

51. Brian Merchant, 'Fully Automated Luxury Communism', *The Guardian*, 18 March 2015 <https://www.theguardian.com/sustainable-business/2015/mar/18/fully-automated-luxury-communism-robots-employment> (accessed 8 December 2017).

52. Trebor Scholz and Nathan Schneider, eds., *Ours to Hack and to Own: The Rise of Platform Cooperativism, a New Vision for the Future of Work and a Fairer Internet* (NewYork: OR Books/Counterpoint, 2017).

53. Francis Fukuyama, *The Origins of Political Order: From Prehuman Times to the French Revolution* (London: Profile Books, 2012), 66.

54. Parmy Olson, 'Meet Improbable, the Startup Building the World's Most Powerful Simulations', *Forbes*, 15 June 2015 <https://www.forbes.com/sites/parmyolson/2015/05/27/improbable-startup-simulations/#6ae2da044045> (accessed 8 December 2017).

55. Kevin Kelly, *The Inevitable: Understanding the 12 Technological Forces that Will Shape Our Future* (NewYork:Viking, 2016), 110.

56. Boyle, *Public Domain*, 38.

57. Richard Susskind and Daniel Susskind, *The Future of the Professions: How Technology Will Transform the Work of Human Experts* (Oxford: Oxford University Press, 2015), 307.

58. See generally Boyle, *Public Domain*.

59. Boyle, *Public Domain*, 5.

60. Boyle, *Public Domain*, 41.

61. Digital Millennium Copyright Act 1998.

62. Boyle, *Public Domain*, 11.

63. Ibid.

64. Yochai Benkler, *The Wealth of Networks: How Social Production Transforms Markets and Freedom* (New Haven and London: Yale University Press, 2006), 96.

65. Benkler, *Wealth of Networks*, 49.

66. Boyle, *Public Domain*, 50; Perzanowksi and Schultz, *End of Ownership*, 135; see also Peter Drahos with John Braithwaite, *Information Feudalism: Who Owns the Knowledge Economy?* (London: Earthscan, 2002).

67. Brynjolfsson, McAfee, and Spence, 'New World Order'.

68. Susskind and Susskind, *Future of the Professions*, 1.

69. Susskind and Susskind, *Future of the Professions*, 307.

70. Arun Sundararajan, *The Sharing Economy: The End of Employment and the Rise of Crowd-Based Capitalism* (Cambridge, Mass: MIT Press, 2017), 3–5.

71. Lauren Goode, 'Delivery Drones Will Mean the End of Ownership', *The Verge*, 8 November 2016 <https://www.theverge.com/a/verge-2021/google-x-astro-teller-interview-drones-innovation> (accessed 8 December 2017). This piece imagines the hammer located in a 'central place', presumably under common ownership. The general point is the same.

72. Allen, *Technology and Inequality*, Kindle Locations 2600–2601.

73. Allen, *Technology and Inequality*, Kindle Locations 2592, 2645–2647.

74. Evgeny Morozov, 'To Tackle Google's Power, Regulators Have to Go After its Ownership of Data', *The Guardian*, 2 July 2017. <https://www.theguardian.com/technology/2017/jul/01/google-european-commission-fine-search-engines?CMP=share_btn_tw> (accessed 8 December 2017).

75. Hamid R. Ekbia and Bonnie A. Nardi, *Heteromation, and Other Stories of Computing and Capitalism* (Cambridge, Mass: MIT Press, 2017), 25.

76. Doug Laney, 'To Facebook, You're Worth $80.95', *CIO Journal: Wall Street Journal Blogs*, 3 May 2012, cited in Ekbia and Nardi, *Heteromation,* 94 (original emphasis).

77. Jaron Lanier, *Who Owns the Future?* (London: Allen Lane, 2014), 15.

78. Lanier, *Who Owns the Future?* 231, 5.

79. Andreas Weigend, *Data for the People: How to Make our Post-Privacy Economy Work for You* (New York: Basic Books, 2017), 24.

80. Pedro Domingos, *The Master Algorithm: How the Quest for the Ultimate Learning Machine Will Remake Our World* (London: Allen Lane, 2015), 275.

Chapter 19

1. Max Weber, 'The Profession and Vocation of Politics' in *Political Writings*, eds. Peter Lassman and Ronald Speirs (Cambridge: Cambridge University Press, 2010), 356.

2. Chris Baraniuk, 'Google Responds on Skewed Holocaust Search Results', *BBC News*, 20 December 2016. <http://www.bbc.co.uk/news/technology-38379453> (accessed 8 December 2017).

3. Plato, *The Republic*, translated by Desmond Lee (London: Penguin, 2003), Book V, Part VII, 474d, 192.

4. Rosa Luxemburg, 'The Russian Revolution', (1918), ch. 6, translated by Bertram Wolfe (New York: Workers Age Publishers, 1940) *Marxists* <https://www.marxists.org/archive/luxemburg/1918/russian-revolution/ch06.htm> (accessed 9 December 2017).

5. See, e.g. Julie E. Cohen, 'The Regulatory State in the Information Age', *Theoretical Inquiries in Law* 17, no. 2 (2016): 369–414.

6. Frank Pasquale, 'From Holocaust Denial to Hitler Admiration, Google's Algorithm is Dangerous', *Huffington Post*, 2 June 2017 <https://www.huffingtonpost.com/entry/holocaust-google-algorithm_us_587e8628e4b0c147f0bb9893> (accessed 8 December 2017).

7. Danielle Keats Citron and Frank Pasquale, 'The Scored Society: Due Process For Automated Predictions', *Washington Law Review* 89, no. 1 (26 March 2014) <https://digital.law.washington.edu/dspace-law/bitstream/handle/1773.1/1318/89WLR0001.pdf?sequence=1> (accessed 1 December 2017).

8. Gerald Dworkin, *The Theory and Practice of Autonomy* (Cambridge: Cambridge University Press, 1989), 88.

9. Ibid.

10. Ibid.

11. Dworkin, *Autonomy*, 87.

12. Meg Leta Jones, *Ctrl + Z: The Right to Be Forgotten* (New York: New York University Press, 2016), 86.

13. See John Rawls, *A Theory of Justice* (Cambridge, Mass: Harvard University Press, 2003), ch. 18.

14. R.B.Friedman,'On the Concept of Authority in Political Philosophy', in *Authority*, ed. Joseph Raz (Oxford: Basil Blackwell, 1990), 58.

15. Adam Entous, Craig Timberg, and Elizabeth Dwoskin, 'Russian Operatives Used Facebook Ads to Exploit America's Racial and Religious Divisions', *Washington Post*, 25 September 2017 <https://www.washingtonpost.com/business/technology/russian-operatives-used-facebook-ads-to-exploit-divisions-over-black-political-activism-and-muslims/2017/09/25/4a011242-a21b-11e7-ade1-76d061d56efa_story.html?utm_term=.8d517bd8e72e> (accessed 8 December 2017).

16. See Regulation (EU) 2016/679 of the European Parliament and of the Council of 27 April 2016 on the protection of natural persons with regard to the processing of personal data and on the free movement of such data, and repealing Directive 95/46/EC (General Data Protection Regulation).

17. Cathy O'Neil, *Weapons of Math Destruction: How Big Data Increases Inequality and Threatens Democracy* (New York: Crown, 2016), 211; Christian Sandvig, Kevin Hamilton, Karrie Karahalios, and Cedric Langbort, 'Auditing Algorithms: Research Methods for Detecting Discrimination on Internet Platforms', paper presented to 'Data and Discrimination: Converting Critical Concerns into Productive Inquiry,' a preconference at the 64th Annual Meeting of the International Communication Association (22 May 2014) Seattle, WA, USA <http://www-personal.umich.edu/~csandvig/research/Auditing%20Algorithms%20--%20Sandvig%20--%20ICA%202014%20Data%20and%20Discrimination%20Preconference.pdf> (accessed 11 December 2017).

18. Andrew Tutt,'An FDA for Algorithms' *Administrative Law Review* 69, no.1 (2017): 83–123.

19. Viktor Mayer-Schönberger and Kenneth Cukier, *Big Data:A Revolution that Will Transform How We Live, Work and Think* (London: John Murray, 2013), 180.

20. Charles de Secondat, baron de Montesquieu, *The Spirit of the Laws*, translated by Anne M. Cohler, Basia Carolyn Miller, and Harold Samuel Stone (Cambridge: Cambridge University Press, 1989) (Kindle Edition).

21. Lawrence Lessig, 'Introduction', in Richard Stallman, *Free Software, Free Society: Selected Essays of Richard M. Stallman* (Boston: GNU Press, 2002), 9.

22. On algorithmic transparency see Frank Pasquale, *The Black Box Society: The Secret Algorithms that Control Money and Information* (Cambridge, Mass: Harvard University Press, 2015).

23. See Mireille Hildebrandt, 'Legal and Technological Normativity: More (and Less) than Twin Sisters', *Techné: Research in Philosophy and Technology* 12, no. 3 (Fall, 2008): 169–83.

24. Carol Gould, *Rethinking Democracy: Freedom and Social Cooperation in Politics, Economy, and Society* (Cambridge: Cambridge University Press, 1990), 26.

25. Michael J. Sandel, *What Money Can't Buy: The Moral Limits of Markets* (London: Penguin, 2012), 10.

26. Karl Marx, *On the Jewish Question*, in *Karl Marx and Frederick Engels: Collected Works Vol. 3* (London: Lawrence & Wishart, 1975), 154.

Chapter 20

1. Yuval Noah Harari, *Homo Deus: A Brief History of Tomorrow* (London: Harvill Secker, 2015), 25.

2. Elizabeth Lopatto, 'Gene Editing Will Transform Cancer Treatment', *The Verge*, 22 November 2016 <https://www.theverge.com/a/verge-2021/jennifer-doudna-crispr-gene-editing-healthcare> (accessed 8 December 2017).

3. See, e.g., Francis Fukuyama, *Our Posthuman Future: Consequences of the Biotechnology Revolution* (London: Profile Books, 2002); Max More and Natasha Vita-More, eds., *The Transhumanist Reader: Classical and Contemporary Essays on the Science, Technology, and Philosophy of the Human Future* (Chichester: John Wiley & Sons, Inc, 2013).

4. See, e.g. Julian Savulescu, Ruud ter Meulen, and Guy Kahane, eds., *Enhancing Human Capacities* (Chichester: Wiley-Blackwell, 2011); Justin Nelson et al., 'The Effects of Transcranial Direct Current Stimulation (tDCS) on Multitasking Throughput Capacity', *Frontiers in Human Neuroscience*, 29 November 2016 <https://www.frontiersin.org/articles/10.3389/fnhum.2016.00589/full> (accessed 8 December 2017); Michael Bess, 'Why Humankind Isn't Ready for the Bionic Revolution', *Ozy*, 24 October 2016 <http://www.ozy.com/opinion/why-humankind-isnt-ready-for-the-bionic-revolution/72555?utm_source=dd&utm_medium=email&utm_campaign=10242016&variable=af3d1702308a23693509dd3317fe68e7> (accessed 8 December 2017).

5. Joel Garreau, *Radical Evolution: The Promise and Peril of Enhancing Our Minds, Our Bodies—and What It Means to Be Human* (New York: Broadway Books, 2005), 7.

6. Yuval Noah Harari, *Sapiens: A Brief History of Humankind* (London: Vintage Books, 2011); Harari, *Homo Deus*; Wendell Wallach, *A Dangerous Master: How to Keep Technology from Slipping Beyond Our Control* (New York: Basic Books, 2015), 172.

7. See Michael J. Sandel, *The Case Against Perfection: Ethics in the Age of Genetic Engineering* (Cambridge, Mass: Harvard University Press, 2007), 5.

8. Eric Schmidt and Jared Cohen, *The New Digital Age: Reshaping the Future of People, Nations and Business* (London: John Murray, 2014), 26–7; Wallach, *Dangerous Master*, 141; Yiannis Laouris, 'Reengineering and Reinventing Both Democracy and the Concept of Life in the Digital Era', in *The Onlife Manifesto: Being Human in a Hyperconnected Era*, ed. Luciano Floridi (Cham: Springer, 2009), 136.

9. See Sandel, *Case Against Perfection*.

10. Steve Fuller and Veronika Lipińska, *The Proactionary Imperative: A Foundation for Transhumanism* (Basingstoke: Palgrave Macmillan, 2014), 122.

11. See e.g. Harari, *Sapiens*, 410; Sandel, *Case Against Perfection*, 15.

12. Jaron Lanier, *Who Owns the Future?* (London: Allen Lane, 2014), 78.

13. Thomas Hobbes, *Leviathan* (Cambridge: Cambridge University Press, 2007), [64], 91.

14. Harari, *Homo Deus*, 44.

15. David Hume, *An Enquiry Concerning the Principles of Morals* (Indianapolis: Hackett Publishing Company, 1983), 25.

16. David Miller, 'Political Philosophy for Earthlings', in *Political Theory: Methods and Approaches*, eds. David Leopold and Marc Stears (Oxford: Oxford University Press, 2010), 37.

17. H. L. A. Hart, *The Concept of Law* (Second Edition) (Oxford: Oxford University Press, 1997), ch. ix.

18. See also the claims in Peter H. Diamandis and Steven Kotler, *Abundance: The Future is Better than You Think* (New York: Free Press, 2014).

19. Nick Bostrom, *Superintelligence: Paths, Dangers, Strategies* (Oxford: Oxford University Press, 2014), 22.

20. Bostrom, *Superintelligence*, 21.

21. Bostrom, *Superintelligence*, 93.

22. Bostrom, *Superintelligence*, 21.

23. See Ray Kurzweil, *The Singularity is Near* (London: Duckworth, 2010); Murray Shanahan, *The Technological Singularity* (Cambridge, Mass: MIT Press, 2015).

24. Karl Marx and Friedrich Engels, *Manifesto of the Communist Party*, in *Karl Marx and Frederick Engels: Collected Works Vol. 6* (London: Lawrence & Wishart, 1976), 489.

25. Thomas Jefferson to John Adams, 4 September 1823, *Library of Congress* <https://www.loc.gov/exhibits/jefferson/202.html> (accessed 8 December 2017).

BIBLIOGRAPHY

Ackerman, Spencer and Sam Thielman. 'US Intelligence Chief: We Might Use the Internet of Things to Spy on You'. *The Guardian*, 9 Feb. 2016 <https://www.theguardian.com/technology/2016/feb/09/internet-of-things-smart-home-devices-government-surveillance-james-clapper> (accessed 1 Dec. 2017).

Affectiva.com. <http://www.affectiva.com/> (accessed 30 Nov. 2017).

Agoravoting.com. <https://agoravoting.com/> (accessed 1 Dec. 2017).

Agüera y Arcas, Blaise, Margaret Mitchell, and Alexander Todorov. 'Physiognomy's New Clothes'. *Medium*, 6 May 2017 <https://medium.com/@blaisea/physiognomys-new-clothes-f2d4b59fdd6a> (accessed 1 Dec. 2017).

Ajunwa, Ifeoma, Kate Crawford, and Jason Schultz. 'Limitless Worker Surveillance'. *California Law Review* 105, no. 3 (2017), 734–76.

Aletras, Nikolaos, Dimitrios Tsarapatsanis, Daniel Preotiuc, and Vasileios Lampos. 'Predicting Judicial Decisions of the European Court of Human Rights: A Natural Language Processing Perspective'. *Peer J Computer Science* 2, e93 (24 Oct. 2016).

Allen, Jonathan P. *Technology and Inequality: Concentrated Wealth in a Digital World*. Kindle Edition: Palgrave Macmillan, 2017.

Ananny, Mike. 'Toward an Ethics of Algorithms: Convening, Observation, Probability, and Timeliness'. *Science, Technology, & Human Values* 41, no. 1 (2016).

Anderson, Berit and Brett Horvath. 'The Rise of the Weaponized AI Propaganda Machine'. *Medium*, 12 Feb. 2017 <https://medium.com/join-scout/the-rise-of-the-weaponized-ai-propaganda-machine-86dac61668b> (accessed 1 Dec. 2017).

Anderson, Elizabeth. *Private Government: How Employers Rule Our Lives (and Why We Don't Talk About It)*. Princeton and Oxford: Princeton University Press, 2017.

Angelidou, Margarita. 'Smart City Strategy: PlanIT Valley (Portugal)'. *Urenio*, 26 Jan. 2015 <http://www.urenio.org/2015/01/26/smart-city-strategy-planlt-valley-portugal/> (accessed 30 Nov. 2017).

Angwin, Julia and Jeff Larson. 'The Tiger Mom Tax: Asians Are Nearly Twice as Likely to Get a Higher Price from Princeton Review'. *ProPublica*, 1 Sep. 2015 <https://www.propublica.org/article/asians-nearly-twice-as-likely-to-get-higher-price-from-princeton-review> (accessed 3 Dec. 2017).

Angwin, Julia, Jeff Larson, Surya Mattu, and Lauren Kirchner. 'Machine Bias'. *ProPublica*, 23 May 2016 <https://www.propublica.org/article/machine-bias-risk-assessments-in-criminal-sentencing> (accessed 1 Dec. 2017).

Aquinas, Thomas. *Political Writings*. Edited and translated by R. W. Dyson. Cambridge: Cambridge University Press, 2002.

Arbesman, Samuel. *Overcomplicated: Technology at the Limits of Comprehension*. New York: Current, 2016.

Arendt, Hannah. *Crises of the Republic*. San Diego and London: Harcourt Brace & Company, 1972.

Arendt, Hannah. *The Human Condition* (Second Edition). Chicago: University of Chicago Press, 1998.

Arendt, Hannah. *Between Past and Future*. London: Penguin, 2006.

Aristotle. *The Politics*. Translated by T. A. Sinclair. London: Penguin, 1992.

Aristotle, *Nichomachean Ethics*. Translated by Terence Irwin (Second Edition). Indianapolis: Hackett Publishing Company, 1999.

Aron, Jacob. 'Revealed: Google's Plan for Quantum Computer Supremacy'. *New Scientist*, 31 Aug. 2016 <https://www.newscientist.com/article/mg23130894-000-revealed-googles-plan-for-quantum-computer-supremacy/> (accessed 28 Nov. 2017).

Assael, Yannis M., Brendan Shillingford, Shimon Whiteson, and Nando De Freitas. 'LipNet: End-to-End Sentence-level Lipreading'. *arXiv*, 16 Dec. 2016 <https://arxiv.org/abs/1611.01599> (accessed 6 Dec. 2017).

Associated Foreign Press. 'Brain-to-Brain "Telepathic" Communication Achieved for First Time.' *The Telegraph*, 5 Sep. 2014 <http://www.telegraph.co.uk/news/worldnews/northamerica/usa/11077094/Brain-to-brain-telepathic-communication-achieved-for-first-time.html> (accessed 30 Nov. 2017).

Associated Press. 'Cockroach-inspired Robots Designed for Disaster Search and Rescue'. *CBC The Associated Press*, 8 Feb. 2016 <http://www.cbc.ca/beta/news/technology/robot-roach-1.3439138> (accessed 30 Nov. 2017).

Atzori, Marcella. 'Blockchain Technology and Decentralised Governance: Is the State Still Necessary?' 1 Dec. 2015 <https://papers.ssrn.com/sol3/papers.cfm?abstract_id=2709713> (accessed 6 Dec. 2017).

Autor, David, David Dorn, Lawrence F. Katz, Christina Patterson, and John van Reenen. 'The Fall of the Labor Share and the Rise of Superstar Firms'. 1 May 2017 <https://economics.mit.edu/files/12979> (accessed 8 Dec. 2017).

Avent, Ryan. *The Wealth of Humans: Work, Power, and Status in the Twenty-First Century*. New York: St. Martin's Press, 2016.

Axelrod, Robert. *The Evolution of Cooperation*. New York: Basic Books, 2006.

Azuma, Hiroki. *General Will 2.0: Rousseau, Freud, Google*. New York: Vertical, Inc, 2014.

Baack, Stefan. 'Datafication and Empowerment: How the Open Data Movement Re-articulates Notions of Democracy, Participation, and Journalism'. *Big Data & Society* 2, no. 2 (2015).

Bachrach, Peter and Morton S. Baratz. 'Two Faces of Power'. *American Political Science Review* 56, no. 4 (Dec. 1962): 947–52.

Bachrach, Peter and Morton S. Baratz. *Power and Poverty: Theory and Practice*. New York: Oxford University Press, 1970.

Baker, Paul and Amanda Potts. '"Why Do White People Have Thin Lips?" Google and the Perpetuation of Stereotypes via Auto-complete Search Forms'. *Critical Discourse Studies* 10, no. 2 (2013).

Ball, James. 'Let's Challenge Google While We Still Can'. *The Guardian*, 16 Apr. 2015 <https://www.theguardian.com/commentisfree/2015/apr/16/challenge-google-while-we-can-eu-anti-trust> (accessed 8 Dec. 2017).

Ball, Terence, James Farr, and Russell L. Hanson, eds. *Political Innovation and Conceptual Change*. New York: Cambridge University Press, 1995.

Baraniuk, Chris. 'Google Responds on Skewed Holocaust Search Results'. *BBC News*, 20 Dec. 2016. <http://www.bbc.co.uk/news/technology-38379453> (accessed 8 Dec. 2017).

Barney, Darin. *Prometheus Wired: The Hope for Democracy in the Age of Network Technology*. Chicago: University of Chicago Press, 2000.

Barr, Alistair. 'Google Mistakenly Tags Black People as "Gorillas", Showing Limits of Algorithms'. *Wall Street Journal*, 1 Jul. 2015 <https://blogs.wsj.com/digits/2015/07/01/google-mistakenly-tags-black-people-as-gorillas-showing-limits-of-algorithms/> (accessed 2 Dec. 2017).

Bartlett, Jamie. *The Dark Net: Inside the Digital Underworld*. London: William Heinemann, 2014.

Bartlett, Jamie and Nathaniel Tkacz. 'Governance by Dashboard'. *Demos*, Mar. 2017 <https://www.demos.co.uk/wp-content/uploads/2017/04/Demos-Governance-by-Dashboard.pdf> (accessed 1 Dec. 2017).

Batt, J. Daniel, ed. *Visions of the Future*. Reno: Lifeboat Foundation, 2015.

BBC. 'Flipped 3D Printer Makes Giant Objects'. *BBC News*, 24 Aug. 2016 <http://www.bbc.co.uk/news/technology-37176662?ocid=socialflow_twitter> (accessed 30 Nov. 2017).

BBC. 'Google Working on "Common-sense" AI Engine at New Zurich Base'. *BBC News*, 17 Jun. 2016 <http://www.bbc.co.uk/news/technology-36558829> (accessed 30 Nov. 2017).

BBC. 'Hack Attack Drains Start-up Investment Fund'. *BBC News*, 21 Jun. 2016 <http://www.bbc.co.uk/news/technology-36585930> (accessed 30 Nov. 2017).

BBC. 'Pokemon Go: Is the Hugely Popular Game a Global Safety Risk?' *BBC News*, 21 Jul. 2016 <http://www.bbc.co.uk/news/world-36854074> (accessed 30 Nov. 2017).

BBC. 'Beijing Park Dispenses Loo Roll Using Facial Recognition'. *BBC News*, 20 Mar. 2017 <http://www.bbc.com/news/world-asia-china-39324431> (accessed 30 Nov. 2017).

BBC. 'German Parents Told to Destroy Cayla Dolls Over Hacking Fears'. *BBC News*, 17 Feb. 2017 <http://www.bbc.co.uk/news/world-europe-39002142> (accessed 1 Dec. 2017).

BBC. 'WhatsApp Must Not Be "Place for Terrorists to Hide"'. *BBC News*, 26 Mar. 2017 <http://www.bbc.co.uk/news/uk-39396578 (accessed 1 Dec. 2017).

Belamaire, Jordan. 'My First Virtual Reality Groping'. *Medium*, 20 Oct. 2016 <https://medium.com/athena-talks/my-first-virtual-reality-sexual-assault-2330410b62ee#.i1o6j1vjy> (accessed 30 Nov. 2017).

Beniger, Andrew J. *Control Revolution: Technological and Economic Origins of the Information Society*. Cambridge, Mass: Harvard University Press, 1986.

Benkler, Yochai. *The Wealth of Networks: How Social Production Transforms Markets and Freedom.* New Haven and London: Yale University Press, 2006.

Benkler, Yochai. 'Networks of Power, Degrees of Freedom'. *International Journal of Communication* 5 (2011): 721–55.

Benkler, Yochai. *The Penguin and the Leviathan: How Cooperation Triumphs over Self-Interest.* New York: Crown Publishing, 2011.

Benkler, Yochai. 'Degrees of Freedom, Dimensions of Power'. *Daedalus* 145, no. 1 (Winter 2016): 18–32.

Berkman Center for Internet and Society. '"DON'T PANIC. Making Progress on the "Going Dark" Debate', 1 Feb. 2016 <https://cyber.harvard.edu/pubrelease/dont-panic/Dont_Panic_Making_Progress_on_Going_Dark_Debate.pdf> (accessed 1 Dec. 2017).

Berlin, Isaiah. *Four Essays on Liberty.* Oxford: Oxford University Press, 1969.

Berlin, Isaiah. *Concepts and Categories.* Ed. Henry Hardy. London: Pimlico, 1999.

Berlin, Isaiah. *The Power of Ideas.* Ed. Henry Hardy. London: Pimlico, 2001.

Berman, Robby. 'New Tech Uses WiFi to Read Your Inner Emotions— Accurately, and From Afar'. *Big Think*, 2016 <http://bigthink.com/robby-berman/new-tech-can-accurately-read-the-emotions-you-may-be-hiding> (accessed 30 Nov. 2017).

Berman, Robby. 'So the Russians Just Arrested a Robot at a Rally'. *Big Think*, 2016 <http://bigthink.com/robby-berman/so-the-russians-just-arrested-a-robot-at-a-rally> (accessed 30 Nov. 2017).

Berners-Lee, Tim with Mark Fischetti. *Weaving the Web: The Original Design and Ultimate Destiny of the World Wide Web.* New York: HarperCollins, 2000.

Bertrand, Romain, Jean-Louis Briquet, and Peter Pels, eds. *The Hidden History of the Secret Ballot.* Bloomington: Indiana University Press, 2006.

Bess, Michael. 'Why Humankind Isn't Ready for the Bionic Revolution'. *Ozy*, 24 Oct. 2016. <http://www.ozy.com/opinion/why-humankind-isnt-ready-for-the-bionic-revolution/72555?utm_source=dd&utm_medium=email&utm_campaign=10242016&variable=af3d1702308a23693509dd3317fe68e7> (accessed 8 Dec. 2017).

BI Intelligence. 'Amazon Accounts for 43% of US Online Retail Sales'. *Business Insider E-Commerce Briefing*, 3 Feb. 2017. <http://uk.businessinsider.com/amazon-accounts-for-43-of-us-online-retail-sales-2017-2?r=US&IR=T> (accessed 9 Dec. 2017).

Bimber, Bruce. *Information and American Democracy: Technology in the Evolution of Political Power*. New York: Cambridge University Press, 2011.

Blake, William. *London*. Poetry Foundation <https://www.poetryfoundation. org/poems/43673/london-56d222777e969> (accessed 7 Dec. 2017).

Blue Brain Project. <https://bluebrain.epfl.ch/page-56882-en.html> (accessed 6 Dec. 2017).

Bock, Gisela, Quentin Skinner, and Maurizio Viroli, eds. *Machiavelli and Republicanism*. Cambridge: Cambridge University Press, 1993.

Boden, Margaret A. *AI: Its Nature and Future*. Oxford: Oxford University Press, 2016.

Bogle, Ariel. 'Good News: Replicas of 16th Century Sculptures Are Not Off-Limits for 3-D Printers'. *Slate*, 27 Jan. 2015 <http://www.slate.com/ blogs/future_tense/2015/01/26/_3_d_printing_and_copyright_replicas_ of_16th_century_sculptures_are_not.html?wpisrc=obnetwork> (accessed 30 Nov. 2017).

Bohman, James, and William Rehg, eds. *Deliberative Democracy: Essays on Reason and Politics*. Cambridge, Mass: MIT Press, 2002.

Boixo, Sergio, Sergi V. Isakov, Vadim N. Smelyanskiy, Ryan Babbush, Nan Ding, Zhang Jiang, Michael J. Bremner, John M. Martinis, and Hartmut Neven. 'Characterizing Quantum Supremacy in Near-Term Devices'. *sarXiv*, 5 Apr. 2017 <https://arxiv.org/abs/1608.00263> (accessed 28 Nov. 2017).

Bollen, Johan, Huina Mao, and Xiao-Jun Zeng. 'Twitter Mood Predicts the Stock Market'. *arXiv*, 14 Oct. 2010 <https://arxiv.org/pdf/1010.3003. pdf> (accessed 1 Dec. 2017).

Bolter, J. David, *Turing's Man: Western Culture in the Computer Age*. London: Duckworth, 1984.

Bolukbasi, Tolga, Kai-Wei Chang, James Zou, and Venkatesh Saligrama. 'Man is to Computer Programmer as Woman is to Homemaker? Debiasing Word Embeddings'. *arXiv*, 21 Jul. 2016 <https://arxiv.org/ pdf/1607.06520.pdf> (accessed 3 Dec. 2017).

Bonchi, Francesco, Carlos Castillo, and Sara Hajian. 'Algorithmic Bias: From Discrimination Discovery to Fairness-aware Data Mining'. *KDD*

2016 Tutorial. <http://francescobonchi.com/tutorial-algorithmic-bias. pdf> (accessed 3 Dec. 2017).

Booth, Robert. 'Facebook Reveals News Feed Experiment to Control Emotions'. *The Guardian*, 30 Jun. 2004 <https://www.theguardian.com/ technology/2014/jun/29/facebook-users-emotions-news-feeds> (accessed 11 Dec. 2017).

Bostrom, Nick. *Superintelligence: Paths, Dangers, Strategies.* Oxford: Oxford University Press, 2014.

Bourzac, Katherine. 'A Health-Monitoring Sticker Powered by Your Cell Phone'. *MIT Technology Review*, 3 Aug. 2016 <https://www. technologyreview.com/s/602067/a-health-monitoring-sticker-powered-by-your-cell-phone/?utm_campaign=socialflow&utm_source=twitter&utm_medium=post> (accessed 29 Nov. 2017).

Boushey, Heather, J. Bradford DeLong, and Marshall Steinbaum, eds. *After Piketty: The Agenda for Economics and Inequality.* Cambridge, Mass: Harvard University Press, 2017.

Boyle, James. 'Foucault in Cyberspace: Surveillance, Sovereignty, and Hardwired Censors'. *University of Cincinnati Law Review* 66 (1997): 177–204.

Boyle, James. *The Public Domain: Enclosing the Commons of the Mind.* New Haven and London: Yale University Press, 2008.

Brabham, Daren C. *Crowdsourcing.* Cambridge, Mass: MIT Press, 2013.

Bradbury, Danny. 'How Block Chain Technology Could Usher in Digital Democracy'. *CoinDesk*, 16 Jun. 2014 <http://www.coindesk.com/block-chain-technology-digital-democracy/> (accessed 1 Dec. 2017).

Braman, Sandra. *Change of State: Information, Policy, and Power.* Cambridge, Mass: MIT Press, 2009.

Bratton, Benjamin H. *The Stack: On Sovereignty and Software.* Cambridge, Mass: MIT Press, 2015.

Bridge, Mark. 'AI Can Identify Alzheimer's Disease a Decade Before Symptoms Appear'. *The Times*, 20 Sep. 2017 <https://www.thetimes. co.uk/article/ai-can-identify-alzheimer-s-a-decade-before-symptoms-appear-9b3qdrrf7> (accessed 1 Dec. 2017).

Brown, Gordon. *My Life, Our Times.* London: Bodley Head, 2017.

Brown, Ian and Christopher T. Marsden. *Regulating Code: Good Governance and Better Regulation in the Information Age*. Cambridge, Mass: MIT Press, 2013.

Brownsword, Roger, and Morag Goodwin. *Law and the Technologies of the Twenty-First Century: Texts and Materials*. Cambridge: Cambridge University Press, 2012.

Brynjolfsson, Erik, Andrew McAfee, and Michael Spence. 'New World Order: Labor, Capital, and Ideas in the Power Law Economy'. *Foreign Affairs*, Jul./Aug. 2014 <https://www.foreignaffairs.com/articles/united-states/2014-06-04/new-world-order> (accessed 8 Dec. 2017).

Brynjolfsson, Erik and Andrew McAfee. *The Second Machine Age: Work, Progress, and Prosperity in a Time of Brilliant Technologies*. New York: W. W. Norton & Company, 2014.

Brynjolfsson, Erik and Andrew McAfee. *Machine Platform Crowd: Harnessing Our Digital Future*. New York: W. W. Norton & Company, 2017.

Burgess, Matt. 'Samsung is Working on Putting AI Voice Assistant Bixby in Your TV and Fridge'. *Wired*, 27 Jun. 2017<https://www.wired.co.uk/article/samsung-bixby-television-refrigerator> (accessed 30 Nov. 2017).

Byford, Sam. 'AlphaGo Beats Ke Jie Again to Wrap Up Three-part Match'. *The Verge*, 25 May 2017 <https://www.theverge.com/2017/5/25/15689462/alphago-ke-jie-game-2-result-google-deepmind-china> (accessed 28 Nov. 2017).

Byrnes, Nanette. 'As Goldman Embraces Automation, Even the Masters of the Universe Are Threatened'. *MIT Technology Review*, 7 Feb. 2017 <https://www.technologyreview.com/s/603431/as-goldman-embraces-automation-even-the-masters-of-the-universe-are-threatened/?set=603585&utm_content=bufferd5a8f&utm_medium=social&utm_source=twitter.com&utm_campaign=buffer> (accessed 1 Dec. 2017).

Cadwalladr, Carole. 'Robert Mercer: The Big Data Billionaire Waging War on Mainstream Media'. *The Guardian*, 26 Feb. 2017 <https://www.theguardian.com/politics/2017/feb/26/robert-mercer-breitbart-war-on-media-steve-bannon-donald-trump-nigel-farage> (accessed 1 Dec. 2017).

Calabresi, Guido, and Philip Bobbit. *Tragic Choices: The Conflicts Society Confronts in the Allocation of Tragically Scarce Resources*. New York: W. W. Norton & Company, 1978.

Calvo, Rafael A., Sidney D'Mello, Jonathan Gratch, and Arvid Kappas, eds. *The Oxford Handbook of Affective Computing*. New York: Oxford University Press, 2015.

Canetti, Elias. *Crowds and Power*. Translated by Carol Stewart. New York: Farrar, Straus and Giroux, 1984.

Campbell, Peter. 'Ford Plans Mass-market Self-driving Car by 2021'. *Financial Times*, 16 Aug. 2016. <https://www.ft.com/content/d2cfc64e-63c0-11e6-a08a-c7ac04ef00aa#axzz4HOGiWvHT> (accessed 28 Nov. 2017).

Carr, Nicholas. *The Big Switch: Rewiring the World from Edison to Google*. New York: W. W. Norton & Company, 2009.

Casanova, Giacomo. *The Story of My Life*. Translated by Sophie Hawkes. London: Penguin, 2000.

Case, Amber. *Calm Technology: Principles and Patterns for Non-Intrusive Design*. Sebastopol, CA: O'Reilly, 2016.

Casey, Anthony J. and Anthony Niblett. 'Self-Driving Laws'. *University of Toronto Law Journal* 429 (Fall 2016) 66: 428–42.

Casey, Anthony J. and Anthony Niblett. 'The Death of Rules and Standards'. *Indiana Law Journal* 92, no. 4 (2017).

Castells, Manuel. *Communication Power*. Oxford: Oxford University Press, 2013.

Cellan-Jones, Rory. ' "Cut!"—the AI Director'. *BBC News*, 23 Jun. 2016. <http://www.bbc.co.uk/news/technology-36608933> (accessed 28 Nov. 2017).

Chadwick, Andrew. *Internet Politics: States, Citizens, and New Communication Technologies*. New York: Oxford University Press, 2006.

Chadwick, Andrew. *The Hybrid Media System: Politics and Power*. New York: Oxford University Press, 2013.

Cheney-Lippold, John, *We Are Data: Algorithms and the Making of Our Digital Selves*. New York: New York University Press, 2017.

Chesterton, G. K. *Orthodoxy*. Cavalier Classics, 2015.

Chin, Josh and Gillian Wong. 'China's New Tool for Social Control: A Credit Rating for Everything'. *Wall Street Journal*, 28 Nov. 2016 <http://www.wsj.com/articles/chinas-new-tool-for-social-control-a-credit-rating-for-everything-1480351590> (accessed 1 Dec. 2017).

Christiano, Thomas. *The Rule of the Many: Fundamental Issues in Democratic Theory*. Colorado and London: Westview Press, 1996.

Chui, Michael, James Manyika, and Mehdi Miremadi. 'Where Machines Could Replace Humans—and Where They Can't (Yet)'. *McKinsey Quarterly*, Jul. 2016 <https://www.mckinsey.com/business-functions/digital-mckinsey/our-insights/where-machines-could-replace-humans-and-where-they-cant-yet> (accessed 8 Dec. 2017).

Churchill, Winston. *My Early Life: A Roving Commission*. London: Reprint Society, 1944.

Cisco. 'VNI Global Fixed and Mobile Internet Traffic Forecasts, Complete Visual Networking Index (VNI) Forecast', 2016 <https://www.cisco.com/c/en/us/solutions/service-provider/visual-networking-index-vni/index.html#~mobile-forecast> (accessed 30 Nov. 2017).

Citron, Danielle Keats. 'Open Code Government'. University of Maryland School of Law Legal Studies Research Paper No. 2008–1 <https://papers.ssrn.com/sol3/papers.cfm?abstract_id=1081689##> (accessed 3 Dec. 2017).

Citron, Danielle Keats and Frank Pasquale. 'The Scored Society: Due Process for Automated Predictions'. *Washington Law Review* 89, no. 1 (26 Mar. 2014) <https://digital.law.washington.edu/dspace-law/bitstream/handle/1773.1/1318/89WLR0001.pdf?sequence=1> (accessed 1 Dec. 2017).

Clark, Justin, Rob Faris, Ryan Morrison-Westphal, Helmi Noman, Casey Tilton, and Jonathan Zittrain, 'The Shifting Landscape of Global Internet Censorship'. *Internet Monitor*, 29 Jun. 2017 <https://thenetmonitor.org/research/2017-global-internet-censorship> (accessed 1 Dec. 2017).

Clarke, Arthur C. *Profiles of the Future: An Inquiry into the Limits of the Possible*. London: Victor Gollancz, 1999.

Clayton, Matthew, and Andrew Williams, eds. *The Ideal of Equality*. New York: Palgrave Macmillan, 2002.

Clayton, Matthew, and Andrew Williams, eds. *Social Justice*. Oxford: Blackwell Publishing, 2005.

Cocco, Federica. 'Most US Manufacturing Jobs Lost to Technology, Not Trade'. *Financial Times*, 2 Dec. 2016 <https://www.ft.com/content/dec677c0-b7e6-11e6-ba85-95d1533d9a62> (accessed 8 Dec. 2017).

Cohen, G. A. *History, Labour, and Freedom: Themes from Marx*. Oxford: Oxford University Press, 1988.

Cohen, G. A. *If You're an Egalitarian, How Come You're So Rich?* Cambridge, Mass: Harvard University Press, 2001.

Cohen, G. A. *Rescuing Justice and Equality*. Cambridge, Mass: Harvard University Press, 2008.

Cohen, Julie E. *Configuring the Networked Self: Law, Code, and the Play of Everyday Practice*. New Haven and London: Yale University Press, 2012.

Cohen, Julie E. 'The Regulatory State in the Information Age'. *Theoretical Inquiries in Law* 17, no. 2 (2016): 369–414.

Coleman, E. Gabriella. *Coding Freedom: The Ethics and Aesthetics of Hacking*. Princeton: Princeton University Press, 2013.

Colvile, Robert. *The Great Acceleration: How the World is Getting Faster, Faster*. London: Bloomsbury, 2016.

Computerscience.org. 'Women in Computer Science: Getting Involved in STEM'. <http://www.computerscience.org/resources/women-in-computer-science/> (accessed 3 Dec. 2017).

Comte, Auguste. *Early Political Writings*. Translated by A. S. Jones. Cambridge: Cambridge University Press, 1998.

Comte, Auguste. *Auguste Comte and Positivism: The Essential Writings*. Ed. Gertrud Lenzer. New York: Transaction, 2006.

Condliffe, Jamie. 'Chip Makers Admit Transistors Are About to Stop Shrinking'. *MIT Technology Review*, 25 Jul. 2016 <https://www.technologyreview.com/s/601962/chip-makers-admit-transistors-are-about-to-stop-shrinking/?> (accessed 28 Nov. 2017).

Conger, Krista. 'Computers Trounce Pathologists in Predicting Lung Cancer Type, Severity'. *Stanford Medicine News Center*, 16 Aug. 2016 <http://med.stanford.edu/news/all-news/2016/08/computers-trounce-pathologists-in-predicting-lung-cancer-severity.html> (accessed 28 Nov. 2017).

Connolly, William E., ed. *Legitimacy and the State*. New York: New York University Press, 1984.

Connolly, William E. *The Terms of Political Discourse* (Third Edition). Oxford: Basil Blackwell, 1994.

Cooper, Daniel. 'These Subtle Smart Gloves Turn Sign Language Into Text'. *Engadget*, 31 May 2017 <https://www.engadget.com/2017/05/31/these-subtle-smart-gloves-turn-sign-language-into-words/?sr_source=Twitter> (accessed 1 Dec. 2017).

Couldry, Nick. *Media, Society, World: Social Theory and Digital Media Practice*. Cambridge: Polity, 2012.

Crawford, Kate. 'Can an Algorithm be Agonistic? Ten Scenes from Life in Calculated Publics'. *Science, Technology & Human Values* 41, no. 1 (2016): 77–92.

Crosland, Anthony. *The Future of Socialism*. London: Constable & Robinson Ltd, 2006.

Crossley, Rob. 'Where in the World is My Data and How Secure is it?' *BBC News*, 9 Aug. 2016 <http://www.bbc.com/news/business-36854292> (accessed 30 Nov. 2017).

Crowdpac. <https://www.crowdpac.co.uk/> (accessed 1 Dec. 2017).

Cukier, Kenneth and Viktor Mayer-Schönberger. 'The Rise of Big Data'. *Foreign Affairs*, May/Jun. 2013 <https://www.foreignaffairs.com/articles/2013-04-03/rise-big-data> (accessed 30 Nov. 2017).

Dahir, Abdi Latif. 'Egypt Has Blocked Over 100 Local and International Websites Including HuffPost and Medium'. *Quartz*, 29 Jun. 2017 <https://qz.com/1017939/egypt-has-blocked-huffington-post-al-jazeera-medium-in-growing-censorship-crackdown/> (accessed 8 Dec. 2017).

Dahl, Robert A. *Who Governs? Democracy and Power in an American City*. New Haven and London: Yale University Press, 1961.

Dahl, Robert A. *A Preface to Economic Democracy*. Cambridge: Polity Press in association with Oxford: Basil Blackwell, 1985.

Dahlberg, Lincoln. 'Re-constructing Digital Democracy: An Outline of Four "Positions"'. *New Media & Society* 13, no. 6 (2011): 855–72.

D'Ancona, Matthew. *Post Truth: The New War on Truth and How to Fight Back*. London: Ebury Press, 2017.

Dandeker, Christopher. *Surveillance, Power and Modernity*. Cambridge: Polity, 1990.

Dayen, David. 'This Budding Movement Wants to Smash Monopolies'. *Nation*, 4 Apr. 2017 <https://www.thenation.com/article/this-budding-movement-wants-to-smash-monopolies/> (accessed 8 Dec. 2017).

De Filippi, Primavera, and Benjamin Loveluck. 'The Invisible Politics of Bitcoin: Governance Crisis of a Decentralized Infrastructure'. *Internet Policy Review* 5, no. 3 (30 Sep. 2016) <http://policyreview.info/articles/analysis/invisible-politics-bitcoin-governance-crisis-decentralised-infrastructure> (accessed 30 Nov. 2017).

De Filippi, Primavera, and Samer Hassan. 'Blockchain Technology as a Regulatory Technology: From Code is Law to Law is Code'. *First Monday* 21, no. 12, 5 Dec. 2016.

De Filippi, Primavera and Aaron Wright, *Blockchain and the Law: The Rule of Code.* Cambridge, Mass: Harvard University Press, 2018.

Dean Sarah. 'A Nation of "Micro-criminals": The 11 Sneaky Crimes We Are Commonly Committing'. *iNews*, 22 Oct. 2016 <https://inews.co.uk/essentials/news/uk/nation-micro-criminals-11-sneaky-crimes-commonly-committing/> (accessed 1 Dec. 2017).

DeDeo, Simon. 'Wrong Side of the Tracks: Big Data and Protected Categories'. *arXiv*, 24 Jun. 2016 <https://arxiv.org/abs/1412.4643> (accessed 5 Dec. 2017).

Delaney, Kevin J. 'The Robot that Takes Your Job Should Pay Taxes, Says Bill Gates'. *Quartz*, 17 Feb. 2017 <https://qz.com/911968/bill-gates-the-robot-that-takes-your-job-should-pay-taxes/> (accessed 8 Dec. 2017).

Deleuze, Gilles. 'Postscript on the Societies of Control'. *October* 59 (Winter, 1992): 3–7.

DemocracyOS. <http://democracyos.org/> (accessed 1 Dec. 2017).

Desert Wolf. 'Skunk Riot Control Copter'. <http://www.desert-wolf.com/dw/products/unmanned-aerial-systems/skunk-riot-control-copter.html> (accessed 1 Dec. 2017).

Desjardins, Jeff. 'How much data is generated each day?' <https://www.weforum.org/agenda/2019/04/how-much-data-is-generated-each-day-cf4bddf29f/> (accessed 5 Jul. 2019).

Desrosières, Alain. *The Politics of Large Numbers: A History of Statistical Reasoning.* Translated by Camille Naish. Cambridge, Mass: Harvard University Press, 1998.

Deuze, Mark, *Media Life.* Cambridge: Polity, 2012.

Devlin, Patrick. *The Enforcement of Morals.* London: Oxford University Press, 1965.

Diamandis, Peter H., and Steven Kotler. *Abundance: The Future is Better Than You Think*. New York: Free Press, 2014.

Doctorow, Cory. 'Riot Control Drone that Fires Paintballs, Pepper-spray and Rubber Bullets at Protesters'. *Boing Boing*, 17 Jun. 2014 <https://boingboing.net/2014/06/17/riot-control-drone-that-paintb.html> (accessed 7 Dec. 2017).

'Domesday Book'. *Wikipedia*, last modified 26 Nov. 2017 <https://en.wikipedia.org/wiki/Domesday_Book> (accessed 28 Nov. 2017).

Domingos, Pedro. *The Master Algorithm: How the Quest for the Ultimate Learning Machine Will Remake Our World*. London: Allen Lane, 2015.

Dommering, E. J. and Lodewijk F. Asscher, eds. *Coding Regulation: Essays on the Normative Role of Information Technology*. The Hague: TMC Asser, 2006.

Dourish, Paul, and Genevieve Bell. *Divining a Digital Future: Mess and Mythology in Ubiquitous Computing*. Cambridge, Mass: MIT Press, 2011.

Drahos, Peter with John Braithwaite. *Information Feudalism: Who Owns the Knowledge Economy?* London: Earthscan, 2002.

Dredge, Stuart. '30 Things Being 3D Printed Right Now (and None of them Are Guns)'. *The Guardian*, 29 Jan. 2014 <https://www.theguardian.com/technology/2014/jan/29/3d-printing-limbs-cars-selfies> (accessed 30 Nov. 2017).

Dryzek, John S., Bonnie Honig, and Anne Phillips, eds. *The Oxford Handbook of Political Theory*. New York: Oxford University Press, 2008.

Duff, Alistair S. *A Normative Theory of the Information Society*. New York: Routledge, 2013.

Dunn, John. *Setting the People Free: The Story of Democracy*. London: Atlantic, 2005.

Dunn, John. *Breaking Democracy's Spell*. New Haven and London: Yale University Press, 2014.

Durkheim, Émile. *The Rules of Sociological Method and Selected Texts on Sociology and its Method*. Translated by W. D. Halls. New York: Free Press, 1982.

Durkheim, Émile. *The Division of Labour in Society*. Translated by W. D. Halls. Basingstoke: Palgrave, 1984.

Dvorsky, George. 'Record-Setting Hard Drive Writes Information One Atom at a Time'. *Gizmodo*, 18 Jul. 2016 <http://gizmodo.com/record-setting-hard-drive-writes-information-one-atom-a-1783740015> (accessed 30 Nov. 2017).

Dworkin, Gerald. *The Theory and Practice of Autonomy*. Cambridge: Cambridge University Press, 1989.

Dwyer, Paula. 'Should America's Tech Giants Be Broken Up?' *Bloomberg Businessweek*, 20 Jul. 2017 <https://www.bloomberg.com/news/articles/2017-07-20/should-america-s-tech-giants-be-broken-up> (accessed 8 Dec. 2017).

Dyer-Witheford, Nick. *Cyber-Proletariat: Global Labour in the Digital Vortex*. London: Pluto Press, 2015.

Dyson, George. *Darwin Among the Machines: The Evolution of Global Intelligence*. Reading, Mass: Addison-Wesley Pub. Co., 1997.

Economist. 'China Invents the Digital Totalitarian State'. 17 Dec. 2016 <http://www.economist.com/news/briefing/21711902-worrying-implications-its-social-credit-project-china-invents-digital-totalitarian> (accessed 1 Dec. 2017).

Economist. 'How Cities Score'. 23 May 2016 <https://www.economist.com/news/special-report/21695194-better-use-data-could-make-cities-more-efficientand-more-democratic-how-cities-score> (accessed 30 Nov. 2017).

Economist. 'Not-so-clever Contracts'. 28 Jul. 2016 <http://www.economist.com/news/business/21702758-time-being-least-human-judgment-still-better-bet-cold-hearted?frsc=dg%7Cd> (accessed 30 Nov. 2017).

Edelman, Benjamin, Michael Luca, and Dan Svirsky. 'Racial Discrimination in the Sharing Economy: Evidence from a Field Experiment'. *American Economic Journal: Applied Economics* 9, no. 2 (Apr. 2017): 1–22.

Edwards, Cory. 'Why and How Chatbots Will Dominate Social Media'. *TechCrunch*, 20 Jul. 2016 <https://techcrunch.com/2016/07/20/why-and-how-chatbots-will-dominate-social-media/?ncid=rss&utm_source=feedburner&utm_medium=feed&utm_campaign=Feed%3A+Techcrunch+%28TechCrunch%29&sr_share=twitter> (accessed 28 Nov. 2017).

Eisenstein, Elizabeth. *The Printing Press as an Agent of Change: Communications and Cultural Transformations in Early-modern Europe, Volumes I and II.* Cambridge: Cambridge University Press, 2009.

Ekbia, Hamid R. and Bonnie A. Nardi. *Heteromation, and Other Stories of Computing and Capitalism.* Cambridge, Mass: MIT Press, 2017.

Electronic Frontier Foundation. 'Free Speech'. <https://www.eff.org/free-speech-weak-link/> (accessed 1 Dec. 2017).

Elster, Jon. *Reason and Rationality.* Princeton: Princeton University Press, 2009.

EMC. 'The Digital Universe of Opportunities: Rich Data and the Increasing Value of the Internet of Things'. Apr. 2014 <https://www.emc.com/leadership/digital-universe/2014iview/executive-summary.htm> (accessed 30 Nov. 2017).

Emerging Technology from the arXiv. 'Racism is Poisoning Online Ad Delivery, Says Harvard Professor'. *MIT Technology Review*, 4 Feb. 2013 <https://www.technologyreview.com/s/510646/racism-is-poisoning-online-ad-delivery-says-harvard-professor/> (accessed 3 Dec. 2017).

Enchassi, Nadia Judith and CNN Wire. 'New Zealand Passport Robot Thinks This Asian Man's Eyes Are Closed'. *KFOR*, 11 Dec. 2016 <http://kfor.com/2016/12/11/new-zealand-passport-robot-thinks-this-asian-mans-eyes-are-closed/> (accessed 2 Dec. 2017).

Engels, Friedrich. *Karl Marx and Frederick Engels: Collected Works Volumes 1–50.* London: Lawrence & Wishart, 1975–2004.

Entous, Adam, Craig Timberg, and Elizabeth Dwoskin. 'Russian Operatives Used Facebook Ads to Exploit America's Racial and Religious Divisions'. *Washington Post*, 25 Sep. 2017. <https://www.washingtonpost.com/business/technology/russian-operatives-used-facebook-ads-to-exploit-divisions-over-black-political-activism-and-muslims/2017/09/25/4a011242-a21b-11e7-ade1-76d061d56efa_story.html?utm_term=.8d517bd8e72e> (accessed 8 Dec. 2017).

Epstein, Robert. 'The New Censorship'. *US News*, 22 Jul. 2016 <http://www.usnews.com/opinion/articles/2016-06-22/google-is-the-worlds-biggest-censor-and-its-power-must-be-regulated> (accessed 1 Dec. 2017).

Esteva, Andre, Brett Kuprel, Roberto A. Novoa, Justin Ko, Susan M. Swetter, Helen M. Blau, and Sebastian Thrun. 'Dermatologist-level

Classification of Skin Cancer with Deep Neural Networks'. *Nature* 542 (2 Feb. 2017): 115–118.

Executive Office of the President. 'Big Data: A Report on Algorithmic Systems, Opportunity, and Civil Rights'. *Obama White House Archives.* May 2016 <https://obamawhitehouse.archives.gov/sites/default/files/microsites/ostp/2016_0504_data_discrimination.pdf> (accessed 2 Dec. 2017).

Facebook Newsroom. 'Facebook, Microsoft, Twitter and YouTube Announce Formation of the Global Internet Forum to Counter Terrorism'. 26 Jun. 2017 <https://newsroom.fb.com/news/2017/06/global-internet-forum-to-counter-terrorism/> (accessed 1 Dec. 2017).

Fairfield, Joshua A. T. *Owned: Property, Privacy, and the New Digital Serfdom.* Cambridge: Cambridge University Press, 2017.

Faris, Robert, Hal Roberts, Bruce Etling, Nikki Bourassa, Ethan Zuckerman, and Yochai Benkler. 'Partisanship, Propaganda, and Disinformation: Online Media and the 2016 U.S. Presidential Election'. *Berkman Klein Center Research Paper* <https://papers.ssrn.com/sol3/papers.cfm?abstract_id=3019414> (accessed 8 Dec. 2017).

Farrand, Benjamin, and Helena Carrapico. 'Networked Governance and the Regulation of Expression on the Internet: The Blurring of the Role of Public and Private Actors as Content Regulators'. *Journal of Information Technology & Politics* 10, no. 4 (2013): 357–68.

Feinberg, Joel. *Harm to Others: The Moral Limits of the Criminal Law.* Oxford: Oxford University Press, 1984.

Feinberg, Joel. *Offense to Others: The Moral Limits of the Criminal Law.* Oxford: Oxford University Press, 1985.

Feinberg, Joel. *Harm to Self: The Moral Limits of the Criminal Law.* Oxford: Oxford University Press, 1986.

Feinberg, Joel. *Harmless Wrongdoing: The Moral Limits of the Criminal Law.* Oxford: Oxford University, 1990.

Finn, Ed. *What Algorithms Want: Imagination in the Age of Computing.* Cambridge, Mass: MIT Press, 2017.

Fischer, Lisa. 'Control Your Phone with These Temporary Tattoos'. *CNN Tech* (undated) <http://money.cnn.com/video/technology/2016/08/15/phone-control-tattoos.cnnmoney/index.html?sr=twCNN091216

phone-control-tattoos.cnnmoney1112PMVideoVideo&linkId= 28654785> (accessed 30 Nov. 2017).

Fiske, Alan Page. *Structures of Social Life: The Four Elementary Forms of Human Relations*. New York: Macmillan, 1991.

Flinders, Matthew. *Defending Politics: Why Democracy Matters in the Twenty-First Century*. Oxford: Oxford University Press, 2013.

Floridi, Luciano. *The 4th Revolution: How the Infosphere is Reshaping Human Reality*. Oxford: Oxford University Press, 2015.

Floridi, Luciano, ed. *The Onlife Manifesto: Being Human in a Hyperconnected Era*. Cham: Springer, 2015.

Foot, Philippa. *Virtues and Vices and Other Essays in Moral Philosophy*. Oxford: Oxford University Press, 2009.

Ford, Martin. *Rise of the Robots: Technology and the Threat of a Jobless Future*. New York: Basic Books, 2015.

Forster, E. M. *The Machine Stops*. London: Penguin, 2011.

Foucault, Michel. *Power/Knowledge: Selected Interviews and Other Writings, 1972–1977*. New York: Vintage Books, 1980.

Foucault, Michel. *Discipline and Punish: The Birth of the Prison*. Translated by Alan Sheridan. New York: Vintage Books, 1995.

Frank, Aaron. 'You Can Ban a Person, But What About Their Hologram?' *Singularity Hub*, 17 Mar. 2017 <https://singularityhub.com/2017/03/17/you-can-ban-a-person-but-what-about-their-hologram/> (accessed 30 Nov. 2017).

Frank, Robert H. *Choosing the Right Pond: Human Behavior and the Quest for Status*. New York: Oxford University Press, 1985.

Frankfurt, Harry. 'Equality as a Moral Ideal'. *Ethics* 98, no. 1 (Oct. 1987): 21–43.

Franklin, Daniel, ed. *Megatech: Technology in 2050*. New York: Profile Books, 2017.

Frayne, David. *The Refusal of Work: The Theory and Practice of Resistance to Work*. London: Zed Books, 2015.

Friedman, Thomas L. *Thank You for Being Late: An Optimist's Guide to Thriving in the Age of Accelerations*. New York: Farrar, Straus, and Giroux, 2016.

Freeden, Michael, *Ideologies and Political Theory: A Conceptual Approach*. Oxford: Oxford University Press, 1998.

Freedom.to. <https://freedom.to/> (accessed 7 Dec. 2017).

Freud, Sigmund. *Civilization and its Discontents*. Oregon: Rough Draft Printing, 2013.

Fromm, Erich. *The Fear of Freedom*. Abingdon: Routledge, 2009.

Fukuyama, Francis. *Our Posthuman Future: Consequences of the Biotechnology Revolution*. London: Profile Books, 2002.

Fukuyama, Francis. *The Origins of Political Order: From Prehuman Times to the French Revolution*. London: Profile Books, 2012.

Fukuyama, Francis. *Political Order and Political Decay: From the Industrial Revolution to the Globalisation of Democracy*. London: Profile Books, 2014.

Full Fact. <https://fullfact.org/> (accessed 1 Dec. 2017).

Fuller, Lon. *The Morality of Law*. New Haven and London: Yale University Press, 1969.

Fuller, Steve, and Veronika Lipińska. *The Proactionary Imperative: A Foundation for Transhumanism*. Basingstoke: Palgrave Macmillan, 2014.

Fullerton, Jamie. 'Democracy Hunters Use Pokémon to Conceal Rallies.' *The Times*, 3 Aug. 2016 <http://www.thetimes.co.uk/article/democracy-hunters-use-pokemon-to-conceal-rallies-j6xrv59jl> (accessed 30 Nov. 2017).

Garreau, Joel. *Radical Evolution: The Promise and Peril of Enhancing Our Minds, Our Bodies—and What it Means to Be Human*. New York: Broadway Books, 2005.

Gartner Newsroom. 'Gartner Says By 2020, a Quarter Billion Connected Vehicles Will Enable New In-Vehicle Services and Automated Driving Capabilities'. 26 Jan. 2015 <http://www.gartner.com/newsroom/id/2970017> (accessed 30 Nov. 2017).

Garton Ash, Timothy. *Free Speech: Ten Principles for a Connected World*. New Haven and London: Yale University Press, 2016.

Gee, Kelsey. 'In Unilever's Radical Hiring Experiment, Resumes Are Out, Algorithms Are In'. *Fox Business*, 26 Jun. 2017 <http://www.foxbusiness.com/features/2017/06/26/in-unilevers-radical-hiring-experiment-resumes-are-out-algorithms-are-in.html> (accessed 1 Dec. 2017).

General Data Protection Regulation. (Regulation (EU) 2016/679 of the European Parliament and of the Council of 27 April 2016 on the protection of natural persons with regard to the processing of personal data and on the free movement of such data, and repealing Directive 95/46/EC).

Gershgorn, Dave. 'Google Has Built Earbuds that Translate 40 Languages in Real Time'. *Quartz*, 4 Oct. 2017 <https://qz.com/1094638/google-goog-built-earbuds-that-translate-40-languages-in-real-time-like-the-hitchhikers-guides-babel-fish/> (accessed 7 Dec. 2017).

Geuss, Raymond. *Philosophy and Real Politics*. Princeton: Princeton University, 2008.

Gleick, James. *The Information: A History, A Theory, A Flood*. London: Fourth Estate, 2012.

Goldman, Alvin I. *Knowledge in a Social World*. Oxford: Oxford University Press, 2003.

Goldman, Bruce. 'Typing With Your Mind: How Technology is Helping the Paralyzed Communicate'. *World Economic Forum*, 1 Mar. 2017 <https://www.weforum.org/agenda/2017/03/this-technology-allows-paralysed-people-to-type-using-their-mind?utm_content=buffer8a986&utm_medium=social&utm_source=twitter.com&utm_campaign=buffer)%20%20(Ref)%20ch.21%20of%20leviathan?> (accessed 1 Dec. 2017).

Goldsmith, Stephen and Susan Crawford. *The Responsive City: Engaging Communities Through Data-Smart Governance*. San Francisco: Jossey-Bass, 2014.

Goldstein, Brett and Lauren Dyson, eds. *Beyond Transparency: Open Data and the Future of Civic Innovation*. San Francisco: Code for America Press, 2013.

Goode, Lauren. 'Delivery Drones Will Mean the End of Ownership'. *The Verge*, 8 Nov. 2016 <https://www.theverge.com/a/verge-2021/google-x-astro-teller-interview-drones-innovation> (accessed 8 Dec. 2017).

Goodman, Marc. *Future Crimes: A Journey to the Dark Side of Technology—and How to Survive It*. London: Bantam Press, 2015.

Gorz, André. *Reclaiming Work: Beyond the Wage-Based Society*. Translated by Chris Turner. Cambridge: Polity Press, 2005.

Gorz, André. *Capitalism, Socialism, Ecology*. Translated by Martin Chalmers. London and New York: Verso, 2012.

Gould, Carol. *Rethinking Democracy: Freedom and Social Cooperation in Politics, Economy, and Society*. Cambridge: Cambridge University Press, 1990.

Graham, Mark and William H. Dutton, eds. *Society and the Internet: How Networks of Information and Communication are Changing Our Lives*. Oxford: Oxford University Press, 2014.

Gramsci, Antonio. *Selections from the Prison Notebooks*. London: Lawrence & Wishart, 2007.

Granka, Laura A. 'The Politics of Search: A Decade Retrospective'. *The Information Society* 26, no. 5 (2010): 364–74.

Gray, John. *The Soul of the Marionette: A Short Enquiry into Human Freedom*. London: Allen Lane, 2015.

Gray, Tim. *Freedom*. Basingstoke: Macmillan Education, 1991.

Green, Harriet. 'Govcoin's Co-founder Robert Kay Explains Why His Firm is Using Blockchain to Change the Lives of Benefits Claimants'. *City AM*, 10 Oct. 2016 <http://www.cityam.com/250993/govcoins-co-founder-robert-kay-explains-why-his-firm-using> (accessed 30 Nov. 2017).

Greenberg, Andy. 'Now Anyone Can Deploy Google's Troll-Fighting AI'. *Wired*, 23 Feb. 2017 <https://www.wired.com/2017/02/googles-troll-fighting-ai-now-belongs-world/?mbid=social_twitter> (accessed 1 Dec. 2017).

Greenfield, Adam, *Everyware: The Dawning Age of Ubiquitous Computing*. Berkley: New Riders, 2006.

Greengard, Samuel. *The Internet of Things*. Cambridge, Mass: MIT Press, 2015.

Grimmelmann, James. 'Regulation by Software', *Yale Law Journal* 114, no. 7 (May 2005): 1719–1758.

Groopman, Jerome. 'Print Thyself'. *New Yorker*, 24 Nov. 2014 <https://www.newyorker.com/magazine/2014/11/24/print-thyself> (accessed 30 Nov. 2017).

Grullon, Gustavo, Yelena Larkin, and Roni Michaely. 'Are U.S. Industries Becoming More Concentrated?' *SSRN*, 2017 <https://papers.ssrn.com/sol3/papers.cfm?abstract_id=2612047> (accessed 8 Dec. 2017).

Guo, Jeff. 'We're So Unprepared for the Robot Apocalypse'. *Washington Post*, 30 Mar. 2017 <https://www.washingtonpost.com/news/wonk/wp/2017/03/30/were-so-unprepared-for-the-robot-apocalypse/?utm_term=.caeece2d19b4> (accessed 8 Dec. 2017).

Gutmann, Amy and Dennis Thompson. *Why Deliberative Democracy?* Princeton: Princeton University Press, 2004.

Haac, Oscar A., ed. *The Correspondence of John Stuart Mill and Auguste Comte.* London: Transaction, 1995.

Habermas, Jürgen. *The Future of Human Nature.* Cambridge: Polity Press, 2003.

Habermas, Jürgen. *The Theory of Communicative Action: Volume 1. Reason and the Rationalization of Society.* Translated by Thomas McCarthy. Cambridge: Polity Press, 2004.

Habermas, Jürgen. *The Structural Transformation of the Public Sphere: An Inquiry into a Category of Bourgeois Society.* Translated by Thomas Burger with the assistance of Frederick Lawrence. Cambridge: Polity Press, 2008.

Habermas, Jürgen. *Between Facts and Norms.* Translated by William Rehg. Cambridge: Polity Press in association with Oxford: Basil Blackwell, 2010.

Hajer, Maarten A. and Hendrik Wagenaar, eds. *Deliberative Policy Analysis: Understanding Governance in the Network Society.* New York: Cambridge University Press, 2003.

Halting Problem. 'Tech Bro Creates Augmented Reality App to Filter Out Homeless People'. *Medium*, 23 Feb. 2016 <https://medium.com/halting-problem/tech-bro-creates-augmented-reality-app-to-filter-out-homeless-people-3bf8d827b0df> (accessed 7 Dec. 2017).

Hamilton, Alexander, James Madison, and John Jay. *The Federalist Papers.* New York: Penguin, 2012.

Hanson, Robin, *The Age of EM: Work, Love, and Life When Robots Rule the Earth.* Oxford: Oxford University Press, 2016.

Harari, Yuval Noah. *Sapiens: A Brief History of Humankind.* London: Vintage Books, 2011.

Harari, Yuval Noah. *Homo Deus: A Brief History of Tomorrow.* London: Harvill Secker, 2015.

Harcourt, Bernard E. *Against Prediction: Profiling, Policing, and Punishing in an Actuarial Age.* Chicago: University of Chicago Press, 2007.

Hare, Ivan and James Weinstein, eds. *Extreme Speech and Democracy.* Oxford: Oxford University Press, 2010.

Harrell, Erika. 'Crime Against Persons with Disabilities, 2009–2015—Statistical Tables'. *Bureau of Justice Statistics*, Jul. 2017 <https://www.bjs.gov/content/pub/pdf/capd0915st.pdf> (accessed 2 Dec. 2017).

Harris, J. W. *Legal Philosophies* (Second Edition). New York: Oxford University Press, 2004.

Hart, H. L. A. *Law, Liberty, and Morality*. Oxford: Oxford University Press, 1991.

Hart, H. L. A. *The Concept of Law* (Second Edition). Oxford: Oxford University Press, 1997.

Hauser, Christine. 'In Connecticut Murder Case, a Fitbit Is a Silent Witness'. *New York Times*, 27 Apr. 2017 <https://www.nytimes.com/2017/04/27/nyregion/in-connecticut-murder-case-a-fitbit-is-a-silent-witness.html?smid=tw-nytimes&smtyp=cur> (accessed 1 Dec. 2017).

Havelock, Eric A. *The Greek Concept of Justice: From its Shadow in Homer to its Substance in Plato*. Cambridge, Mass: Harvard University Press, 1978.

Havelock, Eric A. *The Muse Learns to Write: Reflections on Orality and Literacy from Antiquity to the Present*. New Haven and London: Yale University Press, 1986.

Haven, Douglas. 'The Uncertain Future of Democracy'. *BBC futurenow*, 30 Mar. 2017 <http://www.bbc.com/future/story/20170330-the-uncertain-future-of-democracy?ocid=ww.social.link.twitter> (accessed 1 Dec. 2017).

Hay, Colin, Michael Lister, and David Marsh, eds. *The State: Theories and Issues*. Basingstoke: Palgrave Macmillan, 2006.

Hayek, Friedrich. 'The Use of Knowledge in Society'. *The American Economic Review* 35, no. 4 (Sep. 1945): 519–30.

Hayek, Friedrich. *The Road to Serfdom*. Abingdon: Routledge, 2008.

Hayek, Friedrich. *The Constitution of Liberty*. Abingdon: Routledge, 2009.

Heater, Brian. 'Wilson's Connected Football is a $200 Piece of Smart Pigskin'. *TechCrunch*, 8 Aug. 2016 <https://techcrunch.com/2016/08/08/wilson-x-football/?ncid=rss> (accessed 29 Nov. 2017).

Hegel, G. W. F. *Elements of the Philosophy of Right*. Translated by H. B. Nisbet. Cambridge: Cambridge University Press, 2008.

Held, David, *Models of Democracy* (Third Edition). Cambridge: Polity, 2016.

Hern, Alex. 'Flickr Faces Complaints Over "Offensive" Auto-tagging for Photos'. *The Guardian*, 20 May 2015 <https://www.theguardian.com/technology/2015/may/20/flickr-complaints-offensive-auto-tagging-photos> (accessed 2 Dec. 2017).

Hern, Alex. 'Why Data Is the New Coal'. *The Guardian*, 27 Sep. 2016 <https://www.theguardian.com/technology/2016/sep/27/data-efficiency-deep-learning> (accessed 28 Nov. 2017).

Hern, Alex. 'Vibrator Maker Ordered to Pay Out C\$4m for Tracking Users' Sexual Activity'. *The Guardian*, 14 Mar. 2017 <https://www.theguardian.com/technology/2017/mar/14/we-vibe-vibrator-tracking-users-sexual-habits?CMP=Share_iOSApp_Other> (accessed 1 Dec. 2017).

Hess, Amanda. 'On Twitter, a Battle Among Political Bots'. *New York Times*, 14 Dec. 2016 <https://mobile.nytimes.com/2016/12/14/arts/on-twitter-a-battle-among-political-bots.html?contentCollection=weekendreads&referer=> (accessed 1 Dec. 2017).

Hidalgo, César. *Why Information Grows: The Evolution of Order, from Atoms to Economies*. London: Allen Lane, 2015.

Higgins, Stan. 'IBM Invests \$200 Million in Blockchain-Powered IoT'. *CoinDesk*, 4 Oct. 2016 <https://www.coindesk.com/ibm-blockchain-iot-office/> (accessed 30 Nov. 2017).

Higgs, Eric, Andrew Light, and David Strong. *Technology and the Good Life?* Chicago: University of Chicago Press, 2000.

Hildebrandt, Mireille. 'Legal and Technological Normativity: More (and Less) than Twin Sisters'. *Techné: Research in Philosophy and Technology* 12, no. 3 (Fall, 2008): 169–83.

Hildebrandt, Mireille, and Antoinette Rouvroy, eds. *Law, Human Agency and Autonomic Computing: The Philosophy of Law Meets the Philosophy of Technology*. Abingdon: Routledge, 2013.

Hinchliffe, Emma. 'IBM's Watson Supercomputer Discovers 5 New Genes Linked to ALS'. *Mashable UK*, 14 Dec. 2016 <http://mashable.com/2016/12/14/ibm-watson-als-research/?utm_cid=mash-com-Tw-tech-link%23sd613jsnjlqd#HJziN5r0aGq5> (accessed 28 Nov. 2017).

Hindman, Matthew. *The Myth of Digital Democracy*. Princeton: Princeton University Press, 2009.

Hobbes, Thomas. *Leviathan*. Cambridge: Cambridge University Press, 2007.

Hobsbawm, Eric. *The Age of Revolution: 1789–1848*. New York: Vintage Books, 1996.

Hofstadter, Richard. *The Paranoid Style in American Politics*. New York: Vintage Books, 2008.

Holmes, Oliver Wendell. 'The Path of the Law'. *Harvard Law Review* 10, no. 457 (1897).

Honneth, Axel. *The Struggle for Recognition: The Moral Grammar of Social Conflicts.* Translated by Joel Anderson. Cambridge: Polity Press, 2005.

Hopkins, Nick. 'Revealed: Facebook's Internal Rulebook on Sex, Terrorism and Violence'. *The Guardian*, 21 May 2017 <https://amp. theguardian.com/news/2017/may/21/revealed-facebook-internal-rulebook-sex-terrorism-violence> (accessed 1 Dec. 2017).

Howard, Philip N. *Pax Technica: How the Internet of Things May Set Us Free or Lock Us Up.* New Haven and London: Yale University Press, 2015.

Hudson, Laura. 'Some Like it Bot'. *FiveThirtyEight*, 29 Sep. 2016 <http://fivethirtyeight.com/features/some-like-it-bot/> (accessed 28 Nov. 2017).

Hughes, Thomas P. *Human-Built World: How to Think about Technology and Culture.* Chicago: University of Chicago Press, 2004.

Hume, David. *A Treatise of Human Nature.* London: Penguin, 1969.

Hume, David. *An Enquiry Concerning the Principles of Morals.* Indianapolis: Hackett Publishing Company, 1983.

Hume, David. *Selected Essays.* Oxford: Oxford University Press, 2008.

Hvistendahl, Mara. 'Inside China's Vast New Experiment in Social Ranking'. *Wired*, 14 December 2017 <https://www.wired.com/story/age-of-social-credit/> (accessed 21 Jan. 2018).

IFR. 'World Robotics Report 2016'. *IFR Press Release* <https://ifr.org/ifr-press-releases/news/world-robotics-report-2016> (accessed 30 Nov. 2017).

Innis, Harold. *Empire and Communications.* Lanham: Rowman & Littlefield, 2007.

Intel. 'New Technology Delivers an Unprecedented Combination of Performance and Power Efficiency'. *Intel 22 NM Technology* <http://www.intel.com/content/www/us/en/silicon-innovations/intel-22nm-technology.html> (accessed 28 Nov. 2017).

Internet Live Stats. 'Internet Users'. <http://www.internetlivestats.com/internet-users/> (accessed 30 Nov. 2017).

Internet Live Stats. 'Twitter Users.' <http://www.internetlivestats.com/twitter-statistics/> (accessed 30 Nov. 2017).

Isaacson, Walter. *The Innovators: How a Group of Hackers, Geniuses and Geeks Created the Digital Revolution.* London: Simon & Schuster, 2014.

Issenberg, Sasha. 'How Obama's Team Used Big Data to Rally Voters'. *MIT Techology Review*, 19 Dec. 2012 <https://www.technologyreview.com/s/509026/how-obamas-team-used-big-data-to-rally-voters/> (accessed 1 Dec. 2017).

Jahoda, Marie. *Employment and Unemployment: A Social-psychological Analysis*. Cambridge: Cambridge University Press, 1982.

Jain, Rishabh. 'Charlottesville Attack: Facebook, Reddit, Google and GoDaddy Shut DownHate Groups'. *IBT*, 16 Aug. 2017 <http://www.ibtimes.com/charlottesville-attack-facebook-reddit-google-godaddy-shut-down-hate-groups-2579027> (accessed 1 Dec. 2017).

Jasanoff, Sheila, *The Ethics of Invention: Technology and the Human Future*. New York: W. W. Norton & Company, 2016.

Jay, Anthony, ed. *Lend Me Your Ears: Oxford Dictionary of Political Quotations* (Fourth Edition). Oxford: Oxford University Press, 2010.

Jay, Martin. *The Virtues of Mendacity: On Lying in Politics*. Charlottesville: University of Virginia Press, 2010.

Jefferson, Thomas. Letter to John Adams, 4 Sep. 1823. *Library of Congress*. <https://www.loc.gov/exhibits/jefferson/202.html> (accessed 8 Dec. 2017).

Johnson, Bobby. 'Amazon Kindle Users Surprised by "Big Brother" Move'. *The Guardian*, 17 Jul. 2009. https://www.theguardian.com/technology/2009/jul/17/amazon-kindle-1984 (accessed 8 Dec. 2017).

Johnson, Steven. *Future Perfect: The Case for Progress in a Networked Age*. London: Penguin, 2013.

Jones, Steve. 'Why "Big Data" is the Fourth Factor of Production'. *Financial Times*, 27 Dec. 2012. <https://www.ft.com/content/5086d700-504a-11e2-9b66-00144feab49a> (accessed 9 Dec. 2017).

'Joseph Schumpeter'. *Wikipedia*, last modified 23 Dec. 2017 <https://en.wikipedia.org/wiki/Joseph_Schumpeter> (accessed 21 Jan. 2018).

Jouppi, Norm. 'Google Supercharges Machine Learning Tasks With TPU Custom Chip'. *Google Cloud Platform Blog*, 18 May 2016 <https://cloudplatform.googleblog.com/2016/05/Google-supercharges-machine-learning-tasks-with-custom-chip.html> (accessed 28 Nov. 2017).

Juma, Calestous. *Innovation and its Enemies: Why People Resists New Technologies*. New York: Oxford University Press, 2016.

Kasparov, Garry. *Deep Thinking: Where Machine Intelligence Ends and Human Creativity Begins*. New York: PublicAffairs, 2017.

Kassarnig, Valentin. 'Political Speech Generation'. *arXiv*, 20 Jan. 2016 <https://arxiv.org/abs/1601.03313> (accessed 28 Nov. 2017).

Keen, Andrew. *The Internet is Not the Answer*. London: Atlantic Books, 2015.

Kelion, Leo and Shiroma Silva. 'Pro-Clinton Bots "Fought Back but Outnumbered in Second Debate'. *BBC News*, 19 Oct. 2016 <http://www.bbc.com/news/technology-37703565> (accessed 1 Dec. 2017).

Kellmereit, Daniel and Daniel Obodovski. *The Silent Intelligence: The Internet of Things*. DND Ventures LLC: 2013.

Kelly III, John E. and Steve Hamm. *Smart Machines: IBM's Watson and the Era of Cognitive Computing*. New York: Columbia Business School Publishing, 2014.

Kelly, Kevin. *What Technology Wants*. New York: Penguin, 2010.

Kelly, Kevin. *The Inevitable: Understanding the 12 Technological Forces that Will Shape Our Future*. New York: Viking, 2016.

Kelly, Rick. 'The Next Battle for Internet Freedom Could Be Over 3D Printing'. *TechCrunch*, 26 Aug. 2012 <https://techcrunch.com/2012/08/26/the-next-battle-for-internet-freedom-could-be-over-3d-printing/> (accessed 30 Nov. 2017).

Kelsen, Hans. *Pure Theory of Law*. Translated from the Second (Revised and Enlarged) German Edition by Max Knight. New Jersey: Law Book Exchange, 2009.

Kelty, Christopher M. *Two Bits: The Cultural Significance of Free Software*. Durham & London: Duke University Press, 2008.

Kennedy, John F. Inaugural Address, 20 Jan. 1961.

Khanna, Parag. *Technocracy in America: Rise of the Info-State*. Self-published, 2017.

Khatchadourian, Raffi. 'We Know How You Feel'. *New Yorker*, 19 Jan. 2015 <http://www.newyorker.com/magazine/2015/01/19/know-feel> (accessed 30 Nov. 2017).

Khomami, Nadia. 'Ministers Back Campaign to Give Under-18s Right to Delete Social Media Posts'. *The Guardian*, 28 Jul. 2015 <https://www.theguardian.com/media/2015/jul/28/ministers-back-campaign-under-18s-right-delete-social-media-posts> (accessed 1 Dec. 2017).

Kim, Mark. 'Google Quantum Computer Test Shows Breakthrough is Within Reach'. *New Scientist*, 28 Sep. 2017 <https://www.newscientist.com/article/2148989-google-quantum-computer-test-shows-breakthrough-is-within-reach/> (accessed 6 Dec. 2017).

Kitchin, Rob. *The Data Revolution: Big Data, Open Data, Data Infrastructures and their Consequences*. London: Sage Publications Ltd, 2014.

Kitchin, Rob and Martin Dodge. '"Outlines of a World Coming into Existence": Pervasive Computing and the Ethics of Forgetting'. *Environment and Planning B: Urban Analytics and City Science* 34, no. 3 (2007): 431–45.

Kitchin, Rob and Martin Dodge. *Code/Space: Software and Everyday Life*. Cambridge, Mass: MIT Press, 2014.

Klaas, Brian. *The Despot's Accomplice: How the West is Aiding and Abetting the Decline of Democracy*. Oxford: Oxford University Press, 2016.

Kleinman, Zoe. 'Toyota Launches "Baby" Robot for Companionship.' *BBC News*, 3 Oct. 2016 <http://www.bbc.co.uk/news/technology-37541035> (accessed 30 Nov. 2017).

Kolhatkar, Sheelah. 'The Tech Industry's Gender-Discrimination Problem'. *New Yorker*, 20 Nov. 2017 <https://www.newyorker.com/magazine/2017/11/20/the-tech-industrys-gender-discrimination-problem> (accessed 12 Dec. 2017).

Kollanyi, Bence, Philip N. Howard, and Samuel C. Woolley. 'Bots and Automation over Twitter during the U.S. Election'. *Computational Propaganda Project*, 2016 <http://comprop.oii.ox.ac.uk/2016/11/17/bots-and-automation-over-twitter-during-the-u-s-election/> (accessed 1 Dec. 2017).

Krasodomski-Jones, Alex. 'Talking to Ourselves?' *Demos*, Sep. 2016 <https://www.demos.co.uk/wp-content/uploads/2017/02/Echo-Chambers-final-version.pdf> (accessed 1 Dec. 2017).

Kurzweil, Ray. *The Singularity is Near*. London: Duckworth, 2010.

Kurzweil, Ray. *How to Create a Mind*. London: Duckworth Overlook, 2012.

Kymlicka, Will. *Contemporary Political Philosophy: An Introduction* (Second Edition). Oxford: Oxford University Press, 2002.

Lai, Richard. 'bHaptics' TactSuit is VR Haptic Feedback Done Right'. *Engadget*, 7 Feb. 2017 <https://www.engadget.com/2017/07/02/bhaptics-tactsuit-vr-haptic-feedback-htc-vive-x-demo-day/?sr_source=Twitter> (accessed 30 Nov. 2017).

Landau, Susan. 'Choices: Privacy and Surveillance in a Once and Future Internet'. *Daedalus* 145, no. 1 (Winter 2016): 54–64.

Landemore, Hélène. *Democratic Reason: Politics, Collective Intelligence, and the Rule of the Many*. Princeton: Princeton University Press, 2017.

Lanier, Jaron. *You Are Not a Gadget*. London: Allen Lane, 2010.

Lanier, Jaron. *Who Owns the Future?* London: Allen Lane, 2014.

Lant, Karla. 'Google is Closer than Ever to a Quantum Computing Breakthrough'. *Business Insider*, 24 Jul. 2017 <http://uk.businessinsider.com/google-quantum-computing-chip-ibm-2017-6?r=US&IR=T> (accessed 28 Nov. 2017).

Larson, Selena. 'Research Shows Gender Bias in Google's Voice Recognition'. *Daily Dot*, 15 Jul. 2016 <https://www.dailydot.com/debug/google-voice-recognition-gender-bias/> (accessed 2 Dec. 2017).

Leftwich, Adrian, ed. *What is Politics?* Cambridge: Polity Press, 2015.

Leopold, David and Marc Stears, eds. *Political Theory: Methods and Approaches*. Oxford: Oxford University Press, 2010.

Lessig, Lawrence. *Code Version 2.0*. New York: Basic Books, 2006.

Leta Jones, Meg. *Ctrl + Z: The Right to Be Forgotten*. New York: New York University Press, 2016.

Levy, Steven. *Crypto: How the Code Rebels Beat the Government—Saving Privacy in the Digital Age*. New York: Penguin, 2002.

Lewis, Michael. *Flash Boys: Cracking the Money Code*. London: Allen Lane, 2014.

Linn, Allison. 'Microsoft Creates AI that Can Read a Document and Answer Questions About it As Well as a Person'. *The AI Blog*, 15 Jan. 2018 <https://blogs.microsoft.com/ai/microsoft-creates-ai-can-read-document-answer-questions-well-person/> (accessed 21 January 2018).

Livingston, James. 'Fuck Work'. *Aeon*, 25 Nov. 2016 <https://aeon.co/essays/what-if-jobs-are-not-the-solution-but-the-problem> (accessed 8 Dec. 2017).

Locke, John. *Two Treatises of Government and A Letter Concerning Toleration*. Ed. Ian Shapiro. New Haven and London: Yale University Press, 2003.

Lopatto, Elizabeth. 'Gene Editing Will Transform Cancer Treatment'. *The Verge*, 22 Nov. 2016. <https://www.theverge.com/a/verge-2021/jennifer-doudna-crispr-gene-editing-healthcare> (accessed 8 Dec. 2017).

LSST. <https://www.lsst.org/about> (accessed 30 Nov. 2017).

Lukes, Steven, ed. *Power*. New York: New York University Press, 1986.

Lukes, Steven. *Power: A Radical View* (Second Edition). Basingstoke: Palgrave Macmillan, 2005.

Luxemburg, Rosa. 'The Russian Revolution' (1918). Translated by Bertram Wolfe. New York: Workers Age Publishers, 1940. *Marxists* <https://www.marxists.org/archive/luxemburg/1918/russian-revolution/ch06.htm> (accessed 9 Dec. 2017).

Lynch, Jack. 'For the Price of a Smartphone You Could Bring a Robot Home'. *World Economic Forum*, 7 Jun. 2016 <https://www.weforum.org/agenda/2016/06/for-the-price-of-a-smartphone-you-could-bring-a-robot-home?utm_content=bufferafeb1&utm_medium=social&utm_source=twitter.com&utm_campaign=buffer> (accessed 30 Nov. 2017).

Lyon, David, ed. *Surveillance as Social Sorting: Privacy, Risk and Digital Discrimination*. Abingdon: Routledge, 2003.

Machiavelli, Niccolò, *Discourses on Livy*. Translated by Julia Conaway Bondanella and Peter Bondanella. Oxford: Oxford University Press, 2008.

Machkovech, Sam. 'Marathon Runner's Tracked Data Exposes Phony Time, Cover-up Attempt'. *Ars Technica UK*, 22 Feb. 2017 <https://arstechnica.com/gadgets/2017/02/suspicious-fitness-tracker-data-busted-a-phony-marathon-run/> (accessed 1 Dec. 2017).

MacKinnon, Rebecca. *Consent of the Networked: The Worldwide Struggle for Internet Freedom*. New York: Basic Books, 2013.

Mannheim, Karl. *Ideology and Utopia: An Introduction to the Sociology of Knowledge*. Translated by Louis Wirth and Edward Shils. Connecticut: Martino Publishing, 2015.

Margretts, Helen, Peter John, Scott Hale, and Taha Yasseri. *Political Turbulence: How Social Media Shape Collective Action*. Princeton: Princeton University Press, 2016.

Markoff, John. *Machines of Loving Grace: The Quest for Common Ground Between Humans and Robots*. New York: HarperCollins, 2015.

Markoff, John. 'Automated Pro-Trump Bots Overwhelmed Pro-Clinton Messages, Researchers Say'. *New York Times*, 17 Nov. 2016 <http://www.nytimes.com/2016/11/18/technology/automated-pro-trump-bots-overwhelmed-pro-clinton-messages-researchers-say.html)> (accessed 1 Dec. 2017).

Martinez, Peter. 'Study Reveals Whopping 48M Twitter Accounts Are Actually Bots'. *CBS News*, 10 Mar. 2017 <http://www.cbsnews.com/news/48-million-twitter-accounts-bots-university-of-southern-california-study/?ftag=CNM-00-10aab7e&linkId=35386687> (accessed 1 Dec. 2017).

Marx, Karl. *Karl Marx and Frederick Engels: Collected Works Volumes 1–50*. London: Lawrence & Wishart, 1975–2004.

Mason, Paul. *Postcapitalism: A Guide to Our Future*. London: Allen Lane, 2015.

Mayer-Schönberger, Viktor, and Kenneth Cukier. *Delete: The Virtue of Forgetting in the Digital Age*. Princeton: Princeton University Press, 2009.

Mayer-Schönberger, Viktor, and Kenneth Cukier. *Big Data: A Revolution that Will Transform How We Live, Work and Think*. London: John Murray, 2013.

Mayr, Otto. *Authority, Liberty and Automatic Machinery in Early Modern Europe*. Baltimore: Johns Hopkins University Press, 1989.

McChesney, Robert W. *Digital Disconnect: How Capitalism is Turning the Internet Against Democracy*. New York: The New Press, 2014.

McChesney, Robert W. and John Nichols. *People Get Ready: The Fight Against a Jobless Economy and a Citizenless Democracy*. New York: Nation Books, 2016.

McGinnis, John O. *Accelerating Democracy: Transforming Governance through Technology*. Princeton: Princeton University Press, 2013.

McLuhan, Marshall. *Understanding Media: The Extensions of Man*. Abingdon: Routledge, 2005.

Mearian, Lucas. 'By 2020, there Will Be 5,200 GB of Data for Every Person on Earth'. *ComputerWorld*, 11 Dec. 2012 <http://www.computerworld.com/article/2493701/data-center/by-2020--there-will-be-5-200-gb-of-data-for-every-person-on-earth.html> (accessed 30 Nov. 2017).

Merchant, Brian. 'Fully Automated Luxury Communism'. *The Guardian*, 18 Mar. 2015. <https://www.theguardian.com/sustainable-business/2015/mar/18/fully-automated-luxury-communism-robots-employment> (accessed 8 Dec. 2017).

Metz, Cade. 'Google's AI Wins Fifth and Final Game Against Go'. *Wired*, 15 Mar. 2016 <https://www.wired.com/2016/03/googles-ai-wins-fifth-final-game-go-genius-lee-sedol/> (accessed 28 Nov. 2017).

Metz, Cade. 'Building AI Is Hard—So Facebook is Building AI that Builds AI'. *Wired*, 6 May 2016 <https://www.wired.com/2016/05/facebook-trying-create-ai-can-create-ai/> (accessed 28 Nov. 2017).

Metz, Cade. 'Microsoft Bets its Future on a Reprogrammable Computer Chip'. *Wired*, 25 Aug. 2016 <https://www.wired.com/2016/09/microsoft-bets-future-chip-reprogram-fly/?mbid=social_twitter> (accessed 28 Nov. 2017).

Metz, Cade. 'Elon Musk Isn't the Only One Trying to Computerize Your Brain'. *Wired*, 31 Mar. 2017 <https://www.wired.com/2017/03/elon-musks-neural-lace-really-look-like/?mbid=social_twitter> (accessed 30 Nov. 17).

Metz, Cade. 'Google's Dueling Neural Networks Spar to Get Smarter, No Humans Required'. *Wired*, 11 Apr. 2017 <https://www.wired.com/2017/04/googles-dueling-neural-networks-spar-get-smarter-no-humans-required/> (accessed 28 Nov. 2017).

Metz, Rachel. 'Controlling VR with Your Mind'. *MIT Technology Review*, 22 Mar. 2017 <https://www.technologyreview.com/s/603896/controlling-vr-with-your-mind/> (accessed 1 Dec. 2017).

Mill, John Stuart. 'Thoughts on Parliamentary Reform'. *Collected Works of John Stuart Mill, Volume XIX—Essays on Politics and Society Part 2*. Ed. John M. Robson. Toronto: University of Toronto Press, London: Routledge and Kegan Paul, 1977. <http://oll.libertyfund.org/titles/mill-the-collected-works-of-john-stuart-mill-volume-xix-essays-on-politics-and-society-part-2#lf0223-19_head_002> (accessed 8 Dec. 2017).

Mill, John Stuart. *On Liberty and Other Writings*. Cambridge: Cambridge University Press, 2008.

Mill, John Stuart. *The Autobiography of John Stuart Mill*. US: Seven Treasures Publications, 2009.

Mill, John Stuart. 'Considerations on Representative Government'. *Project Gutenberg* <https://www.gutenberg.org/files/5669/5669-h/5669-h.htm> (accessed 1 Dec. 2017).

Miller, David, ed. *The Liberty Reader*. Edinburgh: Edinburgh University Press, 2006.

Miller, David and Larry Siedentop, eds. *The Nature of Political Theory*. Oxford: Oxford University Press, 1983.

Mills, Laurence. 'Numbers, Data and Algorithms—Why HR Professionals and Employment Lawyers Should Take Data Science and Analytics Seriously'. *Future of Work Hub*, 4 Apr. 2017 <http://www.futureofworkhub.info/comment/2017/4/4/numbers-data-and-algorithms-why-hr-professionals-and-employment-lawyers-should-take-data-science-seriously> (accessed 1 Dec. 2017).

Millward, David. 'How Ford Will create a new generation of driverless cars'. *Telegraph*, 27 Feb. 2017 <http://www.telegraph.co.uk/business/2017/02/27/ford-seeks-pioneer-new-generation-driverless-cars/> (accessed 28 Nov. 2017).

Milton, John. *Paradise Lost*. London: Penguin, 2003.

Misra, Tanvi. '3 Cities Using Open Data in Creative Ways to Solve Problems'. *CityLab*, 22 Apr. 2015 <http://www.citylab.com/cityfixer/2015/04/3-cities-using-open-data-in-creative-ways-to-solve-problems/391035/> (accessed 29 Nov. 2017).

MIT Technology Review Custom, in partnership with Oracle. 'The Rise of Data Capital'. *MIT Technology Review*, 21 Mar. 2016 <https://www.technologyreview.com/s/601081/the-rise-of-data-capital/> (accessed 8 Dec. 2017).

Mitchell, William J. *City of Bits: Space, Place, and the Infobahn*. Cambridge, Mass: MIT Press, 1998.

Mitchell, William J. *E-topia*. Cambridge, Mass: MIT Press, 2000.

Mitchell, William J. *Me ++: The Cyborg Self and the Networked City*. Cambridge, Mass: MIT Press, 2003.

Mittelstand, Brent Daniel, Patrick Allo, Mariarosaria Taddeo, Sandra Wachter, and Luciano Floridi. 'The Ethics of Algorithms: Mapping the Debate'. *Big Data & Society* 3, no. 2 (2016): 1–21.

Mizokami, Kyle. 'The Pentagon Wants to Use Bitcoin Technology to Protect Nuclear Weapons'. *Popular Mechanics*, 11 Oct. 2016 <http://www.popularmechanics.com/military/research/a23336/the-pentagon-wants-to-use-bitcoin-technology-to-guard-nuclearweapons/?utm_content=buffer98698&utm_medium=social&utm_source=twitter.com&utm_campaign=buffer> (accessed 30 Nov. 2017).

Moley. <http://www.moley.com/> (accessed 1 Dec. 2017).

Montesquieu, Charles de Secondat, baron de. *The Spirit of the Laws*. Translated by Anne M. Cohler, Basia Carolyn Miller, and Harold Samuel Stone. Cambridge: Cambridge University Press, 1989 (Kindle Edition).

Moore, Gordon. 'Cramming More Components onto Integrated Circuits'. *Proceedings of the IEEE* 86, no. 1 (Jan. 1998): 82–5.

'Moravec's Paradox', *Wikipedia*, last modified 9 May 2017. <https://en.wikipedia.org/wiki/Moravec%27s_paradox> (accessed 6 Dec. 2017).

More, Max, and Natasha Vita-More, eds. *The Transhumanist Reader: Classical and Contemporary Essays on the Science, Technology, and Philosophy of the Human Future*. Chichester: John Wiley & Sons, Inc, 2013.

Morozov, Evgeny. *The Net Delusion: How Not to Liberate the World*. London: Penguin, 2011.

Morozov, Evgeny. 'The Meme Hustler'. *Baffler* 22, Apr. 2013 <http://thebaffler.com/salvos/the-meme-hustler> (accessed 30 Nov. 2017).

Morozov, Evgeny. *To Save Everything, Click Here: Technology, Solutionism, and the Urge to Fix Problems That Don't Exist*. London: Penguin, 2014.

Morozov, Evgeny. 'To Tackle Google's Power, Regulators Have to Go After its Ownership of Data'. *The Guardian*, 2 Jul. 2017. <https://www.theguardian.com/technology/2017/jul/01/google-european-commission-fine-search-engines?CMP=share_btn_tw> (accessed 8 Dec. 2017).

Mouffe, Chantal. *The Democratic Paradox*. London: Verso, 2009.

Moxley, Lauren. 'E-Rulemaking and Democracy'. *Administrative Law Review* 68, no. 4 (2016): 661–99.

Mumford, Lewis. *The Myth of the Machine*. London: Secker & Warburg, 1967.

Mumford, Lewis. *Technics and Civilization*. Chicago: University of Chicago Press, 2010.

Muoio, Danielle. 'Here's Everything We Know About Google's Driverless Cars'. *Business Insider*, 25 Jul. 2016 <http://uk.businessinsider.com/google-driverless-car-facts-2016-7?r=US&IR=T/#the-cars-have-been-in-a-few-minor-accidents-only-one-of-which-could-be-argued-to-have-been-the-google-cars-fault-11> (accessed 30 Nov. 2017).

Murray, Andrew, and Colin Scott. 'Controlling the New Media: Hybrid Responses to New Forms of Power'. *Modern Law Review* 65, no. 4 (2002): 491–516.

Muse.com. <http://www.choosemuse.com/> (accessed 30 Nov. 2017).

Nagel, Thomas. *The View from Nowhere*. New York: Oxford University Press, 1986.

Negroponte, Nicholas. *Being Digital*. New York: Vintage Books, 1996.

Nelson, Justin, Richard A. McKinley, Chandler Phillips, Lindsey McIntyre, Chuck Goodyear, Aerial Kreiner, and Lanie Monforton. 'The Effects of Transcranial Direct Current Stimulation (tDCS) on Multitasking Throughput Capacity'. *Frontiers in Human Neuroscience*, 29 November 2016 <https://www.frontiersin.org/articles/10.3389/fnhum.2016.00589/full> (accessed 8 Dec. 2017).

Neuman, W. Russell. *The Digital Difference: Media Technology and the Theory of Communication Effects*. Cambridge, Mass: Harvard University Press, 2016.

New York Times. 'Why Not Smart Guns in this High-Tech Era?' Editorial, 26 Nov. 2016. <http://mobile.nytimes.com/2016/11/26/opinion/sunday/why-not-smart-guns-in-this-high-tech-era.html?smid=tw-nytopinion&smtyp=cur&referer=> (accessed 1 Dec. 2017).

Newton, Casey. 'Here's How Snapchat's New Spectacles Will Work'. *The Verge*, 24 Sep. 2016 <http://www.theverge.com/2016/9/24/13042640/snapchat-spectacles-how-to-use> (accessed 28 Nov. 2017).

Nichols, John. 'If Trump's FCC Repeals Net Neutrality, Elites Will Rule the Internet—and the Future'. *Nation*, 24 Nov. 2017 <https://www.thenation.com/article/if-trumps-fcc-repeals-net-neutrality-elites-will-rule-the-internet-and-the-future/> (accessed 1 Dec. 2017).

Nietzsche, Friedrich. *Thus Spoke Zarathustra: A Book for Everyone and No One*. Translated by R. J. Hollingdale. London: Penguin, 2003.

Nietzsche, Friedrich. *Ecce Homo*. Translated by R. J. Hollingdale. London: Penguin, 2004.

Nissenbaum, Helen. *Privacy in Context: Technology, Policy, and the Integrity of Social Life*. Stanford: Stanford University Press, 2010.

Norman, Donald A. *The Design of Future Things*. New York: Basic Books, 2007.

Noveck, Beth Simone. *Wiki Government: How Technology Can Make Government Better, Democracy Stronger, and Citizens More Powerful*. Washington, DC: Brookings Institution Press, 2009.

Noveck, Beth Simone. *Smart Citizens, Smarter State: The Technologies of Expertise and the Future of Governing*. Cambridge, Mass: Harvard University Press, 2015.

Nozick, Robert. *Anarchy, State, and Utopia*. Oxford: Blackwell Publishing, 2008.

NYC Mayor's Office of Technology and Innovation. 'Preparing for the Internet of Everything' (undated) <https://www1.nyc.gov/site/forward/innovations/iot.page> (accessed 6 Dec. 2017).

Ober, Josiah. *Democracy and Knowledge: Innovation and Learning in Classical Athens*. Princeton: Princeton University Press, 2008.

O'Hara, Kieron and David Stevens. *Inequality.com: Power, Poverty and the Digital Divide*. Oxford: Oneworld, 2006.

O'Malley, James. 'Bluetooth Mesh Is Going to Be a Big Deal: Here Are 6 Reasons Why You Should Care'. *Gizmodo*, 18 Jul. 2017 <http://www.gizmodo.co.uk/2017/07/bluetooth-mesh-is-going-to-be-a-big-deal-here-are-6-reasons-why-you-should-care/> (accessed 30 Nov. 2017).

O'Neil, Cathy. *Weapons of Math Destruction: How Big Data Increases Inequality and Threatens Democracy*. New York: Crown, 2016.

Olson, Parmy. 'Meet Improbable, the Startup Building the World's Most Powerful Simulations'. *Forbes*, 15 Jun. 2015. <https://www.forbes.com/sites/parmyolson/2015/05/27/improbable-startup-simulations/#6ae2da044045> (accessed 8 Dec. 2017).

Ong, Walter. *Orality and Literacy*. Abingdon: Routledge, 2012.

Oord, Aäron van den, Sander Dieleman, Heiga Zen, Karen Simonyan, Oriol Vinyals, Alex Graves, Nal Kalchbrenner et al. 'WaveNet: A Generative Model for Raw Audio'. *arXiv* 19 Sep. 2016 <https://arxiv.org/abs/1609.03499> (accessed 6 Dec. 2017).

Orwell, George. *Essays*. London: Penguin, 2000.

Orwell, George. *Nineteen Eighty-Four*. London: Penguin, 2000.

Orwell, George. *Diaries*. London: Penguin, 2009.

Palfrey, John, and Urs Gasser. *Interop: The Promise and Perils of Highly Interconnected Systems*. New York: Basic Books, 2012.

Palmer, Daniel. 'Blockchain Startup to Secure 1 Million e-Health Records in Estonia'. *CoinDesk*, 3 Mar. 2016 <http://www.coindesk.com/blockchain-startup-aims-to-secure-1-million-estonian-health-records/> (accessed 30 Nov. 2017).

Papacharissi, Zizi A. *A Private Sphere: Democracy in a Digital Age*. Cambridge: Polity Press, 2013.

Parijs, Philippe van, and Yannick Vanderborght. *Basic Income: A Radical Proposal for a Free Society and a Sane Economy*. Cambridge, Mass: Harvard, 2017.

Pariser, Eli. *The Filter Bubble: What the Internet is Hiding from You*. London: Penguin, 2011.

Pasquale, Frank. *The Black Box Society: The Secret Algorithms that Control Money and Information*. Cambridge, Mass: Harvard University Press, 2015.

Pasquale, Frank. 'From Holocaust Denial to Hitler Admiration, Google's Algorithm is Dangerous'. *Huffington Post*, 2 Jun. 2017. <https://www.huffingtonpost.com/entry/holocaust-google-algorithm_us_587e8628e4b0c147f0bb9893> (accessed 8 Dec. 2017).

Pasquale, Frank. 'Will Amazon Take Over the World?' *Boston Review*, 20 Jul. 2017 <https://bostonreview.net/class-inequality/frank-pasquale-will-amazon-take-over-world> (accessed 8 Dec. 2017).

Pateman, Carol. *Participation and Democratic Theory*. Cambridge: Cambridge University Press, 1999.

Pearson, Jordan. 'Why an AI-Judged Beauty Contest Picked Nearly All White Winners'. *Motherboard*, 5 Sep. 2016 <https://motherboard.vice.com/en_us/article/78k7de/why-an-ai-judged-beauty-contest-picked-nearly-all-white-winners> (accessed 2 Dec. 2017).

Penney, Jon. 'Internet Surveillance, Regulation, and Chilling Effects Online: A Comparative Case Study'. *Internet Policy Review* 6, no. 2 (2017): 1–38.

Penny, Timothy J. 'Facts Are Facts'. *National Review*, 4 Sep. 2003. <http://www.nationalreview.com/article/207925/facts-are-facts-timothy-j-penny> (accessed 9 Dec. 2017).

Perry, Walter, Brian McInnis, Carter Price, Susan Smith, and John Hollywood. *Predictive Policing: The Role of Crime Forecasting in Law Enforcement Operations*. Santa Monica: RAND Corporation, 2013.

Perzanowksi, Aaron, and Jason Schultz. *The End of Ownership: Personal Property in the Digital Economy*. Cambridge, Mass: MIT Press, 2016.

Peters, John Durham. *Speaking Into the Air: A History of the Idea of Communication*. Chicago: University of Chicago Press, 1999.

Peterson, Andrea. 'Holocaust Museum to Visitors: Please Stop Catching Pokémon Here'. *Washington Post*, 12 Jul. 2016 <https://www.washingtonpost.com/news/the-switch/wp/2016/07/12/holocaust-museum-to-visitors-please-stop-catching-pokemon-here/> (accessed 30 Nov. 2017).

Piketty, Thomas. *Capital in the Twenty-First Century*. Cambridge, Mass: The Belknapp Press of Harvard University Press, 2014.

Plato. *The Republic*. Translated by Desmond Lee. London: Penguin, 2003.

Plato. *Phaedrus*. Translated by Christopher Rowe. London: Penguin, 2005.

Plato. *The Laws*. Translated by Tom Griffith. Cambridge: Cambridge University Press, 2016.

Plaugic, Lizzie. 'Spotify Pulls Several "Hate Bands" from its Service.' *The Verge*, 16 Aug. 2017 <https://www.theverge.com/2017/8/16/16158502/spotify-racist-bands-streaming-service-southern-poverty-law-center> (accessed 1 Dec. 2017).

Pocock, J. G. A. *Politics, Language, and Time: Essays on Political Thought and History*. Chicago: University of Chicago Press, 1989.

Popper, Ben. 'Electrick Lets You Spray Touch Controls Onto Any Object or Surface'. *The Verge*, 8 May 2017 <https://www.theverge.com/2017/5/8/15577390/electrick-spray-on-touch-controls-future-interfaces-group> (accessed 30 Nov. 2017).

Post, David G. 'The "Unsettled Paradox": The Internet, the State, and the Consent of the Governed', *Indiana Journal of Global Legal Studies*, 5(2) (1998), 521–43.

Post, David G. *In Search of Jefferson's Moose: Notes on the State of Cyberspace*. New York: Oxford University Press, 2009.

Prince, Matthew. 'Why We Terminated Daily Stormer.' *Cloudfare*, 16 Aug. 2017 <https://blog.cloudflare.com/why-we-terminated-daily-stormer/> (accessed 1 Dec. 2017).

Pritchard, Tom. 'The EU Wants to Enforce Encryption, and Ban Backdoor Access.' *Gizmodo*, 19 Jun. 2017 <http://www.gizmodo.co.uk/2017/06/the-eu-wants-to-enforce-encryption-and-ban-backdoor-access/> (accessed 1 Dec. 2017).

Raab, Charles and Paul de Hert. 'The Regulation of Technology: Policy Tools and Policy Actors', *TILT Law & Technology Working Paper No. 003/2007*, 15 Nov. 2007 version 1:0.

Radicati Group Inc. 'Email Statistics Report, 2015–2019' (Mar. 2015) <http://www.radicati.com/wp/wp-content/uploads/2015/02/Email-Statistics-Report-2015-2019-Executive-Summary.pdf> (accessed 30 Nov. 2017).

Ratner, Paul. 'Harvard Scientists Create a Revolutionary Robot Octopus'. *Big Think*, 2016 <http://bigthink.com/paul-ratner/harvard-team-creates-octobot-the-worlds-first-autonomous-soft-robot> (accessed 30 Nov. 2017).

Rawls, John. *A Theory of Justice*. Cambridge, Mass: Harvard University Press, 2003.

Rawls, John. *Political Liberalism*. New York: Columbia University Press, 2005.

Raymond, Eric S. *The Cathedral and the Bazaar: Musings on Linux and Open Source by an Accidental Revolutionary*. Cambridge, Mass: O'Reilly Media, 2001.

Raz, Joseph. *The Morality of Freedom*. Oxford: Oxford University Press, 1986.

Raz, Joseph, ed. *Authority*. Oxford: Basil Blackwell Ltd, 1990.

Raz, Joseph. *The Authority of Law* (Second Edition). Oxford: Oxford University Press, 2011.

Remnick, David. 'Obama Reckons With a Trump Presidency'. *New Yorker*, 28 Nov. 2016 <http://www.newyorker.com/magazine/2016/11/28/obama-reckons-with-a-trump-presidency> (accessed 30 Nov. 2017).

Reuters. 'Turkey Blocks Wikipedia Under Law Designed to Protect National Security'. *The Guardian*, 30 Apr. 2017 <https://www.theguardian.com/world/2017/apr/29/turkey-blocks-wikipedia-under-law-designed-to-protect-national-security> (accessed 8 Dec. 2017).

Richards, Neil. *Intellectual Privacy: Rethinking Civil Liberties in the Digital Age*. New York: Oxford University Press, 2015.

Richards, Thomas. *The Imperial Archive: Knowledge and the Fantasy of Empire*. London: Verso, 1993.

Rid, Thomas. *Rise of the Machines: The Lost History of Cybernetics*. London: Scribe, 2016.

Rieff, David. *In Praise of Forgetting: Historical Memory and its Ironies*. New Haven and London: Yale University Press, 2017.

Rifkin, Jeremy. *The Zero Marginal Cost Society: The Internet of Things, the Collaborative Commons, and the Eclipse of Capitalism*. New York: Palgrave Macmillan, 2015.

Riley v. California 134 S. Ct. 2473 Supreme Court 2014.

Rogaway, Phillip, 'The Moral Character of Cryptographic Work', essay written to accompany the 2015 IACR Distinguished Lecture at Asiacrypt 2015, 2 Dec. 2015. <http://web.cs.ucdavis.edu/~rogaway/papers/moral-fn.pdf> (accessed 5 Dec. 2017).

Rorty, Richard. *Contingency, Irony, and Solidarity*. Cambridge: Cambridge University Press, 1997.

Rose, David. *Enchanted Objects: Design, Human Desire, and the Internet of Things*. New York: Scribner, 2014.

Rosen, Michael. *On Voluntary Servitude: False Consciousness and the Theory of Ideology*. Cambridge, Mass: Harvard University Press, 1996.

Rosen, Michael, and Jonathan Wolff, eds. *Political Thought*. Oxford: Oxford University Press, 1999.

Rosenberg, Robin, Shawnee Baughman, and Jeremy Bailenson. 'Virtual Superheroes: Using Superpowers in Virtual Reality to Encourage Prosocial Behaviour'. *PLoS ONE* 8, no. 1 (2013) <https://doi.org/10.1371/journal.pone.0055003> (accessed 5 Dec. 2017).

Rosenbush, Steve. 'The Morning Download: China's Facial Recognition ID's Citizens and Soon May Score Their Behaviour'. *Wall Street Journal*, 27 Jul. 2017 <https://blogs.wsj.com/cio/2017/06/27/the-morning-download-chinas-facial-recognition-ids-citizens-and-soon-may-score-their-behavior/> (accessed 1 Dec. 2017).

Rousseau, Jean-Jacques. *The Social Contract*. Translated by Maurice Cranston. London: Penguin, 1968.

Rousseau, Jean-Jacques. *Discourse on the Origin of Inequality*. Translated by Donald A. Cress. Indianapolis: Hackett Publishing Company, 1992.

Ruan, Lotus, Jeffrey Knockel, Jason Q. Ng, and Masashi Crete-Nishihata. 'One App, Two Systems'. *The Citizen Lab*, 30 Nov. 2016 <https://citizenlab.ca/2016/11/wechat-china-censorship-one-app-two-systems/> (accessed 1 Dec. 2017).

Runciman, David. *Politics*. London: Profile Books, 2014.

Ruparelia, Nayan B. *Cloud Computing*. Cambridge, Mass: MIT Press, 2016.

Rushkoff, Douglas. *Program or be Programmed: Ten Commands for a Digital Age*. New York: Soft Skull Press, 2011.

Rushkoff, Douglas. *Throwing Rocks at the Google Bus: How Growth Became the Enemy of Prosperity*. New York: Portfolio/Penguin, 2016.

Russell, Bertrand. *In Praise of Idleness*. Abingdon: Routledge, 2004.

Russell, Stuart J., and Peter Norvig. *Artificial Intelligence: A Modern Approach* (Third Edition). London: Pearson, 2015.

Ryan, Alan. *On Politics: A History of Political thought from Herodotus to the Present*. London: Penguin, 2013.

Sadowski, Jathan and Frank Pasquale. 'The Spectrum of Control: A Social Theory of the Smart City'. *First Monday* 20, no. 7 (6 Jul. 2015) <http://digitalcommons.law.umaryland.edu/cgi/viewcontent.cgi?article=2545&context=fac_pubs> (accessed 1 Dec. 2017).

Saint-Exupéry, Antoine. *Wind, Sand and Stars*. London: Penguin, 2000.

Sample, Ian. 'Study Reveals Bot-on-Bot Editing Wars Raging on Wikipedia's Pages'. *Guardian*, 23 Feb. 2017 <https://www.theguardian.com/technology/2017/feb/23/wikipedia-bot-editing-war-study> (accessed 1 Dec. 2017).

Sandel, Michael J. *The Case Against Perfection: Ethics in the Age of Genetic Engineering*. Cambridge, Mass: Harvard University Press, 2007.

Sandel, Michael J. *What Money Can't Buy: The Moral Limits of Markets*. London: Penguin, 2012.

Sander, Alison and Meldon Wolfgang. 'The Rise of Robotics'. *BCG Perspectives*, 27 Aug. 2014 <https://www.bcgperspectives.com/content/articles/business_unit_strategy_innovation_rise_of_robotics/> (accessed 30 Nov. 2017).

Sandvig, Christian, Kevin Hamilton, Karrie Karrahalios, and Cedric Langbort. 'Auditing Algorithms: Research Methods for Detecting Discrimination on Internet Platforms'. Paper presented to 'Data and

Discrimination: Converting Critical Concerns into Productive Inquiry', a preconference at the 64th Annual Meeting of the International Communication Association, 22 May 2014, Seattle, WA, USA. <http://www-personal.umich.edu/~csandvig/research/Auditing%20Algorithms%20--%20Sandvig%20--%20ICA%202014%20Data%20and%20Discrimination%20Preconference.pdf> (accessed 11 Dec. 2017).

Sandvig, Christian, Kevin Hamilton, Karrie Karahalios, and Cedric Langbort. 'When the Algorithm Itself is a Racist: Diagnosing Ethical Harm in the Basic Components of Software'. *International Journal of Communications* 10 (2016): 4972–90.

Savulescu, Julian, Ruud ter Meulen, and Guy Kahane, eds. *Enhancing Human Capacities*. Chichester: Wiley-Blackwell, 2011.

Scanlon, Thomas. 'A Theory of Freedom of Expression'. *Philosophy and Public Affairs* 1, no. 2 (1972): 204–26.

Schaff, Kory, ed. *Philosophy and the Problems of Work: A Reader*. Lanham, Maryland: Rowman & Littlefield Publishers, 2001.

Scharff, Robert C. and Val Dusek, eds. *Philosophy of Technology: The Technological Condition: An Anthology* (Second Edition). Oxford: Wiley-Blackwell, 2014.

Schattschneider, E. E. *The Semisovereign People: A Realist's View of Democracy in America*. South Melbourne, Victoria: Wadsworth Thomson Learning, 1975.

Schechner, Sam. 'Why Do Gas Station Prices Constantly Change? Blame the Algorithm'. *Wall Street Journal*, 8 May 2017 <https://www.wsj.com/articles/why-do-gas-station-prices-constantly-change-blame-the-algorithm-1494262674?mod=e2tw> (accessed 1 Dec. 2017).

Schmidt, Eric and Jared Cohen. *The New Digital Age: Reshaping the Future of People, Nations and Business*. London: John Murray, 2014.

Schneier, Bruce. *Data and Goliath: The Hidden Battles to Collect Your Data and Control Your World*. New York: W. W. Norton & Company, 2016.

Scholz, Trebor and Nathan Schneider, eds. *Ours to Hack and to Own: The Rise of Platform Cooperativism, a New Vision for the Future of Work and a Fairer Internet*. New York: OR Books/Counterpoint, 2017.

Schumpeter, Joseph. *Capitalism, Socialism and Democracy*. Abingdon: Routledge, 2010.

Schwab, Klaus. *The Fourth Industrial Revolution*. Geneva: World Economic Forum, 2016.

Scoble, Robert, and Israel Shel. *The Fourth Transformation: How Augmented Reality and Artificial Intelligence Change Everything.* CreateSpace Independent Publishing Platform, 2017.

Scott, Clare. 'Chinese Construction Company 3D Prints an Entire Two-Story House On-Site in 45 Days'. 3Dprint.com, 16 Jun. 2016 <https://3dprint.com/138664/huashang-tengda-3d-print-house/> (accessed 30 Nov. 2017).

Scott, James C. *Seeing Like a State.* New Haven and London: Yale University Press, 1998.

Semayne's Case (1604) 5 Coke Reports 91a 77 E.R. 194.

Sen, Amartya. 'Democracy as a Universal Value'. *Journal of Democracy* 10, no. 3 (1999): 3–17.

Sen, Amartya. *The Idea of Justice.* London: Penguin, 2010.

Sensifall. <http://www.sensifall.com/> (accessed 12 Dec. 2017).

Shakespeare, William. *Hamlet.* Oxford: Oxford University Press, 2008.

Shanahan, Murray. *The Technological Singularity.* Cambridge, Mass: MIT Press, 2015.

Shead, Sam. 'Amazon's Supermarket of the Future Could Operate With Just 3 Staff—and Lots of Robots'. *Business Insider*, 6 Feb. 2017 <http://www.businessinsider.com/amazons-go-supermarket-of-the-future-3-human-staff-2017-2?r=UK&IR=T> (accessed 8 Dec. 2017).

Shiller, Benjamin Reed. 'First-Degree Price Discrimination Using Big Data'. *Brandeis University*, 19 Jan. 2014 <http://benjaminshiller.com/images/First_Degree_PD_Using_Big_Data_Jan_18,_2014.pdf> (accessed 1 Dec. 2017).

Shin, Laura. 'The First Government to Secure Land Titles on the Bitcoin Blockchain Expands Project'. *Forbes*, 7 Feb. 2017 <https://www.forbes.com/sites/laurashin/2017/02/07/the-first-government-to-secure-land-titles-on-the-bitcoin-blockchain-expands-project/#432b8b494dcd> (accessed 30 Nov. 2017).

Shirky, Clay. *Here Comes Everybody: The Power of Organizing Without Organizations.* London: Penguin, 2008.

Shoemaker, Natalie. 'Pilot Earbud Translates Languages in Real-Time'. *Big Think*, 2016 <http://bigthink.com/natalie-shoemaker/pilot-earbud-translates-languages-in-real-time> (accessed 30 Nov. 2017).

Siedentop, Larry. *Inventing the Individual: The Origins of Western Liberalism.* London: Allen Lane, 2014.

Siegel, Eric. *Predictive Analytics: The Power to Predict Who Will Click, Buy, Lie, or Die*. New Jersey: John Wiley & Sons, Inc, 2016.

Sifry, Micah L. *The Big Disconnect: Why the Internet Hasn't Transformed Politics (Yet)*. New York & London: OR Books, 2014.

Silver, David, Julian Schrittwieser, Karen Simonyan, Ioannis Antonoglou, Aja Huang, Arthur Guez, Thomas Hubert, et al. 'Mastering the Game of Go Without Human Knowledge'. *Nature* 550 (19 Oct. 2017): 354–9.

Silver, David, Thomas Hubert, Julian Schrittwieser, Ioannis Antonoglou, Marc Lanctot, Laurent Sifre, Dharshan Kumaran, Thore Graepel, Timothy Lillicrap, Karen Simonyan, Demis Hassabis. 'A general reinforcement learning algorithm that masters chess, shogi, and Go through self-play'. *Science* 362 (7 Dec. 2018): 1140–1144.

Silverman, Craig. 'This Analysis Shows How Viral Fake Election News Stories Outperformed Real News on Facebook'. *BuzzFeed News*, 16 Nov. 2017 <https://www.buzzfeed.com/craigsilverman/viral-fake-election-news-outperformed-real-news-on-facebook?utm_term=.ufqYm8llgv#.sf9JbwppAm> (accessed 1 Dec. 2017).

Simmons, A. John. *Moral Principles and Political Obligations*. Princeton: Princeton University Press, 1981.

Simon, Julie, Theo Bass, Victoria Boelman, and Geoff Mulgan. 'Digital Democracy: The Tools Transforming Political Engagement'. *Nesta*, Feb. 2017. <http://www.nesta.org.uk/sites/default/files/digital_democracy.pdf> (accessed 1 Dec. 2017).

Simonite, Tom. 'Moore's Law Is Dead. Now What?' *MIT Technology Review*, 13 May 2016 <https://www.technologyreview.com/s/601441/moores-law-is-dead-now-what/> (accessed 28 Nov. 2017).

Simonite, Tom. 'Pentagon Bot Battle Shows How Computers Can Fix Their Own Flaws'. *MIT Technology Review*, 4 Aug. 2016 <https://www.technologyreview.com/s/602071/pentagon-bot-battle-shows-how-computers-can-fix-their-own-flaws/?utm_campaign=socialflow&utm_source=twitter&utm_medium=post> (accessed 1 Dec. 2017).

Simonite, Tom. 'Oculus Finally Delivers the Missing Piece for VR'. *MIT Technology Review*, 6 Oct. 2016 <https://www.technologyreview.com/s/602570/oculus-finally-delivers-the-missing-piece-for-vr/?utm_campaign=socialflow&utm_medium=post&utm_source=twitter&set=602564> (accessed 30 Nov. 2017).

Singh Grewal, David. *Network Power: The Social Dynamics of Globalization.* New Haven and London: Yale University Press, 2008.

Skinner, Quentin. *Liberty Before Liberalism.* Cambridge: Cambridge University Press, 2012.

Slee, Tom. *What's Yours is Mine: Against the Sharing Economy.* New York and London: OR Books, 2015.

Smith, Adam. *Wealth of Nations.* Ware: Wordsworth, 2012.

Smith, Bryant Walker. 'Human Error as a Cause of Vehicle Crashes.' *Stanford Center for Internet and Society*, 18 Dec. 2013 <http://cyberlaw. stanford.edu/blog/2013/12/human-error-cause-vehicle-crashes> (accessed 30 Nov. 2017).

Smith, Cooper. 'Facebook Users Are Uploading 350 Million New Photos Each Day'. *Business Insider*, 18 Sep. 2013 <http://www.businessinsider. com/facebook-350-million-photos-each-day-2013-9?IR=T> (accessed 30 Nov. 2017).

Smith, Mat. 'Ralph Lauren Made a Great Fitness Shirt that Also Happens to Be "Smart"'. *Engadget*, 18 Mar. 2016 <https://www.engadget.com/ 2016/03/18/ralph-lauren-polotech-review/> (accessed 6 Dec. 2017).

Smith, Merritt Roe, and Leo Marx, eds. *Does Technology Drive History? The Dilemma of Technological Determinism.* Cambridge, Mass: MIT Press, 1994.

Solon, Olivia. 'World's Largest Hedge Fund to Replace Managers with Artificial Intelligence'. *The Guardian*, 22 Dec. 2016 <https://www. theguardian.com/technology/2016/dec/22/bridgewater-associates- ai-artificial-intelligence-management> (accessed 1 Dec. 2017).

Solove, Daniel J. *The Digital Person: Technology and Privacy in the Information Age.* New York: New York University Press, 2004.

Srnicek, Nick. *Platform Capitalism.* Cambridge: Polity Press, 2017.

Srnicek, Nick and Alex Williams. *Inventing the Future: Postcapitalism and a World Without Work.* London: Verso, 2015.

Stallman, Richard. *Free Software, Free Society: Selected Essays of Richard M. Stallman.* Boston: GNU Press, 2002.

Statista. 'Number of Monthly Active Facebook Users Worldwide as of 3rd Quarter 2017 (in millions). <https://www.statista.com/statistics/ 264810/number-of-monthly-active-facebook-users-worldwide/> (accessed 11 Dec. 2017).

Statista, 'Internet of Things 'IoT connected Devices installed base world-wide from 2015 to 2025 (in billions)'. <https://www.statista.com/statistics/471264/iot-number-of-connected-devices-worldwide/> (accessed 5 Jul. 2019).

Steele, Billy. 'Police Seek Amazon Echo Data in Murder Case'. *Engadget*, 27 Dec. 2016 <https://www.engadget.com/2016/12/27/amazon-echo-audio-data-murder-case/> (accessed 1 Dec. 2017).

Steiner, Christopher. *Automate This: How Algorithms Came to Rule Our World*. London: Portfolio, 2012.

Stephen, James Fitzjames. *Liberty, Equality, Fraternity and Three Brief Essays*. Chicago: University of Chicago Press, 1991.

Sterling, Bruce. *The Epic Struggle for the Internet of Things*. Moscow: Strelka Press, 2014.

Sudha, L.R. and R. Bhavani. 'Biometric Authorization System Using Gait Biometry'. *arXiv*, 2011, <https://arxiv.org/pdf/1108.6294.pdf%3b%20Boden/39–40.pdf> (accessed 30 Nov. 2017).

Sullivan, Danny. 'Google Now Handles at Least 2 Trillion Searches Per Year'. *Search Engine Land* (24 May 2016) <http://searchengineland.com/google-now-handles-2-999-trillion-searches-per-year-250247> (accessed 30 Nov. 2017).

Sullivan, Josh and Angela Zutavern. *The Mathematical Corporation: Where Machine Intelligence and Human Ingenuity Achieve the Impossible*. New York: PublicAffairs, 2017.

Sun, Yiting. 'In China, a Store of the Future—No Checkout, No Staff'. *MIT Technology Review*, 16 Jun. 2017 <https://www.technologyreview.com/s/608104/in-china-a-store-of-the-future-no-checkout-no-staff/> (accessed 8 Dec. 2017).

Sundararajan, Arun. *The Sharing Economy: The End of Employment and the Rise of Crowd-Based Capitalism*. Cambridge, Mass: MIT Press, 2017.

Sunstein, Cass R. *Republic.com 2.0*. Princeton: Princeton University Press, 2007.

Sunstein, Cass R. *The Ethics of Influence: Government in the Age of Behavioral Science*. New York: Cambridge University Press, 2016.

Sunstein, Cass R. *#Republic: Divided Democracy in the Age of Social Media*. Princeton: Princeton University Press, 2017.

Surden, Harry. 'Computable Contracts'. *UC Davis Law Review* 46, no. 629 (2012): 629–700.

Surden, Harry. 'Machine Learning and Law'. *Washington Law Review* 89, no. 1 (2014): 87–115.

Surowiecki, James. *The Wisdom of Crowds: Why the Many are Smarter than the Few*. London: Abacus, 2005.

Susskind, Daniel. 'A Model of Technological Unemployment'. Oxford University Discussion Paper, no. 819, version 6 Jul. 2017 (Jul. 2017) <https://www.danielsusskind.com/research> (accessed 5 Dec. 2017).

Susskind, Daniel. 'Re-thinking the Capabilities of Machines in Economics'. Oxford University Discussion Paper, no. 825, version 1 May 2017 (May 2017) <https://www.danielsusskind.com/research> (accessed 5 Dec. 2017).

Susskind, Jamie. *Karl Marx and British Intellectuals in the 1930s*. Burford: Davenant Press, 2011.

Susskind, Richard. *Expert Systems in Law: A Jurisprudential Inquiry*. Oxford: Oxford University Press, 1987.

Susskind, Richard. *The Future of Law: Facing the Challenges of Information Technology*. Oxford: Oxford University Press, 1998.

Susskind, Richard and Daniel Susskind. *The Future of the Professions: How Technology Will Transform the Work of Human Experts*. Oxford: Oxford University Press, 2015.

Swan, Melanie. *Blockchain: Blueprint for a New Economy*. Sebastopol, CA: O'Reilly, 2015.

Swearingen, Jake. 'Can an Amazon Echo Testify Against You?' *NY Mag*, 27 Dec. 2016 <http://nymag.com/selectall/2016/12/can-an-amazon-echo-testify-against-you.html> (accessed 1 Dec. 2017).

Swift, Adam. *Political Philosophy: A Beginners' Guide for Students and Politicians* (Second Edition). Cambridge: Polity Press, 2007.

Taigman, Yaniv, Ming Yang, Marc' Aurelio Ranzato, and Lior Wolf. 'DeepFace: Closing the Gap to Human-Level Performance in Face Verification'. 2014 IEEE Conference on Computer Vision and Pattern Recognition (CVPR), 2014 <https://www.cs.toronto.edu/~ranzato/publications/taigman_cvpr14.pdf> (accessed 11 Dec. 2017).

Takahashi, Dean. 'Magic Leap Sheds Light on its Retina-based Augmented Reality 3D Displays'. *VentureBeat*, 20 Feb. 2015 <http://venturebeat.com/2015/02/20/magic-leap-sheds-light-on-its-retina-based-augmented-reality-3d-displays/> (accessed 30 Nov. 2017).

Taplin, Jonathan. *Move Fast and Break Things: How Facebook, Google, and Amazon Cornered Culture and Undermined Democracy*. New York: Little, Brown and Company, 2017.

Tapscott, Don and Alex Tapscott. *Blockchain Revolution: How the Technology Behind Bitcoin is Changing Money, Business and the World*. London: Portfolio Penguin, 2016.

Tashea, Jason. 'Courts are Using AI to Sentence Criminals. That Must Stop Now'. *Wired*, 17 Apr. 2017 <https://www.wired.com/2017/04/courts-using-ai-sentence-criminals-must-stop-now/> (accessed 1 Dec. 2017).

Taylor, Astra. *The People's Platform: Taking Back Power and Culture in the Digital Age*. London: Fourth Estate, 2014.

'Teledildonics'. *Wikipedia*, last modified 29 Nov. 2017. <https://en.wikipedia.org/wiki/Teledildonics> (accessed 8 Dec. 2017).

Tennyson, Alfred Lord. *The Major Works*. Oxford: Oxford University Press, 2009.

Tetlock, Philip E. *Expert Political Judgment: How Good Is It? How Can We Know?* Princeton: Princeton University Press, 2006.

Thucydides. *The Peloponnesian War*. Translated by Martin Hammond. New York: Oxford University Press, 2008.

Tibbits, Skylar. *TED*, 2013 <https://www.ted.com/talks/skylar_tibbits_the_emergence_of_4d_printing?language=en> (accessed 30 Nov. 2017).

Time. 'Meet the Robots Shipping Your Amazon Orders'. 1 Dec. 2014 <http://time.com/3605924/amazon-robots/> (accessed 30 Nov. 2017).

Tocqueville, Alexis de. *Democracy in America*. Translated by George Lawrence. New York: HarperCollins, 2006.

Toffler, Alvin. *Future Shock*. New York: Bantam Books, 1990.

Topol, Sarah A. 'Attack of the Killer Robots'. *BuzzFeed News*, 26 Aug. 2016 <https://www.buzzfeed.com/sarahatopol/how-to-save-mankind-from-the-new-breed-of-killer-robots?utm_term=.nm1GdWDBZ#.vaJzgW6va) (http://www.dailystar.com.lb/News/World/2016/Aug-

19/367933-china-eyes-artificial-intelligence-for-new-cruise-missiles. ashx> (accessed 28 Nov. 2017).

Townsend, Anthony M. *Smart Cities: Big Data, Civic Hackers, and the Quest for a New Utopia*. New York: W. W. Norton & Company, 2014.

Trotsky, Leon. 'What is National Socialism?' *Marxists*, last modified 25 Apr. 2007 <https://www.marxists.org/archive/trotsky/germany/1933/330610.htm> (accessed 28 Nov. 2017).

Tucker, Ian. ' "A White Mask Worked Better": Why Algorithms Are Not Colour Blind'. *The Guardian*, 28 May 2017 <https://www.theguardian.com/technology/2017/may/28/joy-buolamwini-when-algorithms-are-racist-facial-recognition-bias> (accessed 2 Dec. 2017).

Tucker, Patrick. *The Naked Future: What Happens in a World that Anticipates Your Every Move?* London: Current, 2015.

Tufekci, Zeynep. 'Engineering the Public: Big Data, Surveillance and Computational Politics'. *First Monday* 19, no. 7 (7 Jul. 2014).

Tufekci, Zeynep. *Twitter and Tear Gas: The Power and Fragility of Networked Protest*. New Haven: Yale University Press, 2017.

Tutt, Andrew. 'An FDA for Algorithms'. *Administrative Law Review* 69, no.1 (2017): 83–123.

Twitter.com. <https://about.twitter.com/company> (accessed 30 Nov. 2017).

UK Government Chief Scientific Advisor. 'Distributed Ledger Technology: Beyond Block Chain.' Crown Copyright, 2016. <https://www.gov.uk/government/uploads/system/uploads/attachment_data/file/492972/gs-16-1-distributed-ledger-technology.pdf> (accessed 5 Dec. 2017).

Useem, Jeremy. 'How Online Shopping Makes Suckers of Us All'. *Atlantic*, May 2017 <https://www.theatlantic.com/magazine/archive/2017/05/how-online-shopping-makes-suckers-of-us-all/521448/?utm_source=nextdraft&utm_medium=email> (accessed 1 Dec. 2017).

Valentino-DeVries, Jennifer, Jeremy Singer-Vine, and Ashkan Soltani. 'Websites Vary Prices, Deals Based on Users' Information'. *Wall Street Journal*, 24 Dec. 2012 <https://www.wsj.com/articles/SB10001424127887323777204578189391813881534> (accessed 1 Dec. 2017).

Van Den Hoven, Jeroen and John Weckert. *Information Technology and Moral Philosophy*. New York: Cambridge University Press, 2008.

Van Reybrouck, David. *Against Elections: The Case for Democracy*. London: Bodley Head, 2016.

Vättö, Kristian. 'Samsung SSD 850 Pro (128GB, 256GB & 1TB) Review: Enter the 3D Era.' *AnandTech*, 1 Jul. 2014 <http://www.anandtech.com/show/8216/samsung-ssd-850-pro-128gb-256gb-1tb-review-enter-the-3d-era> (accessed 28 Nov. 2017).

Wright, Georg Henrik von. *Norm and Action: A Logical Inquiry*. London: Routledge & Kegan Paul, 1963.

Vote for policies.org.uk. <https://voteforpolicies.org.uk/> (accessed 1 Dec. 2017).

Voter.xyz. <http://www.voter.xyz/> (accessed 1 Dec. 2017).

Wakefield, Jane. 'Google, Facebook, Amazon Join Forces on Future of AI'. *BBC News*, 28 Sep. 2016 <http://www.bbc.com/news/technology-37494863> (accessed 30 Nov. 2017).

Wakefield, Jane. 'AI Predicts Outcome of Human Right Cases'. *BBC News*, 23 Oct. 2016 <http://www.bbc.com/news/technology-37727387> (accessed 28 Nov. 2017).

Waldrop, M. Mitchell. 'Neuroelectronics: Smart Connections'. *Nature* 503, no. 7474 (6 Nov. 2013): 22–44.

Waldrop, M. Mitchell. 'The Chips are Down for Moore's Law' *Nature* 530, no. 7589 (9 Feb 2016): 144–7.

Wallach, Wendell. *A Dangerous Master: How to Keep Technology from Slipping Beyond Our Control*. New York: Basic Books, 2015.

Wallach, Wendell and Colin Allen. *Moral Machines: Teaching Robots Right from Wrong*. Oxford: Oxford University Press, 2009.

Walzer, Michael. *Spheres of Justice: A Defense of Pluralism and Equality*. New York: Basic Books, 1983.

Walzer, Michael. *Thinking Politically: Essays in Political Theory.* New Haven & London: Yale University Press, 2007.

Wark, McKenzie. *A Hacker Manifesto*. Cambridge, Mass: Harvard University Press, 2004.

Wassom, Brian D. *Augmented Reality Law, Privacy, and Ethics: Law, Society, and Emerging AR Technologies*. Rockland: Syngress, 2015.

Watkins, Alan and Iman Straitens. *Crowdocracy: The End of Politics*. Rochester: Urbane Publications, 2016.

Waymo. *Google* <https://www.google.com/selfdrivingcar/> (accessed 30 Nov. 2017).

Weber, Max. *Political Writings*. Eds. Peter Lassman and Ronald Speirs. Cambridge: Cambridge University Press, 2010.

Weber, Max. *Economy and Society: An Outline of Interpretive Sociology* (Volume 1). Eds. Guenther Roth and Claus Wittich. Berkley: University of California Press, 2013.

Weber, Max. *Economy and Society: An Outline of Interpretive Sociology* (Volume 2). Eds. Guenther Roth and Claus Wittich. Berkley: University of California Press, 2013.

Weigend, Andreas. *Data for the People: How to Make our Post-Privacy Economy Work For You*. New York: Basic Books, 2017.

Weinberger, David, *Small Pieces Loosely Joined: A Unified Theory of the Web*. New York: Basic Books, 2003.

Weinberger, David. 'Our Machines Now Have Knowledge We'll Never Understand'. *Wired*, 18 Apr. 2017 <https://www.wired.com/story/our-machines-now-have-knowledge-well-never-understand> (accessed 28 Nov. 2017).

Whipple, Tom. 'Nanorobots Could Deliver Drugs by Power of Thought'. *The Times*, 27 Aug. 2016 <http://www.thetimes.co.uk/article/226da2de-6baf-11e6-998d-9617c077f056> (accessed 30 Nov. 2017).

Whitby, Andrew, Audun Jøsang, and Jadwiga Indulska. 'Filtering Out Unfair Ratings in Bayesian Reputation Systems'. *Proceedings of the Workshop on Trust in Agent Societies, at the Autonomous Agents and Multi Agent Systems Conference*, Jul. 2004 <https://www.csee.umbc.edu/~msmith27/readings/public/whitby-2004a.pdf> (accessed 3 Dec. 2017).

Whitehead, Alfred North. *An Introduction to Mathematics*. Milton Keynes: Watchmaker, 2011.

'Who? Whom?' *Wikipedia*, last modified 3 Jun. 2017. <https://en.wikipedia.org/wiki/Who,_whom%3F> (accessed 7 Dec. 2017).

Wiener, Norbert. *Cybernetics or, Control and Communication in the Animal and the Machine* (Second Edition). Connecticut: Martino Publishing, 2013.

Wile, Rob. 'A Venture Capital Firm Just Named an Algorithm to its Board of Directors—Here's What it Actually Does'. *Business Insider*, 13 May 2014 <http://www.businessinsider.com/vital-named-to-board-2014-5> (accessed 28 Nov. 2017).

Winfield, Alan. *Robotics: A Very Short Introduction*. Oxford: Oxford University Press, 2012.

Wittgenstein, Ludwig. *Culture and Value*. Translated by Peter Winch. Chicago: University of Chicago Press, 1980.

Wittgenstein, Ludwig. *Tractatus Logico-Philosophicus*. Abingdon: Routledge, 2001.

Wolff, Jonathan. *An Introduction to Political Philosophy*. Oxford: Oxford University Press, 1996.

Wong, Joon Ian. 'Sweden's Blockchain-powered Land Registry is Inching Towards Reality'. *Quartz Media*, 3 Apr. 2017 <https://qz.com/947064/sweden-is-turning-a-blockchain-powered-land-registry-into-a-reality/> (accessed 30 Nov. 2017).

Wong, Julia. 'Segregated Valley: The Ugly Truth About Google and Diversity in Tech'. *Guardian*, 7 Aug. 2017 <https://www.theguardian.com/technology/2017/aug/07/silicon-valley-google-diversity-black-women-workers> (accessed 28 Nov. 2017).

Wootson Jr, Cleve R. 'A Man Detailed His Escape from a Burning House. His Pacemaker Told Police a Different Story'. *Washington Post*, 8 Feb. 2017 <https://www.washingtonpost.com/news/to-your-health/wp/2017/02/08/a-man-detailed-his-escape-from-a-burning-house-his-pacemaker-told-police-a-different-story/?tid=sm_tw&utm_term=.531d8fabc6d2> (accessed 1 Dec. 2017).

Wright, David, Serge Gutwirth, Michael Friedewald, Elena Vildjiounaite, and Yves Punie, eds. *Safeguards in a World of Ambient Intelligence*. Netherlands: Springer, 2010.

Wright, Ronald. *A Short History of Progress*. London: Canongate Books, 2006.

Wyss Institute. <http://wyss.harvard.edu/viewpage/457> (accessed 30 Nov. 2017).

Wu, Tim. *The Master Switch: The Rise and Fall of Information Empires*. London: Atlantic, 2010.

Wu, Yonghui, Mike Schuster, Zhifeng Chen, Quoc V. Le, Mohammed Norouzi, Wolfgang Macherey, Maxim Krikun, et al. 'Google's Neural

Machine Translation System: Bridging the Gap between Human and Machine Translation'. *arXiv*, 8 Oct. 2016 <https://arxiv.org/abs/1609.08144> (accessed 6 Dec. 2017).

Xiong, Wei, Jasha Droppo, Xupeng Huang, Frank Seide, Michael Seltzer, Andreas Stolcke, Donghan Yu, and Geoffrey Zweig. 'Achieving Human Parity in Conversational Speech Recognition'. *arXiv*, 17 Feb. 2017 <https://arxiv.org/abs/1610.05256> (accessed 28 Nov. 2017).

Yeung, Karen. '"Hypernudge": Big Data as a Mode of Regulation by Design'. *Information, Communication & Society* 20, no. 1 (2017): 118–36.

Yonck, Richard. *Heart of the Machine: Our Future in a World of Artificial Intelligence.* New York: Arcade Publishing, 2017.

Young, Angelo. 'How to Break Up Alphabet, Amazon and Facebook'. *Salon*, 31 May 2017 <https://www.salon.com/2017/05/31/how-to-break-up-alphabet-amazon-and-facebook/> (accessed 8 Dec. 2017).

Young, Iris Marion. *Justice and the Politics of Difference.* Princeton: Princeton University Press, 2011.

YouTube for Press. <https://www.youtube.com/intl/en-GB/yt/about/press/> (accessed 30 Nov. 2017).

Zarsky, Tal. 'Understanding Discrimination in the Scored Society'. *Washington Law Review* 89, no. 4 (2014): 1375–412.

Ziewitz, Malte. 'Governing Algorithms: Myth, Mess, and Methods'. *Science, Technology, & Human Values* 41, no. 1 (2015): 3–16.

Zittrain, Jonathan. *The Future of the Internet and How to Stop It.* London: Allen Lane, 2008.

Zittrain, Jonathan. 'Engineering an Election'. *Harvard Law Review Forum*, 20 Jun. 2014 <https://harvardlawreview.org/2014/06/engineering-an-election/> (accessed 1 Dec. 2017).

Zittrain, Jonathan. 'Apple's Emoji Gun Control'. *New York Times*, 16 Aug. 2016 <https://mobile.nytimes.com/2016/08/16/opinion/get-out-of-gun-control-apple.html?_r=0&referer=https://www.google.com/> (accessed 1 Dec. 2017).

INDEX